Substance Abuse
and Psychopathology

APPLIED CLINICAL PSYCHOLOGY

Series Editors:
Alan S. Bellack, *Medical College of Pennsylvania at EPPI, Philadelphia, Pennsylvania,*
and Michel Hersen, *University of Pittsburgh, Pittsburgh, Pennsylvania*

Substance Abuse and Psychopathology

Edited by

ARTHUR I. ALTERMAN

Veterans Administration Medical Center
Philadelphia, Pennsylvania

1985

PLENUM PRESS • NEW YORK AND LONDON

Library of Congress Cataloging in Publication Data

Main entry under title:

Substance abuse and psychopathology.

(Applied clinical psychology)
Includes bibliographies and index.
1. Substance abuse. 2. Psychology, Pathological. I. Alterman, Arthur I. II. Series.
[DNLM: 1. Psychopathology. 2. Substance Abuse—psychology. WM 270 S9412]
RC564.S8283   1985   616.86   85-6352
ISBN 0-306-41849-5

©1985 Plenum Press, New York
A Division of Plenum Publishing Corporation
233 Spring Street, New York, N.Y. 10013

Printed in the United States of America

Contributors

Arthur I. Alterman, Ph.D., Research Department, VA Medical Center, Philadelphia, Pennsylvania

Sam Castellani, M.D., Department of Psychiatry, School of Medicine, University of Kansas, Wichita, Kansas

Anna Rose Childress, Ph.D., Drug Dependence Unit, Philadelphia VA Medical Center, Philadelphia, Pennsylvania

Nancy Day, Ph.D., Western Psychiatric Institute & Clinic, 3811 O'Hara Street, Pittsburgh, Pennsylvania

Everett Ellinwood, Jr., M.D., Duke University Medical School, Durham, North Carolina

Igor Grant, M.D., San Diego VA Medical Center and University of California, San Diego, School of Medicine, Department of Psychiatry, La Jolla, California

Patricia Ann Harrison, A.B., Department of Psychiatry, Saint Paul-Ramsey Medical Center, St. Paul, Minnesota

Norman G. Hoffmann, Ph.D., Department of Psychiatry, Saint Paul-Ramsey Medical Center, St. Paul, Minnesota

Steven B. Katz, M.D., New York University Medical Center, New York, New York

Edward Kaufman, M.D., Department of Psychiatry and Human Behavior, University of California Irvine, California College of Medicine, Orange, California

David C. Kay, M.D., Department of Psychiatry and Pharmacology, Baylor College of Medicine, Houston, Texas

Kenneth Leonard, Ph.D., Western Psychiatric Institute & Clinic, 3811 O'Hara Street, Pittsburgh, Pennsylvania

A. Thomas McLellan, Ph.D., Drug Dependence Unit, Philadelphia VA Medical Center, Philadelphia, Pennsylvania

Jodi A. Martin, Ph.D., Department of Psychiatry, Saint Paul-Ramsey Medical Center, St. Paul, Minnesota

Demmie Mayfield, M.D., Psychiatry Service, Kansas City VA Medical Center, Kansas City, Missouri

William M. Petrie, M.D., Department of Psychiatry, Vanderbilt University, Nashville, Tennessee

Robert Reed, M.S., San Diego VA Medical Center and University of California, San Diego, School of Medicine, Department of Psychiatry, La Jolla, California

Edwin Robbins, M.D., New York University Medical Center, New York, New York

Marvin Stern, M.D., New York University Medical Center, New York, New York

Vicente B. Tuason, M.D., Department of Psychiatry, Saint Paul-Ramsey Medical Center, St. Paul, Minnesota

W. H. Vogel, Ph.D., Department of Pharmacology, Thomas Jefferson University, Philadelphia, Pennsylvania

Joseph Westermeyer, M.D., Department of Psychiatry, University Hospitals, Minneapolis, Minnesota

George E. Woody, M.D., Drug Dependence Unit, Philadelphia VA Medical Center, Philadelphia, Pennsylvania

Preface

Substance misuse and abuse exist in almost every human society. In our western civilization, the bulk of attention has focused on those individuals who specifically seek treatment or those who have become so disabled by these problems that they require treatment. These individuals usually qualify for a psychiatric diagnosis of alcohol or other substance abuse. However, just as it has been recognized that primary substance abuse is frequently associated with other diagnosable psychiatric disorders, such as sociopathy or attention deficit disorder (residual type) and that the origins of substance abuse are multivariate, we have also begun to become aware that many other individuals in our society with psychiatric or other problems also suffer, to varying degrees, from substance abuse. These problems may be considered secondary by various specialists or treatment personnel; but nevertheless, they are problems, and what disorder is primary or secondary in a given individual may often be very difficult to determine in a meaningful fashion. Thus, within the past decade, research studies have reported significant incidences of substance abuse/or misuse in high school and college-aged populations, in medical populations, and in individuals with other psychiatric disorders such as schizophrenia, depression, and the anxiety and personality disorders. Yet to date little has been done to bring together and systematize this widely scattered data that describes the presence of substance abuse problems in various populations. Furthermore, the practical impact of this information on current treatment practices appears to be negligible. However, my own research, experience, and discussions with care providers has made it clear that substance abuse is a formidable and frequently overlooked or disregarded problem in the treatment of schizophrenia and that it also undoubtedly presents difficulties for persons principally undergoing treatment for other labeled disorders. Yet, there is virtually no literature describing the treatment of psychiatric patients with concurrent substance abuse problems.

The basic objectives in initiating this book, therefore, were to bring together what we presently know about the occurrence of substance abuse/misuse in various psychiatric and nonpsychiatric populations in our society and to attempt to determine, to the extent possible, the causes, roles, effects, and forms of treatment for such persons. What emerges from reading this book is some knowledge, clues, and directions; but what strikes me most profoundly is the preliminary nature of what we have thus far learned. It would therefore seem that a major contribution of this book would be the interest it may stimulate and the questions it will raise about the significance and meaning of substance abuse in various populations in our society. This is a problem that seems to be gradually attracting some attention. Within the past month, I was most encouraged by being contacted by two individuals who were concerned about the treatment of schizophrenic patients with alcohol abuse problems. In this sense, *Substance Abuse and Psychopathology* may be very timely in providing at least a starting point for defining what we currently do know about the relationships between substance misuse/abuse and other problems, the effects and treatment of these problems, and, hopefully, it may stimulate sorely needed systematic inquiries and investigations into this problem area.

The author would like to acknowledge the contribution of several individuals to this endeavor. I would like to express my appreciation to Dr. Michel Hersen, series editor, who was instrumental in encouraging the initiation of this project and provided helpful consultation on numerous occasions. My colleagues, Drs. Gerald Goldstein and Ralph Tarter also provided valuable support, encouragement, and expertise.

ARTHUR I. ALTERMAN

Contents

PART II. PROBLEMS, EFFECTS, AND TREATMENTS

CHAPTER 13

      Patricia Ann Harrison, Jodi A. Martin, Vincente B. Tuason,
      and Norman G. Hoffmann

Relationships between Substance Abuse and Psychopathology

OVERVIEW

Arthur I. Alterman

1. Introduction

My interest in the relationship and role of substance abuse in psychopathology developed over the course of a number of years as I became aware and sensitized to findings from diverse lines of inquiry implicating substance abuse in psychopathology. Some of the findings that suggested this relationship are noted immediately below. For example, McLellan and his colleagues found a high incidence of substance abuse in psychiatric inpatients and demonstrated that the choice of drug(s) by these patients was related to their psychiatric diagnoses (McLellan & Druley, 1977). A subsequent paper by this investigator (McLellan, Woody, & O'Brien, 1979) provided some evidence for the development of schizophrenia in chronic stimulant abusers. My own research has revealed a relatively high incidence of alcohol or drug abuse in schizophrenic inpatients and numerous behavioral, treatment, management, and medical problems associated with such illicit drug use (Alterman & Erdlen, 1983; Alterman, Erdlen, LaPorte, & Erdlen, 1982; Alterman, Erdlen, McLellan, & Mann, 1980; Alterman, Erdlen, & Murphy, 1981). Winokur's research over a decade ago indicated the possibility of genetic

ARTHUR I. ALTERMAN • Research Department, Philadelphia VA Medical Center, Philadelphia, Pennsylvania 19104.

relationships between alcoholism and depression (Winokur, Reich, Rimmer, & Pitts, 1970; Winokur, Rimmer, & Reich, 1971) and more recent work on the treatment of alcoholism with lithium (Reynolds, Merry, & Coppen, 1977) also suggested a relationship with affective disorders. Literature has also begun to cumulate indicating a relationship between substance abuse and anxiety disorders (Mullaney & Trippett, 1979; Quitkin, Rifkin, Kaplan, & Klein, 1972), and it has been established that substance abuse is associated with a number of personality disorders, including sociopathy (American Psychiatric Association, 1980). Finally, a study by Mellinger, Balter, Manheimer, Cisin, and Parry (1978) has shown a relationship between distress and heavy alcohol use among young men in the general population.

Thus, evidence can be marshalled to support the concept that substance abuse and psychopathology are significantly related. However, since the findings are derived from widely varying subject populations and conceptual viewpoints, there has been little effort to systematize and integrate the existing knowledge. The primary purpose of this book, therefore, is to bring together and examine, clarify, and delineate the research findings concerning the relationships between drug/alcohol abuse and psychological and psychiatric symptomatology. A secondary objective is to examine the biomedical consequences and implications of drug/alcohol abuse in the populations under study. Given this framework, a number of questions and considerations can be raised at the outset about the role and relationship of substance abuse to psychopathology. For example:

1. What evidence is there for a relationship between substance abuse and psychological problems?
2. Similarly, what role does substance abuse play in relation to more serious psychiatric symptomatology and disorders? That is, does the chronic abuse of substances lead to psychiatric illness, or conversely, does the presence of psychopathology precede and lead to substance abuse? More generally, what is the etiological and developmental relationship between substance abuse and psychopathology?
3. How does the illicit use of substances affect the care of psychiatric patients? How is the patient with concurrent psychiatric illness and substance abuse problems treated, what are the specific problems associated with his/her treatment, and what is the treatment response of such patients?
4. What are the physical/medical effects and implications, if any, of illicit drug and alcohol use in this population? For example, is

there any evidence of adverse interactions with prescribed medications or other adverse effects—direct or indirect—on the physical health of the patient?

The remaining sections of this introduction provide an overview of the book's chapters, attempting to summarize, describe, and put into perspective some of the more critical findings, given the framework described above.

2. Overview

The chapters of this book were formulated in an attempt to explore and answer questions such as those described above. Because of the preliminary and unsystematic state of knowledge concerning such issues, the knowledge described will, in many instances, only be provisional and suggestive. The book is divided for heuristic purposes into two sections, although the distinction is not entirely complete from a functional viewpoint. The first section of seven chapters attempts primarily to describe the occurrence, interaction, and role of substance abuse in different populations, while the second section of five chapters explores and focusses to a greater extent upon some of the effects and consequences of substance abuse and on efforts to more comprehensively treat problems associated with the conjoint presence of substance abuse and psychopathology.

2.1. Part I

In the initial chapter of the first section, Day and Leonard examine the relationships between alcohol and drug use and psychopathology in the general population. Although many of the findings uncovered at this stage of inquiry are recognizedly tentative, the authors do find evidence for certain personality traits in children being predictive of future drinking problems, a relationship between depression and heavy alcohol use in college-age populations, and a relationship between high distress and heavy alcohol use in young men and nonprescription drug use in young women.

Inasmuch as investigations of the relationship of substance abuse to psychopathology in the general population have been of relatively low priority and are nonintensive by their very nature, the evidence produced would seem to merit serious consideration. It would therefore not be surprising, at some future time, to discover that the findings to date have just been the "tip of the iceberg."

Westermeyer examines the relationship between substance use and psychopathology from a cross-cultural perspective. One significant conclusion drawn from the data reviewed in this chapter is that "genetic and epidemiological data suggest that substance abuse and certain other psychopathological conditions (major depression; generalized anxiety) may be pathoplastic variants of each other." That is, in certain cultures in which substance use is acceptable and prevalent, the problems of anxiety may be relatively low. On the other hand, anxiety is a relatively common phenomenon when individuals of this same cultural group reside in other societal settings where drug use is not so readily accepted. The concept of a pathoplastic relationship between substance abuse and other manifestations of psychopathology appears to be one that may have general utility for our understanding of the development of substance abuse. Other important issues and questions dealt with in this chapter are the relationships between substance abuse and violence and criminality and the negative effects of substance abuse on the individual's use of social networks.

Mayfield reviews the evidence for the occurrence and role of substance abuse in affective disorders. He finds that alcohol use has a profound and pervasive effect on mood. However, it appears to be used more to reduce pain than to cause pleasure. Surprisingly, alcohol intake is not usually increased before the onset of a depressive episode, but prior to the onset of mania. Mayfield concludes that "there is reason to suspect that there is a common underlying process which has not been observed or identified or understood in alcoholism and affective disorders." At the same time, he is able to find little evidence for a relationship between drug abuse and mood disorders.

Ideally, it would have been desirable to have included a chapter in this text on the relationship between anxiety disorders and substance abuse. As suggested earlier, a small body of evidence has begun to accumulate indicating that relatively large numbers of alcoholics and drug abusers are prone to suffer from panic attacks, agoraphobia and social phobias, and vice versa. A decade ago, Quitkin and his colleagues (Quitkin *et al.*, 1972) described an association between phobic anxiety syndrome and severe substance abuse (barbiturates, nonbarbiturate sedatives, minor tranquilizers, alcohol). They found that about 5% to 10% of those suffering from the phobic anxiety syndrome abused drugs in an attempt to self-medicate for chronic anticipatory anxiety. These patients were frequently treated for their substance abuse with the underlying problem of anxiety being overlooked. Long-term treatment with imipramine was found to be relatively effective. Several other studies have now been completed that reveal an association between panic attacks, agoraphobia, social phobia, and alcoholism. Powell and her colleagues (Powell, Penick, Othmer, Bingham, & Rice, 1982) found, for example, that 13% of a large sample of male alcoholic inpatients satisfied the criteria for a diagnosis of panic attacks, while 12% and 10% were diagnosed as obsessive-compulsive and phobic disorder, respectively. Interestingly, the onset of both the panic attacks and the phobic disorders occurred several years prior to the onset of alcoholism, suggesting that the anxiety disorder may have played a role in the instigation of alcoholism. The strongest evidence for the existence of agoraphobia and social phobia in alcoholics has been reported by Mullaney and Trippett (1979). Working with inpatient alcoholics, they found that 13% of the men alcoholics were agoraphobic and another 29% were borderline agoraphobic. Corresponding figures for women alcoholics were 33% and 22%. About one-fourth of both men and women alcoholics were found to be socially phobic and another 33% of both sexes were borderline socially phobic. The fully phobic alcoholics repeatedly experienced anxiety to a disabling degree, either with respect to the agoraphobia or the socially phobic situations. Inasmuch as the age of onset of the phobias occurred on the average about 2 to 3 years before alcohol use became a problem, the authors consider the possibility that there is an interaction between the development of phobias and heavy alcohol consumption.

They argue further that the phobic disorders in alcoholics may not be detected as a rule because alcoholics do not typically receive extensive psychological examination and because the alcoholic himself may not discriminate his phobic anxiety from withdrawal anxiety. The findings on anxiety disorders and substance abuse therefore suggest that alcohol and illicit drugs may be employed by individuals suffering from psychological distress for purposes of self-medication. The evidence thus far accumulated is quite intriguing. One hopes this area will be the object of more intensive inquiry in the near future.

Kay's chapter deals with the relationships between substance abuse and antisocial behavior, psychopathic states, and sociopathy. Several interesting provisional concepts are introduced in this chapter. The author submits that psychopathic or sociopathic states result from an increased need state, such as that seen in a state of protracted abstinence. Sociopathy is then characterized as the chronic manifestation of this problem.

Psychopathic states are dominated by a particular type of unpleasant feeling state termed hypophoria. This negative feeling state is conceptualized as being comparable to depression, but without feelings of unworthiness, sleep disorders, anorexia, decreased libido, or inability to experience joy. Substance abuse may then be associated with the existence of an "affective-antisocial spectrum disorder (Hewett & Martin, 1980) and may therefore represent an effort to treat these feelings.

Psychometric assessment of the substance abuser's current psychopathic state is recommended as a useful procedure for evaluating improvement during the course of treatment.

My own chapter describes the work thus far conducted on alcohol and drug abuse in psychiatric, particularly schizophrenic, inpatients. The relative incidence and family psychiatric history data do not reveal any unique dynamic or developmental relationships between alcoholism and schizophrenia, or vice versa. Nevertheless, alcohol and drug abuse do occur in a significant proportion of the patients with more serious psychiatric disorders, and for a variety of reasons little specific attention has been given to the unique problems of treating these patients. That is, few specialized treatment approaches exist for this population and there is little information concerning how these individuals respond to existing treatment. The available information on this patient population is reviewed. Other problems that could result from illicit substance abuse, such as unknown adverse drug interactions, contamination of laboratory findings, or medical problems resulting from chronic substance abuse, are also considered and described.

The chapter by McLellan, Childress, and Woody describes the findings of a number of studies conducted in recent years by McLellan and his colleagues, which revealed that:

1. Substance-abusing psychiatric patients select substances with similar psychological effects for regular concurrent use;
2. chronic use of the same substances (stimulants) are associated with long-term psychiatric illness, particularly schizophrenia; and
3. chronic use of depressants, such as barbiturates, just as alcohol, are associated with some deficits in cognitive functioning.

The results of a study comparing the relative effectiveness of two forms of treatment for mixed opiate and amphetamine versus mixed opiate and depressant abusers are then described indicating that the former group do better when treated with methadone maintenance, while the latter group appear to fare better when treated in a therapeutic community. It appears clear from McLellan's work that stimulant abusers are much more seriously disturbed individuals. The authors discuss possible explanations for the differential group findings and note, within this context, that positive psychotropic effects have been attributed to methadone, which might explain its effects with stimulant abusers.

The final chapter of Part I of the book, by Castellani, Petrie, and Ellinwood, considers neurobiological mechanisms that underlie both natural and drug-induced psychotic states. The authors consider the effects of hallucinogens, stimulants, phencyclidine, and cannabis in relation to neurotransmitter mechanisms, psychotomimetic drug mechanisms, cortical and limbic electrographic effects, and findings of studies using animal models of psychoses.

Among the conclusions suggested by the literature are that schizophrenic-like symptoms associated with psychotomimetic drug actions would appear to be related primarily to increased central dopamine functioning. This chapter also provides a useful neurobiological perspective for McLellan *et al.*'s clinical findings on chronic stimulant abusers.

2.2. Part II

The second section of this book begins with a review by Vogel on the interactions of drugs of abuse, or illicit drugs, with prescribed drugs. As the author indicates at the outset, substance abusers expose themselves to a considerable degree of risk in terms of organ and tissue damage just by taking drugs or chemicals in large doses and over prolonged periods of time. Although most of the adverse interactions of drugs are not presently known, we would anticipate—based on knowledge of the known mechanisms of drug action—that the concomitant use of medical and nonmedical drugs should result in a variety of adverse effects in most individuals. Some of the more obvious examples of adverse interactions between illicit—and often unknown—drug use and the use of prescribed psychotropic medication (such as the neuroleptics or tricyclic antidepressants) are described. We should emphasize that virtually no direct systematic evidence exists on the actual incidence of such interactions, but related evidence on adverse drug interactions (Miller, 1973) suggests that these undoubtedly occur, and sometimes with serious and life-threatening consequences. One hopes this problem will soon receive the serious attention it deserves.

In line with this section's focus on effects and consequences of substance abuse, the chapter by Robbins, Katz, and Stern is concerned with the identification and treatment of substance-abuse-related problems found in the emergency room. The recommended treatment course for psychiatric and medical symptomatology resulting from a variety of forms of substance abuse are therefore considered. Although the chapter is primarily concerned with problems presented to medical staff in the emergency room setting, it provides a unique contribution to the book in characterizing the spectrum of symptomatology resulting from substance abuse in individuals not necessarily presenting for psychiatric treatment or having diagnosable substance abuse problems.

The chapter by Grant and Reed reviews the existing evidence on the relationship between substance abuse and neuropsychological impairment and brain damage. The effects of a number of substances are considered, including alcohol, opiates, cannabis, phencyclidine, sedative hypnotics, volatile substances, hallucinogens, and stimulants. Contrary to conclusions drawn from earlier research, only limited evidence can be found for permanent neuropsychological deficits in long-term detoxified alcoholics, with the functions most susceptible to impairment being dependent on nonverbal abilities.

In addition, there is no evidence that use of cannabis, hallucinogens, or central stimulants is associated with neuropsychological impairment. On the other hand, it appears that many polydrug abusers exhibit long-term deficits in neuropsychological functioning, a finding of some import since polydrug abuse is becoming increasingly common. The findings on recovery of functioning, of course, presume long-term abstinence and cannot therefore be considered to be representative of the majority of substance abusers. Evidence for neuropsychological impairments in acutely detoxified and chronically abusing individuals is also reviewed in this chapter in some detail. Some of the clinical characteristics associated with this condition are described in the previous chapter by Robbins *et al.* (Chapter 10).

This chapter also deals in some detail with methodological considerations that limit our ability to draw inferences about the specific effects of substance abuse on neuropsychological functioning. For example, biomedical risk factors associated with, but not directly attributable to, substance abuse can result in neuropsychological deficits and brain and other physical injury and it is often difficult to separate these effects from those of the substance abuse. It is important to recognize that such risk factors (e.g., head trauma) may precede or occur concurrently with substance abuse and are therefore significant in their own right when correlates of substance abuse and pathology are considered.

Kaufman's chapter considers the correlates and effects of substance abuse in relation to the health of the family system within which the abuser resides. The effects of alcohol and drug abuse on the family system are considered separately primarily because this has to date been a functional division made by treatment providers. As Kaufman emphasizes, several authors have noted that substance use is essential to maintaining an interactional family equilibrium that resolves a disorganization of the family system which existed prior to consumption of the substance. Given this thesis, the author describes clinical evidence concerning the configurations of pathology that have been found and are considered to exist within the family systems of substance abusers, and describes approaches for treating these problems. Also considered are the family variables that seem to be important for the prevention of substance abuse. The evidence presented must be considered to be tentative at this point in time, awaiting more rigorous investigation. Nevertheless, the chapter presents a different perspective for conceptualizing pathological correlates and effects of substance abuse and one framework for treating them.

The final chapter in the book by Harrison, Martin, Tuason, and Hoffman describes the philosophy, structure, and methods of functioning of one of the few programs in the country designed specifically to conjointly treat persons with dual or multiple psychiatric disorders, one of which is substance abuse. As the authors state at the outset of their chapter, traditional, distinct models of treatment for psychiatric illness and substance abuse are no longer adequate to respond to the clinical reality of coexisting disorders. Often the traditional responses of one subspeciality are not sufficient to meet the needs of patients who cross professional boundaries that are perhaps more distinct and meaningful to the practitioners than to the patients in turmoil. Thus, the multidisciplinary team and departmental support are the cardinal features of the "dual treatment" program. The close working relationships required of staff with different professional backgrounds naturally requires considerable staff training and continuing staff development.

The variety of therapeutic and educational programs provided to different patients seen in the program, staff issues, and treatment problems are presented, as well as a variety of clinical vignettes describing characteristics and problems of patients treated in the program, and approaches to their treatment. Because the "dual treatment" program is just beginning to get beyond the developmental stage of its existence, data evaluating outcome of the patients treated therein are not yet available. The information provided in this chapter on approaches to treating dual and polydiagnosed psychiatric/substance-abusing patients should provide a valuable starting point, nevertheless, for clinicians faced with such problems.

3. Conclusions

Does this book achieve its stated goals? This is, of course, difficult to say. Nevertheless, the chapters do point to a relationship between substance abuse and psychological and psychiatric symptomatology in a variety of populations and societies, although, as indicated, many of the findings are still preliminary. Perhaps this is sufficient, however, to make the point that substance abuse plays a significant role in psychiatric illness and is certainly worthy of careful examination in this respect. Mezzich, Coffman, and Goodpaster (1982) reported recently that substance use disorders are by far the most prevalent second, third, and fourth psychiatric diagnoses among admissions to a comprehensive psychiatric facility, and Treece (1982) has gone so far as to suggest that substance use disorders be assigned a separate DSM-III axis. Mayfield has suggested in this text that understanding alcoholism may be a key to the understanding of affective disorders. I would go further and suggest a similar relation between substance abuse disorders and much of mental disorder. If the nature of these relationships can be articulated, it would seem that at least part of the puzzle of mental illness would be clarified. In this connection, the chapter by Castellani *et al.* provides a perspective on neurobiological theories of the relationships between psychotomimetic drugs (including substances of abuse) and psychosis. Nevertheless, it is clear that only a beginning has been made in delineating the relation of substance abuse to psychopathology, and that much more knowledge of genetic, physiological, psychological, sociological, and other environmental circumstances and articulations has to be obtained before the diverse manifestations of these relationships can be more fully comprehended.

The second section of this book attempted to trace and suggest some of the patient treatment and management implications of substance abuse in persons with other primary psychiatric disorders. To some extent, similar implications would seem to apply to persons with medical disorders. This entire section of the book is even more provisional and tentative than the first section, as direct systematic examination of such problems is virtually nonexistent at the present time. To some extent, these chapters deal more directly with medical/physical consequences of substance abuse than with its interactions with other psychiatric disorders. Readers will therefore have to use their imagination in order to extend the effects described to the substance-abusing individual with other psychiatric disorders. The last two chapters of this section present material on initial efforts to conceptualize and develop modes of treatment for persons with substance abuse problems in the presence of other disorders. Although these efforts are inchoate and incomplete, they nevertheless represent serious attempts to respond to a pressing area of treatment need.

An implicit objective of many books, in addition to communicating knowledge, is to stimulate the attention and interest of the reader. In this particular instance, where there has been little effort to systematize the relevant knowledge, where until now the source material has been put into many separate and independent cubbyholes, an important contribution would be to help the reader recognize the intrinsic relatedness of the subject matter.

If this is accomplished and these individuals—psychiatrists, nurses, physiologists, pharmacologists, physicians, psychologists, students, interns, residents, clinicians, and researchers—are stimulated to further examine and pursue the problems described herein, the book will, in the long run, contribute to a greater understanding of the etiology, causes, functions, and consequences of psychopathology.

4. References

Alterman, A. I., & Erdlen, D. L. Illicit substance abuse in hospitalized psychiatric patients. *Journal of Psychiatric Treatment and Evaluation*, 1983, *5*, 377–380.

Alterman, A. I., Erdlen, F. R., McLellan, A. T., and Mann, S. C. Problem drinking in hospitalized schizophrenic patients. *Addictive Behaviors*, 1980, *5*, 273–276.

Alterman, A. I., Erdlen, F. R., & Murphy, E. Alcohol abuse in the psychiatric hospital population. *Addictive Behaviors*, 1981, *6*, 69–73.

Alterman, A. I., Erdlen, D. L., LaPorte, D. J., & Erdlen, F. R. Effects of illicit drug use in an inpatient psychiatric population. *Addictive Behaviors*, 1982, *7*, 231–242.

American Psychiatric Association. *Diagnostic and statistical manual of mental disorders* (3rd ed.) Washington, D.C.: American Psychiatric Association, 1980.

Hewett, B. B., & Martin, W. R. Psychometric comparisons of sociopathic and psychopathological behaviors of alcoholics and drug abusers versus a low drug control population. *International Journal of Addictions*, 1980, *15*, 77–105.

McLellan, A. T., & Druley, K. A. Non-random relation between drugs of abuse and psychiatric diagnosis. *Journal of Psychiatric Research*, 1977, *13*, 14–27.

McLellan, A. T., Woody, G. E., & O'Brien, C. P. Development of psychiatric disorders in drug abusers. *New England Journal of Medicine*, 1979, *301*, 1310–1314.

Mellinger, G. D., Balter, M. B., Manheimer, D. I., Cisin, I. H., & Parry, H. J. Psychic distress, life crisis, and use of psychotherapeutic medications. *Archives of General Psychiatry*, 1978, *35*, 1045–1052.

Mezzich, J. E., Coffman, G. A., & Goodpastor, S. M. A format for DSM-III diagnostic formulation: Experience with 1,111 consecutive patients. *American Journal of Psychiatry*, 1982, *139*, 591–596.

Miller, R. R. Drug surveillance utilizing epidemiologic methods. *American Journal of Hospital Pharmacy*, 1973, *30*, 584–592.

Mullaney, J. A., & Trippett, C. J. Alcohol dependence and phobias: Clinical description and relevance. *British Journal of Psychiatry*, 1979, *135*, 565–573.

Powell, B. J., Penick, E. C., Othmer, E., Binghan, S. F., & Rice, A. S. Prevalence of additional psychiatric syndromes among male alcoholics. *Journal of Clinical Psychiatry*, 1982, *10*, 404–407.

Quitkin, F. M., Rifkin, A., Kaplan, J., & Klein, D. F. Phobic anxiety syndrome complicated by drug dependence and addiction. *Archives of General Psychiatry*, 1972, *27*, 159–162.

Reynolds, C., Merry, J., & Coppen, A. Prophylactic treatment of alcoholism by lithium carbonate: An initial report. *Alcoholism: Clinical and Experimental Research*, 1977, *1*, 109–111.

Robins, L. N. The diagnosis of alcoholism after DSM-III. In R. Meyer, R. E. Glueck, J. E. O'Brien, T. F. Babor, J. H. Jaffe, & J. R. Stabenau (Eds.), *Evaluation of the alcoholic: Implications for research, theory, and treatment.* Rockville, Md.: United States Department of Health and Human Services, 1981.

Treece, C. DSM-III as a research tool. *American Journal of Psychiatry*, 1982, *139*, 577–583.

Winokur, G., Reich, T., Rimmer, J., & Pitts, F. Alcoholism III: Diagnosis and familial psychiatric illness in 259 alcoholic probands. *Archives of General Psychiatry*, 1970, *23*, 104–111.

Winokur, G., Rimmer, J., & Reich, T. Alcoholism IV: Is there more than one type of alcoholism? *British Journal of Psychiatry*, 1971, *1118*, 525–531.

I

SUBSTANCE ABUSE IN DIFFERENT POPULATIONS

2

Alcohol, Drug Use, and Psychopathology in the General Population

Nancy Day and Kenneth Leonard

1. The Relationship of Alcohol Use and Abuse to Psychopathology in the General Population

This chapter is a review of what we know, or don't know, about the prevalence of alcohol and drug use and abuse, and the relationship of psychopathology to these patterns. Our focus will be on studies of the general population, leaving the discussion of clinical populations to other chapters. Our current state of knowledge will be discussed within the context of definitional and methodologic difficulties encountered in studies of nonclinical populations.

1.1. The Prevalence of Alcohol Use and Abuse in the General Population

In this section we will describe separately the prevalence of use and abuse of alcohol. The methodologic problems of measurement and definition for both alcohol and drug use are very similar and will mainly be explored using alcohol as an example.

Alcohol consumption can be divided into three main components: frequency, quantity, and type of beverage (wine, beer, or liquor). Fre-

NANCY DAY • Western Psychiatric Institute & Clinic, 3811 O'Hara Street, Pittsburgh, Pennsylvania 15213.   KENNETH LEONARD • Western Psychiatric Institute & Clinic, 3811 O'Hara Street, Pittsburgh, Pennsylvania 15213.

quency is measured as a count of the number of times an event happens in a given time period, and as such, is a measure of whether drinking is part of a person's everyday life or a special event. Volume, amount per occasion, and in particular, maximum amount per occasion, are measures of whether the respondent ever drinks to levels of intoxication (Room, 1977). These data are also used as dosage levels, particularly in studies of the biological effects of alcohol consumption.

We know that individual reports of drinking behavior will underestimate the actual amount of drinking, though this bias is not constant across all drinking groups. There is a tendency for the respondent to answer in the most socially desirable manner, which means that heavy drinkers in particular, will be likely to report less drinking than is actually the case, whereas those who drink more moderately are more likely to report their drinking accurately (Jessor, Graves, Hanson, & Jessor, 1968). On balance, there is also some amount of overestimation in the braggadocio of the young male or the nostalgia of the elderly gent, although this certainly does not offset the underreporting.

The national study reported by Cahalan, Cisin, and Crossley (1969) documented the drinking patterns that were found in different demographic groups. As we should expect, religion, ethnicity, region of the country, age and sex, and social class are each important correlates of the type and amount of drinking. Each of these variables is in turn affected by the occupation and educational level of the individuals.

Estimates presented in the National Institute of Alcohol Abuse and Alcoholism's (NIAAA) *Fourth Special Report on Alcohol and Health* (DeLuca, 1981) show that apparent consumption of alcohol averages two drinks per day in the United States for each person 14 years and older. About one third of the population are abstainers, another third drink lightly (.01 to .02 ounces of ethanol per day), 24% drink more than this amount but less than 1 ounce of ethanol per day, and 9% drink more than 1 ounce (two drinks per day) on the average. Women drink less than men and young people more than older people. Drinking is more common in the upper social classes, but among drinkers, heavy drinking is more often found in the lower social status groups (Cahalan *et al.* 1969).

The issue of drinking problems has been variously defined. Cahalan *et al.* (1969) used negative social or health effects as a definition of drinking problems. The NIAAA's most recent *Report on Alcohol and Health* (DeLuca, 1981) has expanded this definition to include not just adverse social or health effects, but also symptoms of alcohol dependence and heavy alcohol consumption in their problem drinkers group. These factors are, of course, highly intercorrelated. Using the latter definition, it is estimated that approximately 10% of U.S. drinkers can be considered problem drinkers (Clark & Midanik, 1982). Earlier studies have shown that young men, particularly those of lower social status who live in large cities, are most likely to have drinking problems (Cahalan & Room, 1972).

The diagnosis of alcoholism is very difficult to make, even in a clinical setting, and, in fact, there is still considerable controversy about which criteria should be used, although the trend is away from the very specific formulation of the National Council on Alcoholism criteria (1972) to a broader definition of dependence. The American Psychiatric Association's *Diagnostic and Statistical Manual of Mental Disorders*, DSM-III (1980) defines a broader group of substance use disorders, separating pathological use into abuse (defined as "impairment in social or occupational functioning due to substance use" for at least one month) and dependence (defined by symptoms of tolerance or withdrawal). For alcohol and marijuana use, both dependence and abuse are required to define dependence. The International Classification of Diseases (ICD) has also dropped the term alcoholism, replacing it with the term alcohol dependence, which is defined as a narrowing of drinking behaviors, increased tolerance, drink-seeking behavior, withdrawal symptoms, compulsion to drink, and returning to drinking after abstinence (Edwards & Gross, 1976).

Keller (1975) and Coakley and Johnson (1978) have estimated that approximately 6% to 7% of the male population of the U.S. may meet criteria for alcohol dependence. Polich (1981) reported that 4.6% of U.S. Air Force personnel met criteria for alcohol dependence, which included tremors, morning drinking, inability to stop drinking, and blackouts. Using the Research Diagnostic Criteria (RDC) for alcoholism (Spitzer, Endicott, & Robins, 1978), Weissman, Myers, and Harding (1980) found that 6.7% of a sample of residents of the New Haven, Connecticut, population had or had ever suffered from alcoholism, and 2.6% met criteria currently. In agreement with other studies, they reported that alcoholism was more common in males, nonwhites, the unmarried, and those of lower social class.

1.2. Relationship of Alcohol Use and Abuse to Psychopathology

This review and analysis will concentrate on what is known about psychopathology as it correlates with alcohol and other drug use as seen in the general population. We have already noted the various alternative definitions of alcohol abuse and alcoholism in the general population. The same definitional problems must now be acknowledged with refer-

ence to psychopathology. Much as in the defining of alcoholism, researchers have used differing criteria, unstandardized instruments, and
varying definitions of psychopathology. Accordingly, we have very few
data that are either reliable or valid. Again, like alcoholism, psychopathology is very difficult to evaluate in the general population, and we
will most often find reports listing symptoms, much like one discusses
alcohol problems or consequences rather than alcoholism itself.

To untangle the concordance between symptoms of alcohol abuse
and symptoms of psychopathology requires a longitudinal study design
and the follow-up of patients over quite a long period of time. Robins
(1974) located individuals who, as children, had been seen at a child
guidance clinic. She found that those who subsequently became alcoholics had high rates of school dropout, more serious antisocial behavior,
and a higher incidence of family problems compared to the childhood of
those who did not display adult psychiatric problems. However, only a
minority of the children with these traits experienced problems with
alcohol as an adult. Thus while the relative risk for adult problems was
high, the actual probability of adult alcoholism problems, given childhood symptomatology, was low. An earlier study by Jones (1968) followed a sample of middle-class children to adulthood and found, similar
to Robins, that males who were adult problem drinkers, were described
as uncontrolled, assertive, hostile, and rebellious as children. Jones also
noted that these male children were often quite gregarious and more
masculine than their counterparts, presenting a self-contained, often
overconfident demeanor. The females displayed inadequate coping skills
and were seen as self-defeating, vulnerable, pessimistic, and withdrawn.
They had fluctuating moods and were often anxious and irritable (Jones,
1971).

Jones' findings agree very well with reports from McCord and Mc-
Cord (1960), the study described earlier by Robins (1974), and studies by
Jessor et al. (1968) and Zucker and DeVoe (1975). This high level of
agreement indicates that these personality traits appear to be rather
reliable predictors of future drinking problems, particularly since these
studies were concerned with very different social class groups from
different regions of the country.

Kammeier, Hoffmann, and Loper (1973) analyzed MMPI scores at
college matriculation of individuals who later presented for treatment of
alcoholism. Those who were future alcoholics were higher on the depression, hysteria, and psychopathologic deviate scales as well as those
of psychasthenia, schizophrenia, and social introversion.

Kandel, Kessler, and Margulies (1978) have shown that different
factors are related to the initiation of behaviors at each point in the

natural history of alcohol use and abuse. The most important predictors of beginning hard-liquor consumption in adolescence were characteristics of the person, such as use of tobacco, use of beer or wine, and minor forms of delinquency. Psychological traits were poor predictors of onset, but were important predictors of the transition from drinker to problem drinker. Studies by Zucker and Barron (1973), Zucker and Devoe (1975), and Jessor et al. (1968) have yielded very similar results. Young drinkers who moved from drinking to problem drinking were more deviant, impulsive, aggressive, and had higher rates of delinquency. Thus, we know that those individuals who are more troubled in their youth are at greater risk of developing alcohol problems as adults. However, it is important to realize that different factors may precede each phase in the transition from nondrinking to problem drinking. Thus, failure to carefully examine the component parts of the progression could lead to a blurring of important associations.

Similar findings have also been obtained in cross-sectional studies. Cahalan and Room (1972) found that those men in the general population who had a drinking problem were most likely to have come from broken homes, and were more impulsive, alienated, and tolerant of deviance. Bell, Schwab, Lehman, Traven, and Warheit (1977) found that heavy drinkers had the highest depression score, and a further report by the same group also found a significant correlation with suicidal ideation (Bell, Lin, Ice, & Bell, 1978). However, whether the suicidal ideation preceded or followed heavy drinking is unclear. Gorenstein (1979), using a college psychology class, demonstrated that depression and the psychopathic deviate scale of the MMPI were significantly associated with levels of consumption and with abusive drinking as measured by the Michigan Alcohol Screening Test (MAST).

Weissman et al. (1980) used a standardized psychiatric interview schedule, the Schedule for Affective Disorder and Schizophrenia (SADS-L) (Endicott & Spitzer, 1978), and RDC to reach a standard diagnosis for each psychiatric entity, on a general population sample of New Haven residents. Among the group diagnosed as alcoholic, 70% had been diagnosed as having at least one other psychiatric disorder during their lifetime. Most frequent of these other diagnoses were depression, depressive personality, minor depression, and drug abuse.

Data are presented below from a study of the Boston Metropolitan area begun in 1976. The study is described in detail elsewhere (Hingson, Scotch, Barrett, Goldman, & Mangione, 1981; Hingson, Scotch, Day, & Culbert, 1980). In this urban population, the rate of consuming three or more drinks per day was 14% for males and 3% for females.

A depression scale was constructed from a series of 11 items, such as "I feel I do not have much to be proud of" and "I certainly feel useless at times," that the respondent ranked on a four-point scale, from "strongly agree" to "strongly disagree." The cumulative total of answers was calculated for each subject, and those subjects who were in the upper 10% of the population on the dpression scale were selected. The relationship between drinking and depression is curvilinear or U-shaped (Table 1). The abstainers and heavy drinkers were most likely to be depressed, and the more moderate drinkers were less depressed. This trend was consistent through all age-sex groups, though the results did not always reach significance when the sample was stratified into small groups.

Measures of alienation, constructed in a similar way to measures of depression, were not significantly associated with the average daily volume of alcohol although, again, the relationship in the total population was U-shaped. There was considerably more variability in the age-sex groupings, however, as young male abstainers and young female heavy drinkers were significantly more alienated.

In a cross-sectional analysis, it is not possible to separate the sequence of associated events. Thus, we don't know whether depression or drinking occurred first in the heavy drinkers. We do, however, have information on whether respondents increased their consumption under certain circumstances. Participants were asked whether they drank more than usual "when you are happy . . . when you want to reward

Table 1
Depression and Average Daily Volume of Alcohol Consumption by Age and Sex[a]

| Current average daily volume | Total population[b] | Proportion in highest level of depression | | | | | |
| | | 17–29 | | 30–44 | | 45–64 | |
		Males	Females	Males	Females	Males	Females
Nondrinker in past year	14.5%	18.2%	15.0%	9.1%	15.4%	14.7%	14.0%
Less than three drinks per day	10.5%	11.1%	12.9%	7.8%	11.4%	9.2%	8.8%
Three or more drinks per day	18.4%	12.1%	33.3%	28.6%	21.4%	12.0%	13.3%
Total	12.0%	11.8%	14.1%	12.0%	12.6%	10.5%	10.3%
N	(196/1630)	(34/287)	(46/326)	(30/251)	(36/286)	(25/237)	(25/242)
P	.002	n.s.	.08	.0003	n.s.	n.s	n.s

[a]Data from Day, Leonard, & Lee, 1983.
[b]Data are unweighted and therefore do not represent the prevalence of use in the sampled population.

yourself for doing something well . . . when you are tense and nervous and want to relax . . . when you are frustrated with people close to you." They answered on a five-point scale from "never" to "very often."

As Table 2 shows, heavier drinkers consume more than their usual amount in response to all four of the occasions presented. Almost one-third of the heavier drinkers indicated that they drank more than usual when they were tense and nervous and wanted to relax, compared to 7.7% of all other drinkers and about 1% of those who drank rarely or not at all.

An even greater proportion of subjects who were both heavy drinkers and depressed drank more than usual when they were tense and nervous or frustrated (Table 3). Of the depressed/heavy drinkers, 57% reported consuming more than their usual amount when they were tense and nervous, compared to 24% of those who were heavy drinkers, but not depressed, and 13% of those who were depressed but not heavy drinkers. Similarly, in response to frustration, 46% of the depressed/heavy drinker group increased their drinking compared to 18% of those who were just heavy drinkers and 14% of those who were just depressed. Although the number who reported drinking more than usual was more dramatic among those who were already heavy drinkers, and especially among those heavy drinkers who were depressed, it was notable that depressed subjects who were not heavy drinkers also increased their consumption significantly in response to negative affect.

Table 2
Relationship between Mood State and Drinking Level[a]

Often or very often drink more than usual when:	Average daily volume[b]		
	Nondrinker in past year	Less than three drinks per day	Three or more drinks per day
Happy	0.4%	5.5%	22.9%
Want to reward yourself	0.4%	5.0%	16.2%
Tense and nervous and want to relax	0.9%	7.7%	30.0%
Frustrated with people close to you	1.3%	5.6%	21.9%

[a]Data from Day *et al.*, 1983.
[b]$p < .0001$.

Table 3
The Proportion of Drinkers Who Increase Drinking in Response to Mood State by Drinking Level and Depression

Often or very often increase drinking when:	Nonheavy drinkers[b]			Heavy drinkers		
	Not depressed[c]	Depressed	Odds ratio	Not depressed	Depressed	Ratio
Happy	5.3%	6.6%	1.2	20.9%	29.7%	1.6
Want reward	4.5%	6.6%	1.5	16.6%	18.9%	1.2
Tense and nervous	7.2%	13.1%	1.9[d]	24.5%	56.8%	4.0[d]
Frustrated	4.7%	13.9%	3.3[d]	17.8%	45.9%	3.9[d]

[a]Data from Day *et al.*, 1983.
[b]Drinking level was dichotomized as three more drinks per day and less than three drinks per day.
[c]Depression was dichotomized to represent the top 10% on the depressive symptomatology scale.
[d]Significant at $p < .05$.

When we compare the proportion of subjects who increase their drinking in each of the four situations, using those who are neither depressed nor heavy drinkers as the base, it is clear that there are independent and significant effects for both depression and drinking, though the increase related to drinking history seems to be larger (Table 4). Further, in the two more negative situations, frustrated and nervous or tense, the effect of being both depressed and a heavy drinker is additive.

Therefore, heavy drinkers significantly increased their drinking in response to both negative and positive stimuli in comparison to non-heavy drinkers. Among heavy drinkers and among non-heavy drinkers, those who were depressed significantly increased their drinking in re-

Table 4
Ratio of Heavy Drinking and Depressed Subjects Who Increase Drinking Relative to Those Who Are Either Not Depressed or Not Heavy Drinkers[a]

Often or very often increase drinking when:	Depressed, nonheavy drinker	Not depressed heavy drinker	Depressed heavy drinker
Happy	1.2	3.9	5.6
Want reward	1.5	3.7	4.2
Tense and nervous	1.8	3.4	7.9
Frustrated	2.9	3.8	9.8

[a]Data from Day *et al.*, 1983.

sponse to negative situations, though there was no significant difference with respect to happier events. Odds ratios presented in Table 3 indicate that heavy drinkers who were depressed were four times as likely to increase drinking in response to negative affect in comparison to heavy drinkers who were not depressed.

Those who were depressed were more likely to report that they currently or had ever had a drinking problem (Table 5). Of the subjects who drank heavily and were depressed, 73% reported ever having had a drinking problem, and 46% reported having one currently. In both cases, the odds ratios indicated a fourfold increase for those heavy drinkers who were depressed compared to heavy drinkers who were not depressed. Non-heavy drinkers who were depressed also reported higher rates of drinking problems. When respondents were asked about the effect of drinking on specific life areas such as work, marriage, finance, health, children, and friendships, heavy drinkers in general, and depressed/heavy drinkers in particular, reported significantly more negative effects. Of the depressed/heavy drinkers, 26% reported that two or more of these life areas were negatively affected by drinking, compared to 16% of other heavy drinkers and 5.7% of other depressed non-heavy drinker subjects. Friendships and children were the only two areas in which negative effects of depression were not found. Younger male drinkers and people who were depressed reported that drinking had a somewhat positive effect on friendships. For these groups it might be that drinking plays a useful role in creating and maintaining social networks.

Heavier drinkers were more likely to be male, but within each drinking group, there was little sex difference in rates of depression. The depressed/heavy drinkers were more likely to be blue-collar workers

Table 5
Respondents' Report of Drinking Problems by Heavy Drinking and Depression[a]

	Nonheavy drinker			Heavy drinker		
	Not depressed ($n = 1071$)	Depressed ($n = 124$)	Odds ratio[b]	Not depressed ($n = 164$)	Depressed ($n = 37$)	Odds ratio[b]
Current problem	1.6%	4.8%	3.1	18.7%	45.9%	3.6
Ever a problem	18.6%	29.6%	1.9	39.0%	73.0%	4.2

[a]Data from day *et al.*, 1983.
[b]Significant at $p < .05$.

(skilled craftsmen, security guards, military, or service workers) and least likely to be in the professional or managerial occupations. Almost 5% of these blue-collar workers fell into the heavy drinker and depressed category compared to 1% for the professionals. The depressed/ heavy drinkers were less likely to have attended or to have graduated from college (13.5%) compared to nondepressed/heavy drinkers (24%). The depressed/heavy drinkers were not less likely to be married.

Two traditional symptoms of drinking problems or alcoholism have been the inability to stop drinking once started and the desire for alcohol. In response to the question, "How hard is it for you to stop drinking once you have started," heavy drinkers found it somewhat or very hard to stop drinking once started, but for depressed/heavy drinkers this was a particular problem (Table 6). Of the depressed/heavy drinkers, 27% said that it was very hard to stop drinking once started, compared to 4.9% of the nondepressed/heavy drinkers. Among the nonheavy drinkers also, those who were depressed had a more difficult time stopping drinking once they had begun.

When they wanted to have a drink but couldn't, 23.4% of the depressed/heavy drinkers reported that they would be very upset, and almost half, 48.6%, said they would be somewhat or very upset (Table 7). Comparable rates for nondepressed/heavy drinkers are 9.2% and 23.3%, less than half that of the depressed/heavy drinker group. For nonheavy drinkers, those who were depressed reported more difficulty than those who were not.

Markers of potential alcoholism include heavy drinking, drinking problems, the inability to stop drinking, and a desire for alcohol. Each of

Table 6
Heavy Drinking, Depression, and Difficulty that Respondents Have Stopping Drinking Once They Have Started[a]

How hard is it to stop drinking once started?	Nonheavy drinkers[b]		Heavy drinkers[c]	
	Not depressed ($n = 1043$)	Depressed ($n = 117$)	Not depressed ($n = 163$)	Depressed ($n = 37$)
Not hard	77.1%	60.7%	49.7%	13.5%
A little hard	17.3%	26.5%	27.0%	32.4%
Somewhat hard	4.4%	10.3%	18.4%	27.0%
Very hard	1.2%	2.6%	4.9%	27.0%

[a]Data from Day et al., 1983.
[b]$\chi^2 = 17.06$; $p < 0.005$.
[c]$\chi^2 = 27.24$; $p < 0.005$.

Table 7

Heavy Drinking, Depression, and the Desire for Alcohol[a]

When you want to drink and can't, how upset are you?	Nonheavy drinkers[b]		Heavy drinkers[c]	
	Not depressed $n = 1043$	Depressed $n = 117$	Not depressed $n = 163$	Depressed $n = 37$
Not at all	68.5%	56.3%	39.3%	8.1%
A little	24.4%	29.7%	37.4%	43.2%
Somewhat	6.1%	7.6%	14.1%	24.3%
Very	1.0%	5.6%	9.2%	24.3%

[a]Data from Day et al., 1983.
[b]$\chi^2 = 16.8$; $p < 0.005$.
[c]$\chi^2 = 15.6$; $p < 0.005$.

these four factors was increased in heavy drinkers, among people who were depressed, and most dramatically, among those who were both heavy drinkers and depressed. Thus, those who are both heavy drinkers and depressed are significantly more likely to have drinking problems, to use alcohol for negative reasons, and to display symptoms of alcoholism, including difficulty stopping and a desire for alcohol.

Thus, from the literature we can demonstrate a relationship between heavy drinking, drinking problems, alcoholism, and psychopathology in the general population. Longitudinal studies indicate that psychological factors precede the development of drinking problems, and alcoholism and cross-sectional studies have repeatedly noted their association.

The data presented from the Boston study of urban drinking practices have substantiated this relationship and also point to the possibility of a spiraling kind of relationship. While heavy drinking is associated with higher levels of depressive symptomatology, the concordance of depression and heavy drinking is more highly predictive of drinking problems and symptoms of alcoholism than either heavy drinking or high levels of depressive symptoms alone. This would point to an additive or interactive relationship between these two factors and problems with alcohol, rather than a simple cause-and-effect model.

2. Drug Use and Abuse

Drug use measures, following the alcohol paradigm, are analyzed in terms of frequency, quantity, and maximum quantity, though often the dimensions of "ever used" versus "never used" and recency of use take on more importance than in alcohol measures.

The problem of accurate measurement, however, is even more difficult when we are measuring drug use. With the exception of prescribed drug use, most subjects are engaging in an illegal activity. In comparison to alcohol consumption, where the tendency is to downplay the amount but not the fact of use, in illicit drug use the social pressure is to deny any use.

The question of use, abuse, and disease states is not unique to the alcohol field, though perhaps more time has been spent on discussion and definition in this area. When we are talking about drugs, these definitions become even more problematic. In general, any use of a drug obtained illicitly is considered abuse, though there may be no adverse consequences. Authorized use of prescribed drugs, on the other hand, is not considered abuse even when there are negative sequelae.

The World Health Organization has proposed a framework for classifying drug and alcohol problems. In this scheme, abuse and misuse are segregated into four categories: (1) unsanctioned use, (2) hazardous use, (3) dysfunctional use or use which can lead to either physical or mental problems, and (4) harmful use which might lead to either physical or mental illness (Edwards, Arif, & Hodgson, 1982). This categorization has unfortunately not yet been applied in drug research, making a description or comparison of rates very difficult.

2.1. Marijuana Use in the General Population

In 1979, over 50 million people in the United States had tried marijuana at least once in their lives. (Relman, 1982). The use of marijuana has increased dramatically in the recent decade, moving marijuana from a deviant to a normative behavior in certain groups. In a 1971 national sample, only 14% of youths aged 12 to 17 and 40% of those aged 18 to 25 reported ever having used marijuana, in contrast to 31% and 68%, respectively, in 1979 (Fishburne, Abelson, & Cisin, 1980). Marijuana use varies by demographic factors; males use it more often than females, urban more than rural residents, and those who live in the Northeast and West more than those in the South. The same parameters predict heavy usage. Race does not seem to differentiate usage, as blacks reported rates similar to whites (Fishburne et al., 1980).

Regular use of marijuana is found in a much smaller number of people. In a sample of 17,000 high school seniors in the Monitoring the Future Project, 9% reported daily or nearly daily consumption (Johnston, Bachman, & O'Malley, 1980). Further data from this study indicate that the rate of daily use has stabilized but that this is occurring at progressively younger ages.

2.2. Psychopathological Correlates of Marijuana Use

Numerous researchers have looked at the personality correlates of marijuana use. Users of marijuana differ from nonusers on attributes that generally reflect nonconventionality, nontraditionality, and nonconformity. They tend to be more critical of society, to have a sense of alienation from and dissatisfaction with general social issues (Jessor, Jessor, & Finney, 1973; Weckowicz and Janssen, 1973), and a more tolerant attitude toward deviance (Jessor & Jessor, 1978; O'Donnell, Voss, Clayton, Slatin, & Room, 1976). Nonusers are more likely to be social conformers (Brook, Lukoff, & Whiteman, 1977) and to be more achievement-oriented (Jessor et al., 1973; Mellinger, Somers, Davidson, & Manheimer, 1976).

Little work has been done looking at the affective and mood states associated with marijuana use. Depressive mood has been reported as a predictor of subsequent marijuana use (National Commission of Marihuana and Drug Abuse, 1972; Paton, Kessler, & Kandel, 1977). Some investigators have reported that marijuana use is related to internal locus of control (Brook et al., 1977) while Jessor and Jessor (1977) found that it was correlated with external locus of control among high school males.

Halikas, Goodwin, and Guze (1972) looked at 100 marijuana users and 50 of their friends and found a high rate of psychopathology in both groups. However, the users were significantly higher in sociopathy. Of the diagnosed psychiatric illnesses, 75% preceded marijuana use, leading to the conclusion that marijuana use may be a symptom of pre-existing psychopathology. It is difficult to generalize from these findings, however, because participants were self-selected.

A similar study in Boston (Harmatz, Shader, & Salzman, 1972) found that psychiatric symptomatology was higher in marijuana users and higher again in users of multiple drugs. Women users were significantly more depressed. Linn (1972) used a symptom checklist with college students and found that students who scored high on the symptom scale were significantly more likely to be marijuana users.

Other researchers have found either equivocal or no correlations between marijuana and psychiatric symptomatology. Richek, Angle, McAdams, and D'Angelo (1975) looked at high school students in the southwest, finding a large difference between polydrug users and nonusers, but no differences between nonusers and those who smoked only marijuana. Hochman and Brill (1973), McAree, Steffhagen, and Zheutlin (1972), and Weckowicz and Janssen (1973) reported few differences between college students who used only marijuana and nonusers.

Thus, marijuana users, and especially heavy users of marijuana, are more likely to be alienated, nonconformist, and perhaps depressed. However, in spite of the rather dramatic reports of the early 1970s, psychopathology is relatively rare among users, and when present, probably predates the marijuana use. Whether the presence of antecedent psychopathology is linked with subsequent marijuana use is not clear.

2.3. Other Illicit Drugs: Prevalence of Use

2.3.1. Hallucinogens

Hallucinogen use, especially use of LSD, appears to be quite frequent. Recent estimates indicate that approximately 20%–25% of young adults aged 18–25 have used LSD at least once in their lifetime (Fishburne et al., 1980; Pope, Ionescu-Pioggia, & Cole, 1981). This lifetime prevalence rate is considerably higher than the rate observed in both younger and older cohorts. High school students have a lifetime prevalence rate of approximately 5%–10%, while older adults have a rate of less than 5% (Fishburne et al., 1980; Lipton, Stephens, Babst, Dembo, Diamond, Spielman, Schmeidler, Bergman, & Uppal, 1977). The lifetime prevalence of LSD use has increased for all age groups. The increase in older adults may be most simply interpreted as a reflection of the high lifetime prevalence of young adults during the early 1970s. However, the increased lifetime prevalence of use among high school individuals indicates that LSD, and probably other hallucinogens as well, are being experimented with at a progressively younger age.

Frequently, studies have examined the recent use of LSD, defined most commonly as any use within the past year. Abelson, Cohen, Schrayer, and Rappeport (1973) reported rates of recent use as 2.6% for adults aged 18 and over and 4% for youths aged 12 to 17. Fishburne et al. (1980) found a similar rate for youths of 4.7%. The rates of recent use among the young adults (18 to 25) and older adults (26+) were 9.9% and 0.6% respectively. Other data reported by Fishburne et al. (1980) suggest that, like lifetime prevalence, recent use is on the increase among young adults. However, this is not true for older adults, suggesting that LSD is first tried during young adulthood, used relatively infrequently, and given up in later years. Support for this hypothesis is provided by Davidson, Mellinger, and Manheimer (1977). Of the 8% of their college sample who were recent users of LSD (previous 6 months) during their freshman year, only 40% were still recent users during their junior year. Fishburne *et al.* (1980) have reported other data suggestive of a relatively short period of usage. For young adults, a sizeable minority of users had used LSD only once or twice, and approximately two-thirds of users had used it ten times or less. Thus, LSD does not appear to be a drug of long-term heavy use in the general population. McGlothlin (1974) suggests that absence of dependence, rapid development of tolerance, inconsistency of effects, and diminished novelty of the drug with increased experience contribute to this pattern of use.

Males are somewhat more likely to have used LSD and considerably more likely to have used it heavily (McGlothlin, 1974). Unlike marijuana and other illicit drugs, LSD is mainly used by whites and those in the upper social strata, though at least one study (O'Donnell *et al.*, 1976) reported slightly higher rates for blacks and lower socioeconomic groups.

2.3.2. Cocaine

Cocaine, a strong, short-lasting stimulant derived from the leaves of the cocoa plant, is currently a very popular drug. Fishburne *et al.* (1980) reported a lifetime prevalence rate of 27.5% for adults aged 18 to 26, while Pope *et al.* (1981) indicated that 30% of college students in 1978 had used cocaine at least once. Recent use is also quite high, with Fishburne *et al.* (1980) finding a rate of nearly 20%. Approximately 60% of users take the drug less than ten times (Fishburne *et al.*, 1980; O'Donnell *et al.*, 1976).

The popularity of cocaine is a relatively recent phenomenon. During the late 1960s and early 1970s, use of cocaine was relatively low. Fishburne *et al.* (1980) reported the lifetime prevalence rate for young adults in 1972 as 9.1%. Pope *et al.* (1981) reported the 1969 rate to be 5%. Thus, the lifetime prevalence rate has at least tripled among young adults (9.1% in 1972 to 27.5% in 1979, according to Fishburne *et al.*, 1980) and may have increased even more than that for some populations (5% in 1969 to 30% in 1978 for college students, according to Pope *et al.*, 1981).

The large general population studies have not reported separate rates of cocaine use for males and females. Thus, it is not known whether there is a sex difference in usage, though as with other illicit drugs, male users probably outnumber female users. O'Donnell *et al.* (1976)

note higher lifetime prevalence and recent usage for blacks than for whites. However, this was true only for older subjects, raising the possibility that black and white rates of cocaine use are converging.

2.3.3. Heroin

Of all the drugs considered, heroin is the least commonly used, perhaps because of a fear of dependency or of other negative effects, or because of the expense or typical method of administration (O'Donnell *et al.*, 1976). For young adults, the lifetime prevalence has been reported to be about 2% to 4% (Davidson *et al.*, 1977; Fishburne *et al.*, 1980; Kopplin, Greenfield, & Wong, 1977; Parry, 1979; Pope *et al.*, 1981). The rates for youth and older adults are usually reported as 1% or less (Abelson *et al.*, 1973; Fishburne *et al.*, 1980).

Recent use of heroin is considerably less than the lifetime prevalence rate. This is, of course, true for all drugs, but is noteworthy considering the assumed addictive potential of heroin. Fishburne *et al.* (1980) reported a recent use rate of about .7%, compared to a lifetime prevalence of 3.5%. Thus, only 20% of heroin users had taken the drug within the preceding year. Similarly, O'Donnell *et al.* (1976) found that 6% of their sample of males had tried heroin, but only 1%–2% had used it within the past year, a recent usage rate of less than 30% for previous users. This suggests that many heroin users refrain from heroin use for considerable periods of time. This interpretation is strengthened by the findings of Robins and Murphy (1967) that, among 22 black addicts, only 3, or 14%, had used heroin in the preceding year.

Over time, the use of heroin appears to have remained relatively stable, despite increases in other drug use. Among young adults in the study by Fishburne *et al.* (1980), lifetime prevalence declined from 4.6% in 1972 to 3.5% in 1979. Similarly, neither Kopplin *et al.* (1977) nor Pope *et al.* (1981) found any increase in heroin use.

Again, as with cocaine usage, none of the general population surveys reported any sex differences in heroin use. The O'Donnell *et al.* (1976) study, however, was conducted with an all-male population; the lifetime prevalence reported by these researchers was 6%, considerably higher than the rates reported for both sexes combined. Thus, it is likely that male use is higher than female use. Racial differences in the use of heroin are considerable. O'Donnell *et al.* (1976) found that 14% of black males, but only 5% of white males, had ever used heroin; recent use of heroin by blacks was 4%, while for whites it was only 1%. This racial difference, as with cocaine, was only apparent among the somewhat older cohort (aged 27–30 at the time of the study). Among the younger cohort (aged 20–21), whites admitted to a higher lifetime prevalence (7%) than blacks (2%). Again, this suggests the possibility of converging rates for blacks and whites with time.

2.4. Psychopathology and Illicit Drug Use

In general, the association between psychopathology and the use of LSD, cocaine, or heroin has not received systematic inquiry. The most sophisticated and systematic research in this area has been conducted by Kandel and her associates (Kandel et al., 1978; Paton et al., 1977). Paton et al. (1977) collected longitudinal data at a 6-month interval on over 5,000 adolescents. During this interval, some individuals began drug use beyond marijuana and alcohol. Thus, it was possible to examine the characteristics of multiple drug users prior to the initiation of their multiple drug use. In addition, some multiple drug users had ceased usage by the second data collection and their characteristics could be compared to multiple drug users who continued their usage. The authors were able to show, using cross-lagged analyses, that depression at Time 1 was related to beginning the use of multiple drugs by Time 2, suggesting that depression may facilitate initiation to drugs stronger than marijuana and alcohol.

2.5. Psychotherapeutic Drug Use

While the use of most of the psychoactive substances discussed thus far is considered illicit and disreputable, certain drugs are condoned by society when obtained and utilized in a specified fashion. These drugs, referred to as psychotherapeutics, include the categories of stimulants, sedatives, hypnotics, major and minor tranqilizers, antidepressants, and antipsychotics. They are typically employed to alleviate symptoms of distress and psychiatric disorders, such as anxiety, tension, fatigue, depression, and sleep difficulties as well as to treat some limited physical disorders, such as obesity and gastrointestinal problems.

Psychotherapeutic drugs may be obtained in mild forms over-the-counter or, in the more potent prescription forms, from a medical or nonmedical source. Regardless of the source, the psychotherapeutics may be used in an appropriate medical, a quasimedical, or a nonmedical (experimental or recreational) fashion. The prevalence of psychotherapeutic drug use and the characteristics of those using these drugs vary as a function of four factors: type of drug, regularity of use, source of acquisition, and type of use.

Recently, a number of studies have examined the epidemiology of psychotherapeutic drug use. These studies vary greatly with respect to their attention to the different patterns of psychotherapeutic drug use, and the classification of drugs is frequently different from one study to the next. As a result, specific prevalence rates are often difficult to ascertain, even though certain generalizations are possible. In the following section, the epidemiology of stimulants, sedatives/hypnotics, and minor tranquilizers will be presented. These three drug classes were chosen since, of all the psychotherapeutics, these are the most widely used in both a medical and nonmedical fashion. The other drug classes are used infrequently in the general population and are almost always used medically and obtained through a physician's prescription.

2.5.1. Stimulants

Stimulants, such as amphetamines or over-the-counter caffeine substances (e.g., No-Doz) are most typically used to combat fatigue or to control appetite, most commonly by young adults. Mellinger, Balter, and Manheimer (1971) estimated the 1-year prevalence of stimulant use to be nearly 25% in the 18–29-year-old age group, but less than 10% in other age groups. These estimates included both ethicals (prescription) and proprietaries (over-the-counter). Most other estimates are restricted in one fashion or another and are, therefore, lower than this prevalence rate. Nevertheless, the pattern remains consistent, with much higher prevalence rates among adults 18–30 years of age, and with rapidly diminishing rates in older adults (Abelson et al., 1973; Fishburne et al., 1980; Parry, Balter, Mellinger, Cisin, & Manheimer, 1973).

The pattern of stimulant use varies considerably as a function of the respondent's sex. Males are somewhat more likely to utilize over-the-counter stimulants than are females (Mellinger et al., 1971); however, this is true mostly for young adults (Parry et al., 1973). In the age group of 18–29 years old, 17% of the men but only 8% of the women had used proprietary stimulants in the past year. In older age groups, approximately 2% of the men and less than 1% of the women used this type of drug.

With prescription stimulants, the situation is reversed. Females are more likely to have used an ethical stimulant in the past year than are males (Abelson et al., 1973; Mellinger et al., 1971; Parry et al., 1973). It is unclear whether this finding is true at all ages because Parry et al. (1973) reported the most pronounced sex difference for younger adults but Mellinger et al. (1971) reported the largest sex difference for older adults.

Although females are more likely to use prescription stimulants than males, males are more likely to acquire prescription stimulants from nonmedical sources (Mellinger *et al.*, 1971) and to use these drugs in a nonmedical fashion (Abelson *et al.*, 1973). Only 30% of young adult male users, compared to nearly 75% of young adult female users, obtained their prescription stimulants from medical sources (Abelson *et al.*, 1973; Mellinger *et al.*, 1971).

In summary, the use of stimulants, irrespective of sex of the user, source of the drug, or type of use, is considerably higher among young adults than among older adults. Males, more so than females, are prone to obtain the drug from a nonmedical source and to use it in a nonmedical fashion.

2.5.2. Sedatives

The class of sedatives/hypnotics consists of both long- and short-acting barbituates as well as a few nonbarbituate drugs prescribed for sedation or sleep. Few studies provide separate data for sedatives and hypnotics; the most common procedure has been to combine these two classes.

The usage pattern for the sedative/hypnotic class of drugs is somewhat different from the pattern for the stimulant class. Abelson *et al.* (1973) reported that 20% of adults and only 4% of youths had used sedatives at least once. For college students, the lifetime prevalence as reported by Davidson *et al.* (1977) was 15%. Abelson *et al.* (1973) estimated 11% of the population had ingested a prescription sedative or hypnotic in the preceding year, and 10% had taken a proprietary of this type.

The data are somewhat discrepant with respect to the use of prescription sedatives and age. Abelson *et al.* (1973) reported a recent use rate of 15% for young adults 18 to 25 years of age and approximately 10% for adults older than 25. Mellinger *et al.* (1971) separated drug use into sedative and hypnotic drug classes and reported a somewhat higher rate of use among older adults for both of these drug classes. Parry *et al.* (1973) found that the rate of recent use of hypnotics increased with age. One way to reconcile these data is to assume that older adults have used both sedatives and hypnotics, while young adults used either one or the other, but not both.

As with stimulants, nonmedical use of prescription sedatives is primarily admitted to by the younger age groups (Abelson *et al.*, 1973; Fishburne *et al.*, 1980). Additionally, the younger age groups are more likely to have obtained their sedative/hypnotic from a nonmedical source (Abelson *et al.*, 1973) and slightly more likely to have used an over-the-counter sedative/hypnotic (sleeping pills) (Parry *et al.*, 1973).

Sex differences in the use of sedatives tend to be weaker than the differences observed in stimulant use. Unlike over-the-counter stimulants, which are used predominantly by males, over-the-counter sedatives are equally likely to have been used by males or females, irrespective of age. However, females are more likely to use prescription sedatives or hypnotics than are males (Parry *et al.*, 1973; Mellinger *et al.*, 1971).

As with stimulants, males are somewhat more likely to have obtained a prescription sedative from a nonmedical source and to have used it in a nonmedical fashion. The difference is not strong for recent use, with 4% of the males and 3% of the females admitting to nonmedical use in the preceding year (Abelson *et al.*, 1973). However, Fishburne *et al.* (1980) note a sizeable difference, especially for young adults, finding that 22% of the young adult males but only 12% of the young adult females have ever used a sedative in a nonmedical fashion. With respect to the source of the prescription sedatives, Abelson *et al.* (1973) estimated that only 68% of the male users, but 82% of the female users, had received their first prescription sedative through a physician. Although males were less likely to obtain sedatives through a physician than females, in fact, both males and females tended to use this source.

2.5.3. Minor Tranquilizers

The most common drugs in this class are the benzodiazepines, such as Valium (diazepam), Librium (chlordiazepoxide), and Serax (oxazepam). The primary use of these drugs is for the alleviation of anxiety or tension.

The use of tranquilizers is fairly common. Abelson *et al.* (1973) report that the lifetime prevalence rates of prescription tranquilizers for adults and youth were 24% and 6%, respectively. These researchers also estimated that 17% had taken a prescription tranquilizer in the past year, but only 5% had used an over-the-counter tranquilizer. The recent use rate reported by Mellinger *et al.* (1971) was considerably less, with 7.5% of males and 12.7% of females acknowledging use of a prescription tranquilizer and 2.6% and 3.8% using a proprietary tranquilizer.

It is difficult to ascertain age trends with respect to the recent use of prescription tranquilizers. However, there does appear to be a tendency for middle-age adults to manifest the highest rate. For adults aged 35–49, Abelson *et al.* (1973) reported a rate of 23%, while for all other ages the rate was approximately 15%. Mellinger *et al.* (1971) found similar results for males, with the highest rate (10%) in the 30–44 age group. In contrast, the highest rate for females (15%) was in the 18–29 age group, and the rates declined steadily with age.

Age trends for nonmedical use of tranquilizers are quite clear and mirror the nonmedical use of stimulants and sedatives/hypnotics. Young adults are more likely both to use prescription tranquilizers in a nonmedical fashion (Abelson *et al.*, 1973; Fishburne *et al.*, 1980) and to obtain their tranquilizers from a nonmedical source (Abelson *et al.*, 1973). Also, young adults have a higher rate of over-the-counter tranquilizer use than older adults (Parry *et al.*, 1973).

Consistent with the findings for stimulants and sedatives/hypnotics, prescription use of tranquilizers is more common in women than in men. As noted earlier, Mellinger *et al.* (1971) found a recent use rate of 12.7% for females and 7.5% for males. Much higher rates were reported by Abelson *et al.* (1973), but the sex difference remains. These authors found that 22% of females and 12% of males had used a prescription tranquilizer in the preceding year. However, unlike stimulants and sedatives/hypnotics, proprietary tranquilizers were also used by females more than by males (Mellinger *et al.*, 1971; Parry *et al.*, 1973), with the most pronounced sex difference occurring among young adults (Parry *et al.*, 1973).

Data concerning the nonmedical utilization of prescription tranquilizers are somewhat contradictory. Fishburne *et al.* (1980) reported that males had a higher lifetime prevalence of nonmedical use of these drugs than did females. Among younger adults, 20% of the males and 11% of the females used prescription tranquilizers in a nonmedical fashion. For older adults, the comparable proportions were 4% of the males and 2% of the females. In contrast, Abelson *et al.* (1973) found that 4% of the males and 7% of the females admitted to recent nonmedical use. Obviously, this discrepancy could be due to several factors, including the years of the studies, the different samples, the different definitions of nonmedical use, or the different measures of prevalence (lifetime versus recent). Finally, males were more likely than females to obtain their first prescription tranquilizer from a nonmedical source, though again, both males (70%) and females (88%) did tend to utilize a medical source (Abelson *et al.*, 1973).

Several trends are apparent from the above review of psychotherapeutic drug use patterns. For stimulants, sedatives/hypnotics, and tranquilizers, females have a higher rate of apparently appropriate use of prescription drugs, while males tend to have a higher rate of nonmedical use, and tend to obtain drugs through nonprescription sources such as over-the-counter outlets or to obtain prescription drugs from a nonmedical source. With respect to age, it appears that nonmedical use and over-the-counter use of any of the psychotherapeutics occurs primarily among young adults. However, for medical and prescription use, the predominant drug is age dependent, with stimulants being used by young adults, tranquilizers being used primarily by middle-aged adults, and sedatives/hypnotics being somewhat more common among older adults. Finally, very few data are available concerning racial or socioeconomic differences in psychotherapeutic drug use.

2.6. Psychopathology and Psychotherapeutic Drug Use

The investigation of the relationship between psychopathology and psychotherapeutic drug use is confronted with one major and rather obvious problem. These drugs, especially the sedative/hypnotics and the minor tranquilizers, are frequently prescribed for complaints such as anxiety, depression, or tension, which are common symptoms of psychopathology. Thus, there is a relationship between use of psychotherapeutics and psychopathology due to the prescribing practices of physicians. However, our concern is not the relationship between psychopathology and the physician's decision to prescribe psychotherapeutics. Rather, it is the role of psychopathology in the individual's decision to use a psychotherapeutic drug.

On the surface, it would appear that the problem (i.e., physicians' prescription practices contaminating the relationships of interest) would be alleviated by simply restricting our focus to the nonmedical use of prescription psychotherapeutics. Unfortunately, this is not so. The distinction between medical and nonmedical use of drugs is often difficult to draw. For example, consider the individual who utilizes an unfinished prescription when several of his symptoms reemerge at a much later time. Additionally, a physician's decision to prescribe a psychotherapeutic drug is, at least sometimes, the result of the dynamic interaction between the physician and the patient. The role of the patient's psychopathology in obtaining medically sanctioned drugs is an issue that needs to be addressed.

These are difficult issues to deal with—explaining why so little research has been undertaken in this area. The few studies that have been done have focused on recent use of any prescription psychotherapeutic without regard to type of drug, type of use, or source of acquisition. The results of these studies, however, are consistent with the hypothesis that drug use and psychopathology are related.

Mellinger, Balter, Manheimer, Cisin, and Parry (1978) examine the relationship between life crises, symptomatology, and the use of psychotherapeutic medications. In a national probability sample, subjects were administered several scales, including a shortened version of the Hopkins Symptom Checklist (SCL), a modified version of the Holmes and Rahe life crisis questionnaire, and a questionnaire measuring alcohol and prescription drug use. Of those subjects scoring in the high range of psychic distress, 30% had used a psychotherapeutic (mostly benzodiazepines) in the preceding year and 12% had used them on a regular basis, defined as daily or almost daily use of the same drug for 2 months or more. In contrast, 8% and 16% of the low- and moderate-distress groups had utilized a psychotherapeutic, with 3% and 4%, respectively, being regular users. This relationship is true for both males and females, though females were more likely to utilize at least one psychotherapeutic than were males at the same distress level. In other words, controlling for the fact that more women reported high levels of distress than did men, more women used a psychotherapeutic drug than did men. A final analysis of just those in the high-distress group suggested that this sex difference in usage occurred primarily in younger adults (aged 18 to 29) and could be attributed to a very low usage rate of 7% in the high-distress males of this age group. Interestingly, many of the members in this latter group (85%) reported moderate to heavy consumption of alcohol, while few of the young high-distress females (8%) reported this level of alcohol use. The appropriate comparisons are not presented, but these data suggest that men, more so than women, appear to choose alcohol as a medicative response to high levels of distress.

There are three other studies of psychotherapeutic drug use and symptoms of psychopathology, each with a quite different focus. Craig and Van Natta (1978) investigated drug intake in the preceding 48 hours and self-reported depression; they reported an association between high levels of depression and the use of four or more drugs. Guttmann (1978) examined drug use in the elderly and discovered that users of psychotropic medication reported lower levels of general life satisfaction. Finally, Uhlenhuth, Balter, and Lipman (1978) attempted to differentiate between different sorts of health problems and minor tranquilizer use. Psychological problems were most associated with tranquilizer use, followed closely by psychosomatic problems.

The above studies tend to support the hypothesis of an association between psychotherapeutic drug use and psychopathology. However, it is clear that there is still much research to be conducted. For example, the above studies do not address whether the observed relationship is simply a function of physicians' prescription practices, nor do they always differentiate drug use with respect to the type of drug. The studies also have not addressed the possibility that psychotherapeutic drug use may influence psychopathology in a detrimental fashion.

3. Summary

There is such a paucity of available literature concerning the relationship between psychopathology and substance abuse in general population samples that drawing substantive conclusions would be misleading. However, it might be useful to pinpoint several issues that need to be addressed in future research.

One of the major issues to be clarified in future studies involves the clarification of substance use variables. Research examining alcohol has received considerable attention in this regard. While a variety of alcohol use measures have been utilized, several conceptually distinct measures have gained some acceptance. Adequate studies currently collect information on current use (abstention versus drinking), regularity and typical amounts of use (quantity-frequency measures), drinking problems, and symptoms of dependence. It is not generally expected, nor should it be, that all of these variables would be related to psychopathology. Indeed, the concern has focused on the variables of excessive use and dependence. Similarly, studies of drug use have sometimes distinguished between lifetime prevalence and current prevalence, age at first use, as well as the current frequency and quantity of use. However, these distinctions have not as a rule been applied to studies of psychopathology and substance abuse. Finally, with respect to the use of psychotherapeutic drugs, though conceptually distinct variables have been utilized that distinguish between different patterns of usage, the practical assessment of these variables is problematic. For example, as mentioned earlier, medical and quasi-medical use of psychotherapeutics are in certain cases difficult to distinguish from each other.

While we must be concerned with distinguishing patterns of usage, we must also be aware that some drugs may be functionally equivalent. This has been a tacit assumption of studies examining multiple drug usage, implying that individuals who progress beyond the more socially acceptable drugs may have certain commonalities, irrespective of which specific drug they choose. The issue is also germane to the study of distress and psychotherapeutic drug and alcohol use reported by Mellinger *et al.* (1978). It will be recalled that high-distress males admitted to high levels of alcohol use, but low levels of psychotherapeutic drug use, while high-distress females evidenced the reverse pattern. This would suggest that alcohol and psychotherapeutic drugs are used by males and females, respectively, for the same purpose—the relief of high levels of distress. Additionally, the possibility of functional equivalence of patterns of usage of the same drug must be recognized. For example, an older woman with a sleep problem may visit her private physician and receive a prescription hypnotic that she utilizes in a manner consistent with medical practice. A younger man with precisely the same problem may obtain the same drug from a nonmedical source and use it in a less medically appropriate fashion. Thus the same symptom may result in different patterns of drug use.

The measurement of psychopathology also requires some clarification. Much of the research has examined either depressive symptomatology or antisocial attitudes or behavior. While these may be indicative of a diagnosable, psychopathological disorder, they are certainly not the same. From a theoretical as well as a practical perspective, it is important to know whether symptoms of a specific nature (e.g., physical, emotional, cognitive, or interpersonal) predict increased substance use or whether only a specific disorder results in such increases.

Even with clarification of substance use and psychopathology variables, the predictive validity of these factors may be relatively weak. As was noted earlier with respect to the Robins (1974) study, there is a low absolute probability that childhood symptoms of disorder result in alcoholism, despite the higher prevalence of these factors noted in alcohol and drug abusers. Clearly, whether psychopathology in childhood results in alcoholism is conditional upon other factors. Similarly, factors noted as strong correlates of abuse in clinical populations may not have the same utility in nonclinical populations. Assessment of these other factors and their effect at each stage of the natural history is critical if we are to better understand the meaning of the relationship between psychopathology and substance abuse.

4. References

Abelson, H., Cohen, R., Schrayer, D., & Rappeport, M. Drug experience, attitudes, and related behavior among adolescents and adults. In *Drug use in America: Problem in perspective. The technical papers of the second report of the National Commission on Marihuana and Drug Abuse. Vol. 1: Patterns and consequences of drug use.* Washington, D.C.: U.S. Government Printing Office, 1973.

American Psychiatric Association, *Diagnostic and statistical manual of mental disorders* (3rd ed.). Washington, D.C.: American Psychiatric Association, 1980.

Bell, R., Schwab, J., Lehman, R., Traven, N., & Warheit, G. An analysis of drinking patterns, social class and racial differences in depressive symptomatology: A community study. In F. Seixas (Ed.), *Currents in Alcoholism IV*. New York: Grune & Stratton, 1977.

Bell, R., Lin, E., Ice, J., & Bell, R. Drinking patterns and suicidal ideation and behavior in a general population. In M. Galanter (Ed.), *Currents in Alcoholism V*. New York: Grune & Stratton, 1978.

Brook, J. S., Lukoff, I. F., & Whiteman, M. Correlates of adolescent marijuana use as related to age, sex, and ethnicity. *Yale Journal of Biology and Medicine*, 1977, *50*, 383–390.

Cahalan, D., & Room, R. Problem drinking among American men aged 21–59. *American Journal of Public Health*, 1972, *62*, 1473–1482.

Cahalan, D., Cisin, I., & Crossley, H. *American drinking practices* (Monograph No. 6). New Brunswick, N.J.: Rutgers Center of Alcohol Studies, 1969.

Clark, W. B., & Midanik, L. Alcohol use and alcohol problems among U.S. adults. In *Alcohol consumption and related problems*. (Alcohol and Health Monograph No. 1). Rockville, Md.: NIAAA, 1982.

Coakley, J., & Johnson, S. *Alcohol abuse and alcoholism in the United States: Selected recent prevalence estimates* (Working Paper 1). Washington, D.C.: NIAAA, 1978.

Craig, T. J., & Van Natta, P. A. Current medication use and symptoms of depression in a general population. *American Journal of Psychiatry*, 1978, *135*, 1036–1039.

Davidson, S. T., Mellinger, G. D., & Manheimer, D. I. Changing patterns of drug use among university males. *Addictive Disease: An International Journal*, 1977, *3*, 215–234.

Day, N., Leonard, K., & Lee, S. W. *Alcohol use and psychiatric symptomatology in a general population*. Unpublished manuscript, 1983.

DeLuca, J. R. (Ed.), *Fourth special report to the U.S. Congress on alcohol and health*. Washington, D.C.: NIAAA, U.S. Government Printing Office, 1976.

Edwards, G., & Gross, M. Alcohol dependence: Provisional description of a clinical syndrome. *British Medical Journal*, 1976, *1*, 1058–1061.

Edwards, G., Arif, A., & Hodgson, R. Nomenclature and classification of drug- and alcohol-related problems: A shortened version of a WHO memorandum. *British Journal of Addiction*, 1982, *77*, 3–20.

Endicott, J., & Spitzer, R. L. A diagnostic interview: The Schedule for Affective Disorders and Schizophrenia. *Archives of General Psychiatry*, 1978, *35*, 837–844.

Fishburne, P., Abelson, H., & Cisin, I. *National survey on drug abuse: Main findings: 1979* (DHHS Publication No. (ADM) 80–976). Rockville, MD.: National Institute on Drug Abuse, 1980.

Gorenstein, E. Relationships of subclinical depression, psychopathy and hysteria to patterns of alcohol consumption and abuse in males and females. In M. Galanter (Ed.), *Currents in Alcoholism VII*, New York: Grune & Stratton, 1979.

Guttmann, D. Patterns of legal drug use by older Americans. *Addictive Diseases: An International Journal*, 1978, *3*, 337–356.

Halikas, J., Goodwin, D., & Guze, S. Marihuana use and psychiatric illness. *Archives of General Psychiatry*, 1972, *27*, 162–165.

Harmatz, J., Shader, R., & Salzman, C. Marihuana users and nonusers. *Archives of General Psychiatry*, 1972, *26*, 108–112.

Hingson, R., Scotch, N., Day, N., & Culbert, A. Recognizing and seeking help for drinking problems. A study in the Boston metropolitan area. *Journal of Studies on Alcohol*, 1980, *41*, 1102–1117.

Hingson, R., Scotch, N., Barrett, J., Goldman, E., & Mangione, T. Life satisfaction and drinking practices in the Boston metropolitan area. *Journal of Studies on Alcohol*, 1981, *41*, 24–37.

Hochman, J. S., & Brill, N. Q. Chronic marijuana use and psychosocial adaptation. *American Journal of Psychiatry*, 1973, *130*, 132–140.

Jessor, R., & Jessor, S. *Problem behavior and psychosocial development: A longitudinal study of youth.* New York: Academic Press, 1977.

Jessor, R., & Jessor, S. Theory testing in longitudinal research on marijuana use. In D. B. Kandel (Ed.), *Longitudinal research on drug use: Empirical findings and methodological issues.* Washington, D.C.: Hemisphere (Halstead-Wiley), 1978.

Jessor, R., Graves, T., Hanson, R., & Jessor, S. *Society personality, and deviant behavior.* New York: Holt, Rinehart and Winston, 1968.

Jessor, R., Jessor, S. L., & Finney, J. A social psychology of marijuana use: Longitudinal studies of high school and college youth. *Journal of Personality and Social Psychology*, 1973, *26*, 1–15.

Johnston, L. D., Bachman, J. G., & O'Malley, P. M. *Drug use among high school students 1975–1977.* National Institute on Drug Abuse. Washington, D.C.: U.S. Government Printing Office, 1977.

Johnston, L. D., Bachman, J. G., & O'Malley, P. M. *Highlights from Student Drug Use in America, 1975–1980.* (DHHS Publication No. (ADM) 81-1066). Washington, D.C.: U.S. Government Printing Office, 1980.

Jones, M. C. Personality correlates and antecedents of drinking patterns in adult males. *Journal of Consulting and Clinical Psychology*, 1968, *1*, 2–12.

Jones, M. C. Personality antecedents and correlates of drinking patterns in women. *Journal of Consulting and Clinical Psychology*, 1971, *36*, 61–69.

Kammeier, M., Hoffmann, H., & Loper, R. Personality characteristics of alcoholics as college freshmen and at time of treatment. *Quarterly Journal of Studies on Alcohol*, 1973, *34*, 390–399.

Kandel, D. B., Kessler, R. C., & Margulies, R. Z. Antecedents of adolescent initiation into stages of drug use: A developmental analysis. In D. B. Kandel (Ed.), *Longitudinal research on drug use: Empirical findings and methdological issues.* Washington, D.C.: Hemisphere (Halstead-Wiley), 1978.

Keller, M. Problems of epidemiology in alcohol problems. *Journal of Studies on Alcohol*, 1975, *36*, 1442–1451.

Kopplin, D. A., Greenfield, T. K., & Wong, H. Z. Changing patterns of substance use on campus: A four-year follow-up study. *International Journal of Addictions*, 1977, *12*, 73–94.

Linn, L. Psychopathology and experience with marihuana. *British Journal of Addiction*, 1972, *67*, 55–64.

Lipton, D. S., Stephens, R. C., Babst, D. V., Dembo, R., Diamond, S. C., Spielman, C. R., Schmeidler, J., Bergman, P. J., & Uppal, G. S. A survey of substance use among junior and senior high school students in New York State, Winter 1974–75. *American Journal of Drug and Alcohol Abuse*, 1977, *4*, 153–164.

McAree, C. P., Steffhagen, R. A., & Zheutlin, L. S. Personality factors and patterns of drug usage in college students. *American Journal of Psychiatry*, 1972, *128*, 890–893.

McCord, W., & McCord, J. *Origins of alcoholism.* Stanford, Calif.: Stanford University Press, 1960.

McGlothlin, W. H. The epidemiology of hallucinogenic drug use. In E. Josephson & E. E. Carroll (Eds.), *Drug use: Epidemiological and sociological approaches.* New York: John Wiley and Sons, 1974.

Mellinger, G. D., Balter, M. B., & Manheimer, D. I. Patterns of psychotherapeutic drug use among adults in San Francisco. *Archives of General Psychiatry*, 1971, *25*, 385–394.

42   NANCY DAY AND KENNETH LEONARD

Mellinger, G. D., Somers, R. H., Davidson, S. T., & Manheimer, D. I. The amotivational syndrome and the college student. *Annals of the New York Academy of Science*, 1976, *282*, 37–55.

Mellinger, G. D., Balter, M. B., Manheimer, D. I., Cisin, I. H., & Parry, H. J. Psychic distress, life crisis, and use of psychotherapeutic medications. *Archives of General Psychiatry*, 1978, *35*, 1045–1052.

National Commission on Marihuana and Drug Abuse. *Marihuana: Signal of misunderstanding*. Appendix: Volume I. Washington, D.C.: U.S. Government Printing Office, 1972.

National Council on Alcoholism. Criteria for the diagnosis of alcoholism. *American Journal of Psychiatry*, 1972, *129*, 127–135.

O'Donnell, J. A., Voss, H. L., Clayton, R. R., Slatin, G. T., & Room, R. *Young men and drugs: A nationwide survey*. NIDA Research Monograph No. 5. Washington, D.C.: U.S. Government Printing Office, 1976.

Parry, H. J. Sample surveys of drug abuse. In R. Dupont, A. Goldstein, J. O'Donnell, & B. Brown (Eds.), *Handbook on drug abuse*. Rockville, MD.: METROTEC Research Associates, National Institute on Drug Abuse, 1979, 381–394.

Parry, H. J., Balter, M. B., Mellinger, G. D., Cisin, I. H., & Manheimer, D. I. National patterns of psychotherapeutic drug use. *Archives of General Psychiatry*, 1973, *28*, 769–783.

Paton, S., Kessler, R., & Kandel, D. B. Depressive mood and illegal drug use: A longitudinal analysis. *Journal of Genetic Psychology*, 1977, *131*, 267–289.

Polich, J. M. Epidemiology of alcohol abuse in military and civilian populations. *American Journal of Public Health*, 1981, *71*, 1125–1132.

Pope, H. G., Ionescu-Piogga, M., & Cole, J. O. Drug use and life style among college undergraduates. *Archives of General Psychiatry*, 1981, *38*, 588–591.

Relman, A. (Ed.), *Marijuana and health* (Report of a Study by a Committee of the Institute of Medicine, Divison of Health Sciences Policy). Washington, D.C.: National Academy Press, 1982.

Richek, H. G., Angle, J. F., McAdams, W. S., & D'Angelo, J. Personality/mental health correlates of drug use by high school students. *Journal of Nervous and Mental Disease*, 1975, *160*, 435–442.

Robins, L. *Deviant children grown up*. Huntington, New York: Krieger, 1974.

Robins, L. N., & Murphy, G. E. Drug use in a normal population of young Negro men. *American Journal of Public Health*, 1967, *57*, 1580–1596.

Room, R. The measurement and distribution of drinking patterns and problems in general populations. In G. Edwards, M. Gross, M. Keller, J. Moser, & R. Rooms (Eds.), *Alcohol-related disabilities* (Offset Publication No. 32). Geneva: World Health Organization, 1977.

Spitzer, R. L., Endicott, J., & Robins, E. Research diagnostic criteria: Rationale and reliability. *Archives of General Psychiatry*, 1978, *35*, 773–782.

Uhlenhuth, E. H., Balter, M. B., & Lipman, R. S. Minor tranquilizers: Clinical correlates of use in an urban population. *Archives of General Psychiatry*, 1978, *35*, 650–655.

Weckowicz, T. E., & Janssen, D. V. Cognitive functions, personality traits, and social values in heavy marihuana smokers and nonsmoker controls. *Journal of Abnormal Psychology*, 1973, *81*, 264–269.

Weissman, M., Myers, J., & Harding, P. Prevalence and psychiatric heterogeneity of alcoholism in a United States urban community. *Journal of Studies on Alcohol*, 1980, *41*, 672–681.

Zucker, R., & Barron, F. Parental behaviors associated with problem drinking and antisocial behavior among adolescent males. In M. Chafetz (Ed.), *Research on alcoholism I:*

Clinical problems and special populations (DHEW publication 74-675). Washington, D.C.: U.S. Government Printing Office, 1973.

Zucker, R., & DeVoe, C. Life history characteristics associated with problem drinking and antisocial behavior in adolescent girls: A comparison with male findings. In R. Wirt, G. Winokur, & M. Roff (Eds.), *Life history research in psychopathology* (Vol. 4). Minneapolis: University of Minnesota Press, 1975.

3

Substance Abuse and Psychopathology

SOCIOCULTURAL FACTORS

JOSEPH WESTERMEYER

1. Definitions and Methodologies

Comparing the psychopathology of substance abuse across cultural boundaries involves certain problems of definition and research methodology. The definitions of substance abuse and psychopathology vary both within as well as across cultural boundaries. Both of these concepts—substance abuse and psychopathology—are not all-or-none phenomena, like cancer or pregnancy. Rather, they tend to vary over a spectrum, more like hypertension or depression, from mild to severe cases. Both substance abuse and psychopathology also overlap with normal behavior and socially acceptable substance use, moral, ethical, and religious considerations, social learning and deviance, and law, so that the problems of definition within a culture become even greater with cross-cultural comparisons (MacAndrew & Edgerton, 1969; Westermeyer, 1976a). The definition of a sociocultural group is also problematic. Culture involves such characteristics as language, religion, political organization, social class, technology, aesthetics, symbol, role, and other behavioral concerns. Any one element of culture (say, the Moslem religion, or use of the horse for transportation, or polygamy) occurs among

JOSEPH WESTERMEYER • Department of Psychiatry, University Hospitals, Box 393 Mayo Memorial Building, 420 Delaware Street S.E., Minneapolis, Minnesota.

cultures that may differ widely in other respects. The concept of culture (a learned entity) is often interwoven in the lay person's thinking with notions of race (an inherited entity). To further complicate matters, there are increasingly shadings and gradations among ethnic groups, rather than the distinct boundaries, as was more often the case in the past (Keyes, 1976).

A second dilemma involves sampling methods, both within and across cultures. It is difficult to replicate substance abuse samples in the same society. Replicating them across cultures can be virtually impossible.

Third, research instruments such as questionnaires or self-rating scales may become distorted as they are translated or applied without contextual changes in different groups. Seldom can the same investigator, sensitive to sociocultural matters, conduct precisely the same studies in two or more ethnic groups.

Despite these problems, some investigators have conducted cross-cultural studies using similar samples and methods (Carstairs, 1954; Chegwidden & Flaherty, 1977; Glad, 1947; Kane, 1981; Kunitz, Levy, Odoroff, & Bollinger, 1971; Vitols, 1968; Westermeyer, 1972b, 1977a). Valuable cross-cultural studies and comparisons of substance abuse can be undertaken. In part, this owes to the cross-cultural similarity of the biopharmacological factors involved (e.g., opium can produce tolerance and addiction in people from any culture). Substance abuse syndromes also have fairly consistent psychosocial effects across various cultures. And, while different sampling methods can produce drastically different findings, they may reveal startling similarities (Westermeyer, 1981a, 1981b).

2. Historical Perspectives

2.1. Cultural Selection

Drug-using and drug-abusing behaviors are strongly influenced by social learning and the social expectations of the culture (Bunzel, 1940; Westermeyer, 1971). Many societies have assimilated certain drugs into the warp and woof of their cultural fabric in such a way as to have few or no problems associated with drug use, even with episodic mild to moderate intoxication. Prominent examples include alcohol use by Jewish people (Bacon, 1951; Glad, 1947; Knupfer & Room, 1967) and hallucinogen use by American Indian peoples—in both North and South America—to commune with the spirit world (Bergman, 1971; Furst,

1972; LaBarre, 1964; Opler, 1942). It has been suggested that certain cultural groups select a particular drug for its cultural-congruent pharmacological effects. For example, groups that value emotional and behavior control may reinforce opium use and addiction as compared to alcohol use and addiction (Bailey, 1967; Singer, 1974; Westermeyer, 1971).

Unfortunately, cultures do not necessarily transfer the integrated or controlled use of a substance from one drug to another drug. Whereas the Chinese have traditionally had little trouble with alcohol, they have had considerable problems with opium (Barnett, 1955; Singer, 1974). Some American Indian groups have had few problems with stimulants and hallucinogens, but devastating problems with alcohol (Kuttner & Lorincz, 1967; Kunitz et al., 1971; Westermeyer & Brantner, 1972). While Jewish people have traditionally had few alcohol-related problems, they have had serious problems associated with use of sedatives, hallucinogens, and other drugs (Deutsch, 1975; Klepfisz & Ray, 1973; Rosenbloom, 1959; Parry, 1968; Westermeyer & Walzer, 1975; Zimberg, 1977).

In recent years, some groups that have traditionally not had problems with a particular substance have begun to develop them. This is particularly true for alcohol. Cultural groups with traditional, controlled drinking now having problems include the Chinese (Wang, 1968), Japanese (Sargent, 1967), Greek (Karayannis & Kelepouris, 1967), and Jewish (Hes, 1970) peoples.

2.2. Social Uses of Psychotropic Substances

Studies of tribal people during recent times and archeological data indicate that people in most times and places have taken psychoactive substances. Such drugs have sometimes been taken to preserve health, and at other times to treat illness. They have come primarily from plants, including roots, bark, leaves, flowers, berries, and fruits.

Psychoactive plant compounds developed over the centuries have included digitalis, belladonna, reserpine, and opium. Several drugs of animal origin discovered in the last century or so (such as thyroid extract, insulin, antiserum to certain toxics, and vitamin B_{12}) have not been directly psychoactive. Over the last few decades, biochemists have begun to target psychoactive compounds from the laboratory, at first accidentally and later for specific therapeutic purposes.

Drugs of abuse long used for medicinal purposes included opium, cannabis, alcohol, and various naturally occurring stimulants and hallucinogens. Recent synthetic drugs of abuse have usually been marketed first as medications, such as the synthetic sedatives (e.g., barbiturates), opioids (e.g., meperidine), and even certain hallucinogens with anesthetic properties (e.g., phencyclidine or PCP). These compounds relieve cough, diarrhea, pain, nausea, vomiting, asthma or other respiratory problems, insomnia, agitation, anxiety, and melancholia. While they have greatly benefitted mankind, they have also caused great mischief.

Ritual drug and alcohol use have occurred since recorded times, and probably earlier. High priests of the Mayan and Aztec societies took stimulant and hallucinogenic drugs to commune with the gods. Central and South Americans still use similar compounds for the same purpose (DuToit, 1977; Furst; 1972, 1977). The celebrants at Jewish high holidays consume wine. Besides facilitating relationships between mankind and the gods, drugs have also served to facilitate relationships among people. Middle Eastern tribesmen have traditionally honored friends and relatives by sharing meat during a sacrificial meal, along with alcohol or cannabis (Smith, 1965). Peoples of Asia proffer opium, alcohol and/or tobacco to anyone invited as a guest into their homes (Westermeyer, 1982).

Some drugs attain a near-sacred or sacramental status. Examples include wine consumed during Catholic mass and the hallucinogen peyote consumed during Native American church rituals (LaBarre, 1964). Ritual feasting, which has origins in animal sacrifice and the subsequent consumption of the sacrificed animal, often accompanies alcohol or cannabis intoxication. Such animal sacrifices, along with intoxication, are done to bring spiritual blessings on a new marriage or friendship, or to announce a new or special relationship to the ancestors or the gods.

2.3. Individual Use of Psychotropic Substances

Medical and ritual use imply that the decision to take a psychoactive drug is made by more than one person. In the medical context, the physician, patient, and often the family or others (such as a nurse or pharmacist) participate in the drug-giving/drug-taking event. With ritual use, the society approves and limits the drug taking. With both forms, professional or social mores determine dose, frequency, pattern of use, and expected drug response (e.g., pain, relief, euphoria, rest, increased energy, and so forth).

Alternatively, the individual decides to take a psychoactive drug without necessarily sharing that decision with another person. For example, a person may choose to purchase an over-the-counter (i.e., nonprescription) drug to relieve a particular symptom. Or one may decide to join friends in taking alcohol, cannabis, or opium. In contrast, the hospital patient with acute cholecystitis must rely on the physician, nurse, and pharmacist to administer, say, morphine sulfate. And the Catholic priest must drink wine in prescribed amounts during daily mass.

Of course, social norms and even legal norms do govern individual substance use to some extent. Participants at a cocktail party are not expected to take eight or ten drinks. Opiate compounds are not sold without a prescription in most countries today. However, social and legal norms are not as binding on the individual as the medical use governed by practitioners or the ritual use determined by cultural role and ritual. Whether for recreation or self-treatment, individuals may choose to take the drug often or not at all, and to vary the dosage of the drug from small to large amounts.

Individual drug use serves many purposes. It can relieve minor or self-limited illnesses, aid social conviviality, or enhance food. In all of these cases, the objective is to enjoy life, or at least to tolerate continued existence. Therapeutic and social benefit accrues from the use of psychoactive substances, especially when they are modulated by medical or ritual boundaries. They can also pose a considerable danger when idiosyncratic use leads to excessive dose or excessive frequency in taking these drugs.

2.4. Production of Psychotropic Substances

There has been wide variation in methods of producing alcohol. Sources of carbohydrate to unleash ethanol have included mammalian milk, grain, fruits, sugar cane, honey, and even tubers such as the potato. Production methods have varied from beer making, which takes only days or weeks, to making wines over months or years, to the chemical distillation process and aging of whiskies. While methods for alcohol production were most widely investigated in the Old World, they were also pursued in certain areas of the New World. Pulque, a fermented drink from a fleshy agave plant, is still popular in Mexico.

Opium was probably first developed in the Middle East, as was cannabis. The origin of betel is not clear, but its continued widespread use from South Asia, to Southeast Asia, to the Malay Archipelago, to Oceania suggests its possible origin somewhere in the region (Burton-Bradley, 1977; Westermeyer, 1982). Africa has a long history of beer production, widespread cannabis use, and some opium consumption. Stimulants and mild hallucinogens also originated there; chat, yohimbine, and coca are only a few examples (DuToit, 1977; Getahun & Krikorias, 1973; Rubin, 1975). Prior to European contact during the sixteenth century, the tribes as well as the highly civilized societies of the Americas refined hundreds of plant stimulants and hallucinogens. Many are still consumed only locally; others are consumed around the world, such as coffee and tobacco (DuToit, 1977; Furst, 1972; Goddard, deGoddard, & Whitehead, 1969; Hanna, 1976).

2.5. Consumption of Psychoactive Substances

Oral ingestion may be the oldest route of administration for psychoactive substances. Cannabis cakes in the Middle East, opium spherules in South Asia, coca leaf in the Andes Mountains, and beverage tea, coffee, or alcohol almost everywhere are ingested *per os*. This method offers ease of administration, no need of special paraphernalia, and general medical safety as compared to other routes. Some drugs, such as heroin, are either poorly absorbed in the gut or are inactivated by gastric juice. Since absorption into the blood stream tends to occur in 20 minutes to 1–2 hours, the onset tends to be slow. While that may be desirable from a medical perspective in order to reduce side effects or intoxication, substance abusers typically seek rapid onset of drug effect, since the latter is more desirable as compared to slow or gradual onset.

Drug chewing leads to absorption through the oral mucosa. It was employed in both the Old World (e.g., for betel) and the New World (e.g., for coca leaf). Onset of action is more rapid than ingestion, and this method avoids the acidic environment of the stomach. Certain drugs that produce gastrointestinal irritation (e.g., tobacco, the betel cud) are better tolerated by chewing.

Snorting, snuffing, or nasal insufflation also rely on absorption through the mucosa, but in the nasal region. A small pinch of powdered drug is held to the external nares, followed by a sudden inhalation through the nose (generally one side of the nose is held shut). Absorption and onset of action are much like smoking or parenteral administration. Perforation of the nasal septum may occur as a complication. While this method has waxed and waned in popularity over the centuries, it has largely been replaced by chewing, smoking, and parenteral use.

Smoking seems to have evolved in the New World. It leads to rapid onset of drug effect, much like snorting. Special preparation or paraphernalia are needed for pipes or other means of smoking. Since fire must be applied to ignite combustion, accidental fires and burns can be a problem. Originally only New World drugs such as tobacco and kinnikanick were smoked. The method was adapted to opium and cannabis several centuries ago, and more recently to barbitone, heroin, cocaine freebase, phencyclidine (PCP), and various mixtures of these and other drugs.

Rectal clysis was first discovered in the New World. Administration of drugs by rectum is recorded from pre-Columbian times (Furst, 1977). Subsequently adopted by medical practitioners throughout the world, it continues to be used effectively today. It has not proven popular, however, among those who abuse drugs and become dependent on them. A parallel method, vaginal clysis, has recently appeared among drug users, thereby linking sexual and drug experiences.

Parenteral administration first appeared in the mid-nineteenth century as the product of modern biomedical development. Morphine addicts first employed it for self-administration. None of the drug is lost to volatilization (as in smoking) or to the atmosphere (as in smoking and snuffing) or is inactivated in the gut (as with ingestion), so this is an efficient use of the entire dose. Specific problems associated with parenteral injections involve allergic, inflammatory, and infectious complications: cardiac arrest, hypotension, pulmonary edema, hepatitis, septicemia, endocarditis, nephritis, thrombophlebitis, and abscesses—to name but a few.

3. Research of Psychopathological Conditions

3.1. Drug Type

Choice of intoxicant influences the type of substance-associated psychopathology manifested in a society. For example, plant stimulants such as betel chewing and tobacco smoking do not produce typical psychopathological conditions—although both can produce cancers after prolonged usage. On the other hand, synthetic stimulants (such as amphetamine), concentrated stimulants of plant origin (such as cocaine), and either synthetic or plant-origin hallucinogens (such as peyote, cannabis, PCP, LSD) can produce short-lived stimulant psychosis or—in sensitive individuals—precipitate long-lasting psychoses resembling schizophrenia or mania (Chopra & Smith, 1974). Such psychopathological problems can be negligible or virtually absent in societies not using these drugs widely, or they can assume epidemic proportions, as they have with amphetamine epidemics in Japan, Scandinavia, North America, Ireland (Moorehead, 1968), and, as I have recently observed, in areas of Thailand.

The opiate compounds, including heroin, tend to be unique in producing little or none of the traditional psychopathological states besides substance abuse *per se*. This is not to say that psychopathological states are not sometimes associated with opiates; indeed, they are. But reports of depression, panic attacks, psychosis, dementia, and other conditions can generally be ascribed to either concomitant disorders (e.g., affective disorder) or to complications of substance abuse (e.g., nutritional deficiencies, septicemia, cerebral embolus).

Over the last few decades inhalation of volatile hydrocarbons has been observed among children in several areas of the world (Cockerham, Forslund, & Roboin, 1976; Eastwell, 1979; Kaufman, 1975). Short-term effects include irritability, drop-off in school performance, and cognitive impairment. Acute confusional states may occur. Accidental deaths are infrequent, but do occur. Long-term outcome has not been extensively studied, but occasional dementia cases of varying degree have been reported. It is a matter of public health concern that, in communities with hydrocarbon epidemics, many children may experience mild brain damage which—while not clinically evident—may limit their intellectual potential and maturation.

Alcohol is a special case, since alcohol abuse is associated with high rates of psychopathology in some cultures and low rates in other cultures.

3.2. Alcoholic Psychoses

Among some ethnic groups a pattern is manifested whereby episodic, heavy drinking alternates with periods of sobriety. This binge behavior has been observed among the Irish, Chicanos, and Chippewa (Kelleher, 1976; Knupfer & Room, 1967; Paine, 1977; Walsh & Walsh, 1973; Westermeyer, 1972a). Other groups have a tradition for drinking lower amounts per day, but without periods of abstinence. This latter form, which may be called "titer" drinking, predominates among such peoples as Italians, Hopi, Spanish, Greeks, and most middle-class Americans (Bonfiglio, Falli, & Pacini, 1977; Fernandez, 1976; Hartocollis, 1966; Kunitz *et al.*, 1971). To be sure, no distinct line separates these two patterns; considerable overlap exists within cultures, and even within individuals over time (Tomsovic, 1974). Still, it is possible to discern these two modal patterns of drinking behavior.

Binge drinking among the Irish and American Indians is associated with high rates of alcoholic psychosis. These include alcoholic hallucinosis, delirium tremens, alcoholic paranoia, and Korsakoff's psychosis (Westermeyer, 1972a,b). Conversely, the "titer" drinkers have less alcoholic psychosis, but more hepatic cirrhosis.

3.3. Alcoholic Amnesia

Memory lapse or blackout while drinking is generally presumed to have an organic basis, although functional or hysterical factors may sometimes be present (Goodwin, Crane & Guze, 1969). Rimmer, Pitts, and Reich (1971) in St. Louis observed that black and white alcoholics reported similar rates of blackout when they were matched for social class. However, lower-class alcoholics from both groups reported higher rates of blackout. Women also reported more blackouts than men.

Negrete (1973) found differences across ethnic boundaries in Canada. Franco-Catholic alcoholics had the lowest blackout rate (i.e., 49%, n = 29); Anglo-Protestants and Anglo-Catholics had the highest rates (74%, n = 34, and 86%, n = 28, respectively). Negrete suggests that the ethnic groups with the highest need for "time out" behavior (i.e., behavior free of their usual social constraints) are more apt to report blackout. This writer's work with Chippewa alcoholics supports his hypothesis. Virtually all of this latter group report blackout in association with un-Chippewayan (i.e., shameful) behavior while drinking (Westermeyer, 1972a).

3.4. Alcoholism, Depression, and Other Disorders

Genetic and family studies over the last 15 years have pointed strongly toward a genetic link between alcoholism and depression (Freed, 1970. Morrison, 1975; Pitts & Winokur, 1966; Taylor & Abrams, 1973; Winokur, Cadoret, & Dorzab, 1971; Winokur & Clayton, 1967). Recent work has indicated that alcoholism and anxiety neurosis also possess genetic links to each other (Noyes, Clancy, & Crowe, 1978). Clinical studies going back three decades support the pathoplastic nature between these disorders. Roberts and Meyers (1954) observed that in a clinical population, Jewish patients of that time had high neurosis rates, but no alcoholism, while Irish patients had high alcoholism rates, but no neurosis. Most other ethnic groups had various admixtures of both problems in the Roberts and Meyers study.

Epidemiological studies also support the inverse relationship between alcoholism and neurosis. In community surveys Leighton (1969) found high lifetime prevalence rates of psychoneurosis in areas of Africa and North America with infrequent alcoholism (only a few percent or absent). Shore, Kinzie, and Hampson (1973) found 31 cases of alcoholism and 18 cases of psychoneurosis among 100 American Indians in the northwestern United States. In both the Leighton and Shore et al.

studies, the lifetime prevalence of psychoneurosis and alcoholism suggests that increased prevalence of alcoholism replaces psychoneurosis, and vice versa, rather than their being mutually exclusive. In an area of Asia with very high opiate addiction, Westermeyer (1982) found neurotic disorders to be distinctly uncommon, especially in comparison to high rates of neurotic disorders among the same Asians in the U.S.

3.5. Violence and Criminality

The rates of death from accidents, homicide, and suicide vary widely across ethnic groups. Groups with high rates of alcoholism tend to have high rates of violent death. For example, among American Indians in Minnesota, violence has been the most common cause of death over the last few decades (Westermeyer & Brantner, 1972). Of interest, the percentage of alcohol-related deaths and the mean blood alcohol levels (BAL) observed at autopsy in Minneapolis differed according to the type of death (i.e., vehicular accidents, nonvehicular accidents, and homicide. However, the mean BAL's were remarkably similar with type of death across ethnic boundaries. (See Table 1.)

Some societies tolerate extremely high rates of accidental death in association with alcohol abuse (Stull, 1972; Westermeyer & Brantner, 1972). Other societies, especially in Northern Europe, have passed stringent laws to reduce this type of violence.

Arrest rates for alcohol-related offenses were found to vary widely among three ethnic groups in the southwestern United States (Biegel, Hunter, Tamerin, Chapin, & Lowery, 1974). American Indians had the

Table 1
Alcohol-Related Deaths among Minnesota American Indians and General Population in Relation to Mean Blood Levels at Autopsy

Category of death	Percentage with positive BAL	Mean level of BAL in positive cases
Nonvehicular accidents		
Indian deaths ($n = 21$)	62%	0.27 Gm%
All deaths ($n = 297$)	63%	0.21 Gm%
Vehicular accidents		
Indian deaths ($n = 11$)	73%	0.18 Gm%
All deaths ($n = 163$)	67%	0.18 Gm%
Homicide victims		
Indian deaths ($n = 7$)	86%	0.16 Gm%
All deaths ($n = 40$)	77%	0.18 Gm%

Source: Westermeyer & Brantner, 1972.

highest rates, Hispanics had intermediate rates, and Anglos had the lowest rates. Similarly, alcohol-related child abuse was very high among Indian people in Minnesota (Westermeyer, 1977b,c). Social class may again intersect here, as the lowest classes had the highest rates of alcohol-related criminality.

Criminality is also related to the legal status and cost of a drug in any society. For example, while criminality commonly attends opiate addiction in North America (Suffet & Brotman, 1976), it is rare among poppy-growing addicts in Asia (Westermeyer, 1971).

3.6. Psychosocial and Medical Concomitants

Park and Whitehead (1973) have compared male alcoholic Finns and Americans with regard to the type and sequence of problems that alcoholics encounter in these two countries. They found that the psychological, behavioral, and social concomitants of alcoholism developed in about the same sequence in both countries. Moreover, the duration of time over which these problems developed was about the same. Black and white alcoholics have also been compared in the United States (Cahalan, Cisin, & Crossley, 1969; King, Murphy, Robbins, & Darvish, 1969). Major social and medical complications in these two racial-ethnic groups occur with approximately the same prevalence.

Paradoxically, intoxicants that enhance sexuality when used episodically and moderately are generally detrimental if used heavily and chronically. Alcoholics and drug addicts typically report less interest in sex, less frequent sexual activity, and increased complications such as impotence and amenorrhea (Bell & Trethowan, 1961; DeLeon & Wexler, 1973). The existence of these phenomena across cultural boundaries suggests that they are probably biological, and not related to sociocultural factors.

3.7. Substance Abuse, Social Networks, and Ethnic Affiliation

The work of Pattison (1973, 1977), Speck and Attneave (1973), and others has underlined the importance of social networks in maintaining psychological well-being and social competence. People entering treatment for substance abuse (as well as other psychiatric disorders, including schizophrenia and depression) have markedly reduced and impaired social networks. Their relationships with others tend to be nonreciprocal; that is, they involve much receiving and little giving. By contrast, most normal adult relationships are fairly reciprocal over time, with mutual giving and receiving. In addition, most people in the average network know other members of the network—about 80% in most cases. But in the networks of substance abusers, any one person in the network may only know, on the average, less than 60% of the other network members. Some network members may only know the substance abuser and no other members of the network—a most unusual circumstance among most social networks. Whereas social networks are usually quite stable, those of substance abusers tend to change at a more rapid rate.

Besides reduction in their social networks, substance abusers interact less with ethnic peers in family rituals, ethnic celebrations, and religious activities (Westermeyer, 1976b, 1982). Their new social contacts revolve increasingly around drug usage (Finestone, 1957). While their behavior becomes increasingly nonethnic and drug-centered, their verbally reported norms still remain traditional. Miller, Sensenig, Stocker, and Campbell (1973) found that American black and white addicts, males as well as females, retained idealized norms and values similar to nonaddicts of their own race and sex, despite the fact that their behavior differed considerably from that of nonaddicts.

4. Research on Sociocultural Factors in Treatment and Treatment Outcome

Research on the sociocultural context of treatment involves not only awareness of treatment factors, but also awareness of the sociocultural context within which treatment occurs. Few researchers have so far been able to bring both of these factors into focus at the same time.

Conducting treatment outcome research in the field of substance abuse is also a particularly difficult endeavor. In order to obtain relevant data, the follow-up must be done 1 or 2 years following treatment since most recurrence has occurred by that time. Of course it is not always easy to locate people whose lives are marked by family alienation, marital disruption, episodic unemployment, and financial crises. Despite the difficulties and expense, some perspicacious investigators have conducted studies of treatment and its outcome.

4.1. Access to Treatment

Access to treatment varies with many factors, including geographic distance between the patient's residence and treatment, education, income, and occupation (Edwards, Kyle, & Nicholls, 1974; Hoffman, 1974; Kosa, Antonovsky, & Zola, 1969; Mellsop, 1969; Pattison, Coe, &

Rhodes, 1969). Since ethnicity often covaries with these demographic characteristics, it cannot be assumed that ethnicity *per se* influences access to treatment. However, studies in which these other variables show no or minor difference still indicate that the ethnic composition of the treatment staff influences access (Kane, 1981).

For example, black and American Indian alcoholics have been markedly underrepresented at treatment facilities in North Carolina (Vitols, 1968), Georgia (Lowe & Alston, 1973; Lowe & Hodges, 1972), and Minnesota (Westermeyer, 1976a). Black, Chicano, and American Indian alcoholics seeking treatment for their alcoholism in majority institutions tend to have more severe problems (Paine, 1977; Rimmer *et al.*, 1971; Westermeyer, 1972b), suggesting that those minority alcoholics with less severe problems either stay away or wait until the problems exacerbate. In another study, when severity of withdrawal was considered, American Indian alcoholics went to a less expensive, short-term detoxification center staffed by less trained people, while non-Indian alcoholics tended to go to more expensive, longer-stay hospital facilities staffed by better trained people (Westermeyer & Lang, 1975). Social alienation among certain ethnic groups is associated with decreased willingness to enter treatment (Muller & Brunner-Orne, 1967). Further, it has been suggested that the stigma against alcoholism in Jewish communities (Zimberg, 1977) and the continued support of heavy drinking in some black communities (Lowe & Hodges, 1972) discourage entry into treatment.

4.2. Treatment Modalities and Outcome

Biomedical treatment varies little around the world. For example, withdrawal regimens are essentially identical (Favazza & Martin, 1974; Hartocollis, 1966). Disulfiram has been used as an adjunct to alcoholism treatment with good effect in many cultures (Beaubrun, 1967; Ferguson, 1970; Gumede, 1972; Gerrein, Rosenberg, & Manohar, 1973; Savard, 1968). Even the psychotherapies and sociotherapies for alcoholism greatly resemble one another in areas around the world (Beaubrun, 1967; Gumede, 1972; Hartocolis, 1966; Suwaki, 1975).

Certain culture-bound therapies (which usually contain religious and quasi-religious components) have had good effect with some cases of substance abuse. Among American Indians, the Native American Church (Albaugh & Anderson, 1974) and pre-Columbian guardian spirit religion (Jilek, 1976, 1982) have employed healing rituals for alcoholism. Folk treatment for opium addiction has likewise been used in Asia (Westermeyer, 1973, 1979). Unfortunately, there has been almost no outcome evaluation of these methods. One uncontrolled comparison of medical and Buddhist treatment programs for opium addicts in Asia showed a significantly higher short-term mortality in the latter form of treatment; but long-term outcomes in both programs were the same (Westermeyer & Bourne, 1978; Westermeyer, 1979).

Economic, historical, and social class issues related to ethnicity may influence the need for special treatment modalities for minority groups. For example, given their limited job skills and financial resources, Aborigine alcoholics in Australia have greater need for aftercare and rehabilitation than other Australian alcoholics (Chegwidden & Flaherty, 1977). The same is true for many American Indian alcoholics in Minnesota (Westermeyer & Lang, 1975).

Minority members, once they choose to enter majority treatment programs, usually have outcomes similar to majority members of similar marital and employment status. Treatment follow-up data at a white-run program were the same for blacks and whites (Lowe & Hodges, 1972). Studies have been done comparing black and white alcoholics in Georgia (Lowe & Alston, 1973), Aborigine and white alcoholics in Australia (Chegwidden & Flaherty, 1977), American Indian and white alcoholics in Minnesota (Hoffman & Noem, 1975), and Lao and minority addicts in Laos (Westermeyer, 1982). They have compared several aspects of treatment, including possible racial or ethnic bias in type of treatment modalities administered to patients, length of stay, or postdischarge referral patterns. There has been no demonstrated difference in these studies *once patients sought and were admitted for treatment* (an important caveat). Of course, this absence of demonstrated difference has obviously not encouraged many minority substance abusers to enter majority-staffed programs. And if they fail to enter treatment, then they are unlikely to benefit from it (although they may benefit indirectly from the influence of ethnic peers who have entered treatment and returned to the ethnic community with new attitudes toward drugs and alcohol).

4.3. Ethnicity, Social Class, and Prognosis

Data comparing American whites and blacks (Lowe & Hodges, 1972) and American whites and Indians (Kuttner & Lorincz, 1967; Westermeyer & Brantner, 1972) suggest that social class—that is, education, occupation, employment, residence, and material resources—affects outcome from alcoholism. Middle- and upper-class people with drinking problems may be more apt to appear in a treatment facility. Conversely, lower-class people with drinking problems may be more apt to surface in prisons and morgues.

It has also been demonstrated that the social characteristics of the substance abuser when entering treatment are powerful predictors of outcomes (Crawford, 1976; Goldfried, 1969; Rosenblatt, Gross, & Chartoff, 1969; Suwaki, 1975; Wanberg & Jones, 1973)—probably more powerful than treatment factors *per se*. For example, residence within a family and active employment or active school attendance favor a better prognosis. The poor to fair treatment outcomes among lower-class alcoholics may not be unusual in view of their frequent alienation from family and unemployment upon entering treatment (Westermeyer & Peake, 1983). Unemployment is often due to limited job skills and access to employment as well as to alcohol-related problems.

5. Efforts at Prevention

5.1. Production and Importation of Psychoactive Substances

About two centuries ago various countries began to pass laws regarding the production or importation of psychoactive substances. Currently every nation in the world has laws regarding both medical and recreational drugs. The Aztecs created laws governing the amount and frequency of alcohol use several centuries ago (Paredes, 1975).

Such laws are most effective when the governed concur with the right of the government to pass such laws. However, individuals or groups do not accept central-government rule in many areas of the world. Illicit production, sale, or smuggling of psychoactive substances can result from these circumstances.

Laws are most apt to be useful before widespread drug abuse has appeared. After it has appeared, laws alone are rarely successful without concomitant antidrug efforts by police, health officials, educators, welfare officials, religious leaders, and others (Westermeyer, 1974).

5.2. Medical Prescribing

Pharmaceutical drugs in many areas of the world today can be obtained without a physician's prescription. Even in areas requiring a prescription, pharmacy records may be lax. Regardless of pharmacy practice, some physicians prescribe addicting psychoactive substances too readily, or for chronic conditions in which drug dependence may be an imminent risk. Prescriptions can be easily forged or stolen, if careful surveillance is not maintained.

Prescribing laws and practices are an increasingly important preventive activity against drug dependence. The most common means of
obtaining illicit stimulants, sedatives, or opiates in some regions is
through forging prescriptions or bribing pharmacists and physicians to
supply these drugs.

5.3. Taxation

Taxes on tea, coffee, and distilled alcohol were established over 200
years ago (Greden, 1976). While originally these taxes were meant only to
raise funds for government expenditures, another purpose has evolved
in recent years. Increased cost limits the public consumption of recreational intoxicants, such as tobacco or alcohol. However, there are limits to
the effect of increasing these taxes. Eventually, at a high enough tax,
people begin to produce their own drug—such as the homebrew *poteen* in
Ireland (Connell, 1961)—or purchase it through illicit channels. Not only
does this lose revenue for the government; it also undermines public
respect for the law.

5.4. Prohibition

A society may prohibit a psychoactive substance for any recreational use. This presents little or no problem if the substance is not used
or is used only by a few. However, prohibition becomes much more
difficult if many people traditionally use the substance (Musto, 1973).
Prohibition sometimes has succeeded, as exemplified by Chinese and
Japanese laws against opium during the 1940s and 1950s. Prohibition
has sometimes been a mixed success, such as the 1930s prohibition
against alcohol in the United States. The American law was successful in
reducing some alcohol-related problems (such as cirrhosis death), but it
caused increasing police corruption, bootlegging of illegal beer and liquor, the rise of organized criminal gangs, and general disrespect for the
law. Similarly, anti-opium prohibition in countries of Southeast and
South Asia has had disastrous consequences in some places and times
(Westermeyer, 1974), and relatively beneficial outcomes in other places
and times.

5.5. Culture-Wide Attitudinal Change

Nonlegal methods have been applied to stem widespread drug
abuse. Cartoons, literature, and other applications of the mass media
have changed people's attitudes toward heavy alcohol or drug use. Such

an approach was effective in stemming the English "Gin Epidemic" several hundred years ago. A mass media campaign against heroin use in Hong Kong has had highly beneficial results over the last several years.

Certain religious groups, such as the Mormons, have induced people to abstain from personal or recreational drug use—a primary preventive technique. Other religious groups have been active in helping people with alcohol or drug problems to substitute religious affiliation for their dypsomania or narcotomania (Kearny, 1970). Self-help groups or drug rehabilitation groups have met in Islamic mosques, Christian churches, and Judaic synagogues.

Social network support for sobriety can help to prevent drug abuse, or intervene in its early stages. Widespread knowledge about the early signs of pathological drug use, and skills to confront the person firmly but supportively, can be a powerful means to reduce the prevalence of drug abuse (Swed, 1966).

6. Discussion

6.1. Historical Trends across Cultures

History demonstrates that the acute and chronic effects of psychoactive drugs have been known for over 2,000 years. Literature from the Middle East to South and East Asia describes such drug problems as opiate addiction and delirium tremens. Current observations of tribal people have documented some very high drug dependence prevalence rates. At times, these rates exceed those among urban people. Thus, substance abuse is not just a phenomenon of twentieth century urban societies.

Sporadic cases of substance abuse have been known throughout historical times, but wodespread or epidemic drug dependence is a comparitively recent event. About 300 years ago, governments began to express alarm at the people's increasing use of one or another substance, including not just alcohol and opium, but also tea and coffee. Countries either passed laws against production or importation of certain psychoactive substances or levied taxes on their importation. Patients began being treated for alcoholism in the asylums of 200 years ago. These early epidemics of drug and alcohol abuse spread despite efforts to contain them.

Over the last 100 years, social efforts to deal with alcohol and drug epidemics have intensified. These have included special drug laws, governmental commissions, police bureaus, privately and publicly supported treatment, mass media campaigns, concerned citizens' groups, government-supervised distribution of opiates to registered addicts, new drugs and treatments, and various law enforcement approaches. Two new events have further exacerbated the problem of substance abuse. First, new psychoactive substances have been synthesized in chemical laboratories. And second, subgroups or subcultures of people have appeared, whose common unifying feature is the consumption of one or another psychoactive substance.

The post-Columbian era has seen a diffusion of drug use technology around the world. More sophisticated navigation and shipbuilding, increased travel and commerce, and the explosive transfer of previously unknown technologies between the Old and New Worlds have contributed to this diffusion. Drugs have spread rapidly from one region to another. Routes of administration have also spread. People in the Old World not only imported methods of consumption for imported drugs (e.g., tobacco smoking), but they also began to use the imported methods of consumption for their own drugs (e.g., cannabis and opium smoking). These exchanges have wrought considerable mischief (along with the good effects of technology transfer). North Americans who had no drug problems with their tobacco smoking developed severe problems with alcohol drinking. Similarly, East Asian peoples who had used oral opium for medicinal purposes were soon smoking opium as a recreational pastime. And Europeans acquired compulsive tobacco habits, with all of its attendant problems.

6.2. The Role of Affluence

The post-Columbian era also ushered in a period of relative affluence as new agricultural and animal products became available. Transportation and commercial advances and the Industrial Revolution contributed further. Even the most remote peasant and tribal people today have some share in this labor-saving progress, if only in the form of matches, a tin pot, or an iron machete. The subsequent increase in disposable income facilitates the production or purchase of psychoactive substances. Entrepreneurs are always available to meet an increased social appetite for one or another drug. Of course, affluence is not always associated with increased use of psychoactive compounds. Some religious groups, and some nations, have been able to eliminate certain forms of widespread drug abuse through broad-based societal efforts (Lowinger, 1977).

7. Comment

Despite definitional and methodological problems, sociocultural comparisons can be made so long as limitations are kept in mind and conclusions are not overdrawn. The relationship between culture and drugs is a dynamic one. A given society may have no difficulty with one drug and considerable trouble with another drug. Or the society may have no problems with a drug at one point in time and subsequently develop problems.

Even psychopathological conditions with probable pharmacological and neurophysiological bases (such as alcoholic psychosis and alcoholic amnesia) can vary considerably across cultural boundaries. This variance appears to be related to such factors as different drinking patterns as well as different expectations or secondary gain from intoxication.

Genetic and epidemiological data suggests that substance abuse and certain other psychopathological conditions (e.g., major depression, generalized anxiety) may be pathoplastic variants of each other. The violence and criminality associated with substance abuse are probably related to substance abuse *per se*, at least to some extent. But violence and criminality associated with substance abuse are also influenced by social class, legal codes, the cost of drugs and their licit or illicit status in the society, and cultural norms and mores.

Sociocultural factors often exert considerable influence on access to treatment. However, once treatment is undertaken, cultural treatment factors appear less important while the extent of the patient's social resources (e.g., job, family) have a strong influence on outcome.

8. References

Albaugh, B. J., & Anderson, P. O. Peyote in the treatment of alcoholism among American Indians. *American Journal of Psychiatry*, 1974, *131*, 1247–1250.

Bacon, S. D. Studies of drinking in Jewish culture: 1. General introduction. *Quarterly Journal Studies of Alcohol*, 1951, *12*, 444–450.

Bailey, W. Primary drug addiction. *Medical Opinion and Review*, 1967, *3*, 82–91.

Barnett, M. L. Association in the Cantonese of New York City: An anthropological study. In O. Diethelm (Ed.), *Etiology of chronic alcoholism*. Springfield, Ill.: Charles C Thomas, 1955.

Beaubrun, M. H. Treatment of alcoholism in Trinidad and Tobago, 1956–65. *British Journal of Psychiatry*, 1967, *113*, 643–658.

Bell, D. S. & Trethowan, W. H. Amphetamine addiction and disturbed sexuality. *Archives of General Psychiatry*, 1961, *4*, 74–78.

Bergman, R. L. Navaho peyote use: Its apparent safety. *American Journal of Psychiatry*, 1971, *128*, 695–699.

Biegel, A., Hunter, E. J., Tamerin, J. S., Chapin, E. H., & Lowery, M. J. Planning for the development of comprehensive community alcoholism services: I. The prevalence survey. *American Journal of Psychiatry*, 1974, *131*, 112–116.

Bonfiglio, G., Falli, S. & Pacini, A. Alcoholism in Italy: An outline highlighting some special features. *British Journal of Addiction*, 1977, *72*, 3–12.

Bunzel, R. Role of alcoholism in two Central American cultures. *Psychiatry*, 1940, *3*, 361–387.

Burton-Bradley, B. G. Some implications of betel chewing. *Medical Journal of Australia*, 1977, *2*, 744–746.

Cahalan, D., Cisin, I. H., & Crossley, H. M. *American drinking practices: A national study of drinking behavior and attitudes.* New Brunswick, N.J.: Rutgers Center of Alcohol Studies, 1969.

Carstairs, G. M. Daru and bhang: Cultural factors in the choice of intoxicant. *Quarterly Journal Studies of Alcohol*, 1954, *15*, 220–237.

Chegwidden, M., & Flaherty, B. J. Aboriginal versus non-Aboriginal alcoholics in an alcohol withdrawal unit. *Medical Journal of Australia*, 1977, *1*, 299–703.

Chapra, G. S., & Smith, J. W. Psychotic reactions following cannabis use in East Indians. *Archives of General Psychiatry*, 1974, *30*, 24–27.

Cockerham, W. C., Forslund, M. A., & Roboin, R. M. Drug use among White and American Indian high school youth. *International Journal of Addictions*, 1976, *11*, 209–220.

Connell, K. H. Illicit distillation: An Irish peasant industry. *Historical Studies of Ireland*, 1961, *3*, 58–91.

Crawford, R. J. M. Treatment success in alcoholism. *New Zealand Medical Journal*, 1976, *84*, 93–96.

DeLeon, G., & Wexler, H. K. Heroin addiction: Its relation to sexual behavior and sexual experience. *Journal of Abnormal Psychology*, 1973, *81*, 36–38.

Deutsch, A. Observations of a sidewalk ashram. *Archives of General Psychiatry*, 1975, *32*, 166–175.

DuToit, B. M. *Drugs, rituals and altered states of consciousness.* Rotterdam: Balkema, 1977.

Eastwell, H. D. Petrol-inhalation in Aboriginal towns. *Medical Journal of Australia*, 1979, *2*, 221–224.

Edwards, G., Kyle, E., & Nicholls, P. Alcoholics admitted to four hospitals in England: I. Social class and the interaction of alcoholics with the treatment system. *Quarterly Journal Studies of Alcohol*, 1974, *35*, 499–522.

Favazza, A. R., & Martin, P. Chemotherapy of delerium tremens: A survey of physicians' preferences. *American Journal of Psychiatry*, 1974, *131*, 1031–1033.

Ferguson, F. N. A treatment program for Navaho alcoholics: Results after four years. *Quarterly Journal Studies of Alcohol*, 1970, *31*, 898–919.

Fernandez, F. A. The state of alcoholism in Spain covering its epidemiological and aetiological aspects. *British Journal of Addiction*, 1976, *71*, 235–242.

Finestone, H. Cats, kicks, and color. *Social Problems*, 1957, *5*, 3–13.

Freed, E. X. Alcoholism and manic depressive disorders: Some perspectives. *Quarterly Journal Studies of Alcohol*, 1970, *31*, 62–89.

Furst, P. T. (Ed.). *Flesh of the gods: The ritual use of hallucinogens.* New York: Praeger, 1972.

Furst, P. T., & Coe, M. D. Ritual enemas. *Natural History*, 1977, *86*, 88–91.

Gerrein, J. R., Rosenberg, C. M., & Manohar, V. Disulfiram maintenance in outpatient treatment of alcoholism. *Archives of General Psychiatry*, 1973, *28*, 798–802.

Getahun, A., & Krikorias, A. D. Chat: Coffee's rival from Harar, Ethiopia. *Economic Botany*, 1973, *27*, 353–389.

Glad, D. D. Attitudes and experiences of American–Jewish and American–Irish male youth as related to differences in adult rates of inebriety. *Quarterly Journal Studies of Alcohol*, 1947, *8*, 406–472.

Goddard, D., deGoddard, S. N., & Whitehead, P. C. Social factors associated with coca use in the Andean region. *International Journal of Addictions*, 1969, *4*, 577–590.

Goldfried, M. R. Prediction of improvement in an alcoholism outpatient clinic. *Quarterly Journal Studies of Alcohol*, 1969, *30*, 129–139.

Goodwin, D. W., Crane, J. B., & Guze, S. R. Alcoholic "blackouts": Review and clinical study of 100 alcoholics. *American Journal of Psychiatry*, 1969, *126*, 191–198.

Greden, J. The tea controversy in colonial America. *Journal of the American Medical Association* 1976, *236*, 63–65.

Gumede, M. V. The treatment of African alcoholics: A review of the first 400 cases seen at Kwasimama Clinic. *South African Medical Journal*, 1972, *46*, 430–433.

Hanna, J. M. Coca leaf in Southern Peru: Some biosocial aspects. *American Anthropologist*, 1976, *76*, 281–286.

Hartocollis, P. Alcoholism in contemporary Greece. *Quarterly Journal Studies of Alcohol*, 1966, *27*, 721–727.

Hes, J. P. Drinking in a Yemenite rural settlement in Israel. *British Journal of Addiction*, 1970, *65*, 293–296.

Hoffman, H. County characteristics and admission to state hospital for treatment of alcoholism and psychiatric disorder. *Psychological Reports*, 1974, *35*, 1275–1277.

Hoffman, H., & Noem, A. A. Adjustment of Chippewa Indian alcoholics to a predominantly White treatment program. *Psychological Reports*, 1975, *37*, 1284–1286.

Jilek, W. G. "Brainwashing" as a therapeutic technique in contemporary Canadian Indian spirit dancing: A case in theory building. In J. Westermeyer (Ed.), *Anthropology and mental health*. The Hague: Mouton, 1976.

Jilek, W. G. *Indian healing: Shamanistic ceremonialism in the pacific northwest today*. Surrey, Canada: Hancock House, 1982.

Kane, G. *Inner city alcoholism: An ecological and cross cultural study*. New York: Human Sciences Press, 1981.

Karayannis, A. D., & Kelepouris, M. B. Impressions of the drinking habits and alcohol problem in modern Greece. *British Journal of Addiction*, 1967, *62*, 71–73.

Kaufman, A. Gasoline sniffing among children in a Pueblo Indian village. *Pediatrics*, 1975, *51*, 1060–1063.

Kearny, M. Drunkenness and religious conversion in a Mexican village. *Quarterly Journal Studies of Alcohol*, 1970, *31*, 248–249.

Kelleher, M. J. Alcohol and affective disorder in Irish mental hospital admissions. *Journal of the Irish Medical Association*, 1976, *69*, 140–143.

Keyes, C. F. Towards a new formulation of the concept of ethnic group. *Ethnicity*, 1976, *3*, 202–312.

King, L. J., Murphy, G. E., Robbins, L. N., & Darvish, H. Alcohol abuse: A crucial factor in the social problems of Negro men. *American Journal of Psychiatry*, 1969, *125*, 1682–1690.

Klepfisz, A. & Ray, J. Homicide and LSD. *Journal of the American Medical Association*, 1973, *223*, 429–430.

Knupfer, G., & Room, R. Drinking patterns and attitudes of Irish, Jewish and White Protestant men. *Quarterly Journal Studies of Alcohol*, 1967, *28*, 676–699.

Kosa, J., Antonovsky, A., & Zola, I. K. *Poverty and health*. Cambridge, Mass.: Harvard University Press, 1969.

Kunitz, S. J., Levy, J. E., Odoroff, C. L., & Bollinger, J. The epidemiology of cirrhosis in two southwestern Indian tribes. *Quarterly Journal Studies of Alcohol*, 1971, *32*, 706–720.

Kuttner, R., & Lorincz, A. Alcoholism and addiction in urbanized Soiux Indians. *Mental Hygiene*, 1967, *51*, 530–542.

LaBarre, W. *The peyote cult*. Hamden, Conn.: The Shoe String Press, 1964.

Leighton, A. H. A comparative study of psychiatric disorder in Nigeria and rural North America. In S. C. Plog & R. B. Edgerton (Eds.), *Changing perspectives in mental illness*. New York: Holt Rinehart Winston, 1969.

Lowe, G. D., & Alston, J. P. An analysis of racial differences in services to alcoholics in a southern clinic. *Hospital Community Psychiatry*, 1973, *24*, 547–551.

Lowe, G. D., & Hodges, H. E. Race and the treatment of alcoholism in a southern state. *Social Problems*, 1972, *20*, 240–252.

Lowinger, P. The solution to narcotic addiction in the People's Republic of China. *American Journal of Drug and Alcohol Abuse*, 1977, *4*, 165–178.

MacAndrew, C., & Edgerton, R. B. *Drunken comportment*. Chicago: Aldine, 1969.

Mellsop, G. W. The effect of distance in determining hospital admission rates. *Medical Journal of Australia*, 1969, *2*, 814–817.

Miller, J. S., Sensenig, J., Stocker, R. B., & Campbell, R. Value patterns of drug addicts as a function of race and sex. *International Journal of Addiction*, 1973, *8*, 589–598.

Moorehead, N. C. Amphetamine consumption in Northern Ireland. *Journal of the Irish Medical Association*, 1968, *61*, 80–84.

Morrison, J. R. The family histories of manic-depressive patients with and without alcoholism. *Journal of Nervous and Mental Disease*, 1975, *160*, 227–229.

Muller, J. J., & Brunner-Orne, M. Social alienation as a factor in the acceptance of outpatient psychiatric treatment by the alcoholic. *Journal of Clinical Psychology*, 1967, *23*, 517–518.

Musto, D. *The American disease*. New Haven, Conn.: Yale University Press, 1973.

Negrete, J. C. Cultural influences on social performance of alcoholics: A comparative study. *Quarterly Journal Studies of Alcohol*, 1973, *34*, 905–916.

Noyes, R., Clancy, J., & Crowe, R. The familial prevalence of anxiety neurosis. *Archives of General Psychiatry*, 1978, *35*, 1057–1059.

Opler, M. Fact and fancy in Ute peyotism. *American Anthropologist*, 1942, *44*, 151–159.

Paine, H. J. Attitudes and patterns of alcohol use among Mexican Americans: Implications for service delivery. *Journal Studies of Alcohol*, 1977, *38*, 544–553.

Paredes, A. Social control of drinking among the Aztec Indians of Meso-America. *Journal Studies of Alcohol*, 1975, *36*, 1139–1153.

Park, P., & Whitehead, P. C. Developmental sequence and dimensions of alcoholism. *Quarterly Journal Studies of Alcohol*, 1973, *34*, 887–904.

Parry, H. J. Use of psychotropic drugs by U.S. adults. *Public Health Reports*, 1968, *83*, 799–810.

Pattison, E. M. Social system psychotherapy. *American Journal of Psychotherapy*, 1973, *27*, 396–409.

Pattison, E. M. Clinical social systems interventions. *Psychiatry Digest*, 1977, *38*, 25–33.

Pattison, E. M., Coe, R., & Rhodes, R. J. Evaluation of alcoholism treatment: A comparison of three facilities. *Archives of General Psychiatry*, 1969, *20*, 478–488.

Pitts, F. N., & Winokur, G. Affective disorder: VIII. Alcoholism and affective disorder. *Journal of Psychiatric Research*, 1966, *4*, 37–50.

Rimmer, J., Pitts, F. N., & Reich, T. Alcoholism: II. Sex, socioeconomic status and race in two hospitalized samples. *Quarterly Journal Studies of Alcohol* 1971, *32*, 942–952.

Roberts, B., & Meyers, J. Religion, national origin, immigration, and mental illness. *American Journal of Psychiatry*, 1954, *110*, 759–764.

Rosenblatt, S., Gross, M., & Chartoff, S. Marital status and multiple psychiatric admissions for alcoholism. *Quarterly Journal Studies of Alcohol*, 1969, *30*, 445–447.

Rosenbloom, J. R. Notes on Jewish addicts. *Psychological Reports*, 1959, *5*, 769–882.

Rubin, V. (Ed.). *Cannabis and culture*. The Hague: Mouton, 1975.

Sargent, M. J. Changes in Japanese drinking patterns. *Quarterly Journal Studies of Alcohol*, 1967, *28*, 709–722.

Savard, R. J. Effects of disulfiram therapy on relationships within the Navaho drinking group. *Quarterly Journal Studies of Alcohol*, 1968, *29*, 909–916.

Shore, J., Kinzie, J. D., & Hampson, J. L., Pattison, E. M. Psychiatric epidemiology in an Indian village. *Psychiatry*, 1973, *36*, 70–81.

Singer, K. The choice of intoxicant among the Chinese. *British Journal of Addiction*, 1974, *69*, 257–268.

Smith, W. R. Sacrifice among the Semites. In W. A. Lessa, & E. Z. Vogt (Eds.), *Reader in comparative religion*, New York: Harper & Row, 1965.

Speck, R. V., & Attneave, C. L. *Family networks*, New York: Pantheon, 1973.

Stull, D. D. Victims of modernization: Accident rates and Papago Indian adjustment. *Human Disorganization*, 1972, *31*, 227–240.

Suffet, F., & Brotman, R. Employment and social disability among opium addicts. *American Journal of Drug and Alcohol Abuse*, 1976, *3*, 387–395.

Suwaki, H. A follow-up study of alcoholic patients. *Psychiatria and Neurologica Japonica*, 1975, *77*, 89–106.

Swed, J. F. Gossip, drinking and social control: Consensus and communication in a Newfoundland parish. *Ethnology*, 1966, *5*, 434–441.

Taylor, M., & Abrams, R. Manic states: A genetic study of early and late onset of affective disorders. *Archives of General Psychiatry*, 1973, *28*, 656–658.

Tomsovic, M. Binge and continuous drinkers. *Quarterly Journal Studies of Alcohol*, 1974, *35*, 558–564.

Vitols, M. M. Culture patterns of drinking in Negro and White alcoholics. *Diseases of the Nervous System*, 1968, *29*, 391–392.

Walsh, B. M., & Walsh, D. Validity of indices of alcoholism: A comment from the Irish experience. *British Journal of Preventive Social Medicine*, 1973, *27*, 18–26.

Wanberg, K. W., & Jones, E. Initial contact and admission of persons requesting treatment for alcohol problems. *British Journal of Addiction*, 1973, *68*, 281–285.

Wang, R. P. A study of alcoholism in Chinatown, *International Journal of Social Psychiatry*, 1968, *14*, 260–267.

Westermeyer, J. Use of alcohol and opium by the Meo of Laos. *American Journal of Psychiatry* 1971, *127*, 1010–1023.

Westermeyer, J. Options regarding alcohol use among the Chippewa. *American Journal of Orthopsychiatry*, 1972, *42*, 398–403. (a)

Westermeyer, J. Chippewa and majority alcoholism in the Twin Cities: A comparison. *Journal of Nervous and Mental Disease*, 1972, *155*, 322–327. (b)

Westermeyer, J. Folk treatments for opium addiction in Laos. *British Journal of Addiction*, 1973, *68*, 345–349.

Westermeyer, J. The pro-heroin effects of anti-opium laws in Asia. *Archives of General Psychiatry*, 1974, *33*, 1135–1139.

Westermeyer, J. Use of a social indicator system to assess alcoholism among Indian people in Minnesota. *American Journal of Drug and Alcohol Abuse*, 1976, *3*, 447–456. (a)

Westermeyer, J. Models for chemical dependency. *Primer on chemical dependency*. Baltimore, Md.: Williams and Wilkins, 1976. (b)

Westermeyer, J. Narcotic addiction in two Asian cultures: A comparison and analysis. *Drug Alcohol Dependence*, 1977, *2*, 273–285. (a)

Westermeyer, J. Cross-racial foster home placement among Native American psychiatric patients. *Journal of the National Medical Association,* 1977, *69,* 231–236 (b)

Westermeyer, J. The ravages of Indian families in crisis. In S. Unger (Ed.), *The destruction of Indian families.* New York: Association of American Indian Affairs, 1977. (c)

Westermeyer, J. Medical and nonmedical treatment programs for opium addicts: A comparison in Asia. *Journal of Nervous and Mental Disease,* 1979, *167,* 205–211.

Westermeyer, J. A comparison of three case finding methods for opiate addicts. A study among the Hmong in Laos. *International Journal of Addictions,* 1981, *16,* 173–183. (a)

Westermeyer, J. Influence of opium availability on addiction rates in Laos. *American Journal of Epidemiology,* 1981, *109,* 550–562. (b)

Westermeyer, J. *Poppies, pipes and people.* Los Angeles: University of California Press, 1982.

Westermeyer, J., & Bourne, P. Treatment outcome and the role of the community in narcotic addiction. *Journal of Nervous and Mental Disease,* 1978, *166,* 51–58.

Westermeyer, J. & Brantner, J. Violent death and alcohol use among the Chippewa in Minnesota. *Minnesota Medicine,* 1972, *55,* 749–752.

Westermeyer, J., & Lang, G. Ethnic differences in use of alcoholism facilities. *International Journal of Addictions,* 1975, *10,* 513–520.

Westermeyer, J., & Peake, E. A ten year follow up of alcoholic Native Americans in Minnesota. *American Journal of Psychiatry,* 1983, *140,* 189–194.

Westermeyer, J. & Walzer, V. Drug usage: An alternative to religion? *Disorders of the Nervous System,* 1975, *36,* 492–495.

Winokur, G., & Clayton, P. J. Family history studies: II. Sex difference and alcoholism in primary affective illness. *British Journal of Psychiatry,* 1967, *113,* 973–979.

Winokur, G., Cadoret, R., & Dorzab, J. Depressive disease: A genetic study. *Archives of General Psychiatry,* 1971, *24,* 135–144.

Zimberg, S. Sociopsychiatric perspectives on Jewish alcohol abuse: Implications for the prevention of alcoholism. *American Journal of Drug and Alcohol Abuse,* 1977, *4,* 571–579.

4

Substance Abuse in the Affective Disorders

Demmie Mayfield

Common sense suggests that there might be a relationship between affective disorder and substance abuse. That drugs might be sought and abused by individuals who are experiencing unhappiness or dysphoria is consistent with popular notions about drug-taking behavior. There is, in fact, a great deal of evidence for a rather strong association between alcoholism and affective disorder, though by no means as simple or as well understood as the popular commonsense notion would suggest. On the other hand, there is very little evidence of any particular association between other drugs and affective disorder. This may in part be due to lack of study or to the relative newness of other drugs of abuse, but the discrepancy is striking and is in itself an interesting finding. In any event, there is good reason to consider alcohol and other drugs of abuse separately with respect to their relationship to affective disorder.

1. Alcohol

A relationship between alcoholism and affective disorder has been noted by clinicians since the terms manic-depressive and dipsomania were coined. In fact the term dipsomania, the alcoholism of the old nomenclature, was based on the assumption that the mania to drink was

Demmie Mayfield • Psychiatry Service, Kansas City VA Medical Center, 4801 Linwood Boulevard, Kansas City, Missouri 64128.

a periodic phenomena, and was thus closely related to manic-depressive disorder.

The first systematic studies that suggested an unusual association between alcoholism and affective disorder were those appearing about the turn of the century and showing a high incidence of alcoholism among suicides and suicide attempters (East, 1913; Sullivan, 1900). Subsequently a high incidence of alcoholism has repeatedly been found in studies of individuals who commit suicide (Robins, Murphy, Wilkinson, Gassner, & Kayes, 1959), and a high incidence of suicide has consistently been reported in follow-up studies of alcoholics (Lemere, 1953; Norwig & Borge, 1956). Alcoholism thus shares with affective disorder this common cause of death (Pitts & Winokur, 1966; Robins et al., 1959).

A high coincidence of alcoholism and affective disorder is more recent but is well established as a consistent finding by investigators who have sought to demonstrate affective disorder in alcoholism as well as alcoholism among patients with affective disorder. Cassidy, Flanagan, Spellman, and Cohen (1957) found that 8% of their manic-depressive patients were excessive drinkers, while Parker, Meiller, and Andrews (1960) found that 32.8% of their manic-depressives were alcoholic. Mayfield and Coleman (1968) found alcoholism in 20% of their bipolar patients. The figures for the incidence of affective disorder among alcoholics are less striking but still impressively high, with Amark (1951) reporting 9% among his chronic alcoholics and Sherfey (1955) identifying 6.8% affective disorder among her alcoholics. Practically without exception, appropriate studies have shown a remarkable coincidence of the two disorders.

1.1. Family Studies

Though there had been incidental reports of a high occurrence of affective disorder in the lineage of alcoholics and an extraordinary incidence of alcoholism in the families of patients with affective disorder, it was Pitts and Winokur (1966) who first brought attention to the curious familial relationship of the two disorders. They systematically gathered family history data from consecutive admissions to the psychiatric hospital and compared the rates of psychiatric disorders in relatives with that found in medical and surgical controls. They found a high incidence of alcoholism in male first-degree relatives of all categories of psychiatric patients, but this incidence was especially high in patients with affective disorder. They also found a high incidence of affective disorder, but no other psychiatric illness, in relatives of patients with alcoholism.

In similar studies, the group at St. Louis and then Iowa have further defined family histories in alcoholism and affective disorder. Winokur, Reich, Rimmer, and Pitts (1970) found that 41% of fathers and brothers of male alcoholics and 48% of fathers and brothers of female alcoholics were themselves alcoholic. Only 6% of mothers and sisters of both male and female alcoholics were themselves alcoholic. However, 33% of the mothers and sisters of male alcoholics and 48% of the mothers and sisters of female alcoholics had suffered from depression, while 8% of the fathers and brothers of male alcoholics and 11% of the fathers and brothers of female alcoholics had depressive illness.

These studies also looked in detail (Winokur, 1979) at alcoholism in the families of depressed patients, finding that mothers of affectively ill patients had a 22.9% incidence of affective disorder and that fathers had a 13.6% incidence of affective disorder, both significantly higher than that found in a control group. They further found that fathers of affectively ill patients had a 9.5% incidence of alcoholism, which was significantly higher than that of the control group; the mothers had a 1.1% incidence of alcoholism, which was not significantly different from that for the control group.

In a similar study Angst (1972) investigated psychiatric illness in parents and siblings of patients with unipolar affective disorder. He found that the incidence of alcoholism (14.5%) and affective disorder (9.6%) among the fathers and brothers of the patients totalled the same (25.1%) as the incidence of alcoholism (1.0%) and affective disorder (24.1%) among their mothers and sisters.

These findings strongly suggest a linkage of some sort between alcoholism and affective disorder. It has been suggested that the two disorders are manifestations of the same disease—expressed as alcoholism in males and depression in females.

Winokur and his co-workers (Behar & Winokur, 1979) have been led by these findings to devise a hypothetical subdivision of unipolar affective disorder into: (1) "depressive spectrum disease," a serious unipolar depression occurring in a person with a first-degree relative suffering from either alcoholism or antisocial personality with or without first-degree relatives with unipolar depression; (2) "familial pure depressive disease," occurring in an individual with a first-degree relative suffering only from unipolar depression; and (3) "sporadic depressive disease," occurring in the absence of any psychiatric illness among first-degree relatives. It is in "depression spectrum disease" that there seems to be some common underlying familial attribute that may be expressed as either unipolar affective disorder or alcoholism.

Winocur and his colleagues are continuing their efforts at defining these groups, although to date no clinical differences have been noted among the patients in the diffeent depressive subgroups. Interestingly enough, bipolar affective disorder does not seem to show this familial relationship to alcoholism (Dunner, Hensel, & Fieve, 1979; Morrison, 1975). Genetic linkage studies thus far have not yielded any markers.

1.2. Alcohol Use and Affective Disorder

For years the high incidence of excessive drinking in manic-depressive disorder was noted and accepted as the expected effect of mood state on drinking behavior (i.e., that patients drank during periods of depression to ameliorate their dysphoria). Increased drinking during mania was inconsistent with this formulation and was largely ignored and unreported or, when acknowledged, dismissed as an unexplainable curiosity.

In 1957, Cassidy et al. inquired into changes in drinking concomitant with mood swings in patients with manic-depressive disorders. They found that both increase and decrease in drinking was common, but failed to distinguish between manic and depressive episodes in their inquiry.

It was not until 1968 that Mayfield and Coleman reported the first detailed study of changes in drinking concomitant with episodes of cyclic affective disorder. Such a study was of interest not only to establish the exact relationship of affective disorder and alcoholism, but because it presented one of the few opportunities to examine the relationship between affective state and drinking.

The role of affect in the cause of excessive drinking had been difficult to evaluate because alcoholics are usually available for study immediately after prolonged intoxication, when their affective state is more relevant to the psychological and physiological effects of drinking. The identification of affects pertinent to the onset of drinking tends to be unreliable because this requires the recall of an affective state that occurred either long ago or immediately prior to or during a period of intoxication. The evaluation of the role of affect in drinking should be more clear-cut in manic-depressive illness where a persistent, profound disturbance in a variety of affects occurring over a circumscribed period of time alternates with more normal mood states.

With this strategy in mind, Mayfield and Coleman (1968) studied the changes in drinking concomitant with episodes of cyclic affective disorder. As expected, they found a high incidence (20%) of excessive drinking in the group of 59 patients with cyclic affective disorders. They also found that a change in drinking with mood swing was quite common. Contrary to expectations, they found that increased drinking was associated with elation and not depression. The findings of their study were that, in individuals alternating between elation and depression, the change in drinking was predominantly one of increase with elation. A decrease in drinking with elation, if it occurred at all, was rare. The change in drinking with depression was more likely to be a decrease than an increase. More drinking when depressed than when elated was uncommon, and an increase to excessive proportions while depressed was exceptional. An attempt at drinking during depression tended to be quickly abandoned.

This association of drinking with mania has since been noted by other investigators (Freed, 1969, 1970). Reich, Davies, and Himmelhoch (1974) felt that a number of their manic patients were attempting to control their mania with alcohol.

Mayfield and Coleman (1968) also studied a group of patients with histories of excessive drinking and unipolar affective disorders to determine the correlation between depression and drinking. In most patients there was no correlation whatever, and in no patient was drinking delimited to depressive episodes. Similarly, Pauleikoff (1953) found only 2 cases of excessive drinking restricted to depression among almost 900 patients with "cyclic depression." Campanella and Fossi (1963) found only one case of excessive drinking delimited to depressive episodes, but found that four of five chronic alcoholics ceased drinking during depressive episodes.

Their findings seem to cast doubt on the notion that the drinking seen in association with affective disorder is a result of the attempt to alleviate pathologically depressed mood. They also cast some doubt on certain notions about the use and excessive use of alcohol generally. Another line of inquiry, experimental studies of the influence of intoxication on affect, has been pursued in an effort to clarify these questions.

1.3. Alcohol and Affect

The quest for intoxication with alcohol is as old as recorded history and so too, probably, are notions about why people drink. It has generally been assumed that people drink because it makes them feel better, and, logically enough, it has been assumed that people drink more and then "too much" for the same reason.

What is the basis for this commonsense explanation for drinking? Social drinkers say that they feel mildly better when they drink to the low evel of intoxication described as a "glow" or "buzz." This pleasant experience is apt to be described in terms such as cheerful, mellow, relaxed, carefree—occasionally with an element of stimulation or arousal. It is a phenomenon of low dose, subject to careful titer. These moderate social drinkers regularly stop when they have "enough," and they do not value the experience so highly that they go to extraordinary effort to seek out drink.

When these moderate social drinkers are brought into the laboratory and experimentally intoxicated, they behave pretty much as expected. They do undergo mild improvement in a variety of mood factors at low levels of intoxication. They undergo deterioration in mood with progression to slightly higher levels of definite intoxication (Williams, 1966). This mild affective improvement has regularly been demonstrated in healthy social drinkers experimentally intoxicated with different beverages, different routes of administration (Warren & Raynes, 1972) and in a variety of settings (Smith, Parker, & Noble, 1975). There is improvement in affect, but the change is not impressive, suggesting that "euphoria" is perhaps too strong a term to use in describing it. It would appear that alcohol has a weaker pharmacological signature in this regard than cocaine or amphetamine.

Thus, the findings in social drinkers seem entirely consistent with the properties of the drug and the behavior of the users with respect to it. More important to theories of alcoholism is what intoxication does to or for alcoholics, and how they differ from nonalcoholics. Common sense would suggest that alcoholics get more affective benefit from alcohol than do those who drink it less and seem to like it less. When detoxified, asymptomatic alcoholics have been studied while acutely intoxicated, they differ little from nonalcoholics. If there is any difference, it is in the direction of slightly less improvement in mood for alcoholics. The modest "euphoria" alcoholics experience is, as in nonalcoholics, an early, low-dose phenomenon reversed at levels they typically achieve in the drinking they do in their natural habitat (Mayfield, 1968b).

These findings are not consistent with the drinking behavior of alcoholics and are at odds with theories that hold that alcoholics drink heavily simply because they find the experience more affectively gratifying. Acute low-level experimental intoxication, however, bears little resemblance to most of the drinking of alcoholics, who are apt to be heavily intoxicated for extended periods. It was not until the early 1960s that Mendelson (1964) and co-workers first studied alcoholics during chronic experimental intoxication. Much to everyone's surprise, these alcoholics experienced *deterioration* in mood with chronic heavy drinking. They became progressively more anxious, more depressed, and less friendly as drinking continued. These findings have been replicated in numerous similar studies by Mendelson's group (McGuire, Mendelson,

& Stein, 1966), and a number of other investigators (Allman, Taylor, & Nathan, 1972; Goldman, 1974). Chronic intoxication is regularly accompanied by affective discomfort that is progressive as intoxication continues. This deterioration in affective state has been observed to proceed to the point of a severe depressive syndrome with suicidal ideation (Tamerin & Mendelson, 1969). With cessation of drinking, this mood disturbance promptly disappears, even in the presence of withdrawal symptoms.

Mayfield and Montgomery (1972) noted a similar phenomenon that they labeled "depressive syndrome of chronic intoxication" in more than one-third of alcoholics admitted to the hospital for treatment of injuries resulting from suicide attempts. These patients had made their attempts during a clearly defined severe depressive syndrome that developed after 3 weeks or more of continuous heavy drinking. As in the case of experimentally intoxicated subjects, the depressive syndrome cleared promptly after hospitalization. These patients had practically no recall for their depressive symptoms, a feature prominent in experimentally intoxicated subjects as well.

The findings of experimental intoxication in alcoholics, as we find them in hospitals, suggest that if these heavy drinkers are seeking "euphoria" with their drinking, they are probably not finding it. They may, however, be finding relief from dysphoria—and this is indeed what is proposed as an explanation for a high incidence of excessive drinking among patients with affective disorder. The study of the palliative effects of alcohol had been hindered by the difficulty in inducing, in the laboratory, the appropriate state of affective disorder against which to test the effects of the drug.

To overcome this difficulty Mayfield (1968a) and Mayfield and Allen (1967) studied patients with acute severe depressive syndromes. They found that subjects with disordered mood underwent profound improvement in most mood factors after mild, acute intoxication with intravenous alcohol. These findings indicated that alcohol does have a profound and pervasive palliative effect on mood. They also found that depressed subjects who had past histories of excessive drinking showed significantly less improvement than subjects who never used alcohol excessively.

Alcohol did have a marvelous effect on the mood of profoundly depressed patients, whether they had been excessive drinkers or not. The experimental findings predict that these individuals would be likely to seek out and abuse alcohol when they were depressed. Such, however, was not the case. Though they had the potential for immediate relief with alcohol, the subjects did not manifest increased drinking with depressed mood. The behavior with respect to alcohol was found to be paradoxical with respect to laboratory response. In fact the alcohol–affect relationship seems at present to be a complete set of paradoxes. In other words, if you feel very bad, alcohol will make you feel a lot better. But if you drink a lot of alcohol, it will make you feel very bad. Feeling very bad from drinking a lot does not seem to encourage people to stop drinking. Feeling a lot better from drinking does not seem to encourage people to continue drinking.

The evidence does not support the oft cited alcoholism formulation: depression—drinking—relief—repetition—addiction. The data do support uncertainty, and this confusion, combined with the experience of clinicians confronted with dysphoria in excessive drinkers presenting for treatment, has encouraged widespread use of antidepressant drugs.

1.4. Alcoholism and Antidepressant Drug Treatment

The two factors of, first, the frequent appearance of depressive symptoms in alcoholics presenting for treatment, and second, the commonly held notion among clinicians that alcoholism was a symptomatic expression of affective disorder, both served to encourage the rather unselective use of antidepressant drugs in the treatment of alcoholism. Most early reports of results of antidepressant treatment in alcoholics were favorable, but tightly designed, controlled studies were quite the exception—as was the case in most areas of psychopharmacology at the time.

After antidepressant drugs had been in use for about a decade, Ditman (1966) reviewed the evidence and concluded that antidepressant usage had been somewhat indiscriminate and that evidence was lacking to support general efficacy of antidepressants in alcoholism. Subsequently Viamontes (1972) reviewed 16 studies, finding that 7 of 9 uncontrolled studies reported favorable results in the use of antidepressants for alcoholism, while none of the 7 controlled studies reported a beneficial effect.

More recent reports of clinical studies have been more selective in their use of antidepressant drugs in alcoholic populations (although clinical practice may have changed very little). There is now a general consensus in the literature that presenting alcoholics are a heterogeneous population—even more heterogeneous than the depressive population (Pottenger, McKernon, Patrie, Weissman, Ruben, & Newberry, 1978). Rates of depression among alcoholics as high as 98% (Shaw, Donley, Morgan, & Robinson, 1975) or as low as 3% (Winokur, Rimmer, & Reich, 1971) have been reported, depending upon the rigor of application of the label "depression." Keeler, Taylor, and Miller (1979) reported rates of 8.6%, 28%, 43%, and 66% in the same presenting alcoholic population, depending upon whether clinical diagnosis was employed or upon which the three depression rating scales was used as criteria. Hamm, Major, and Brown (1979) found their young, healthy, navy alcoholism-program referrals to be relatively free of depression. Tyndel (1973), on the other hand, found that 35% of a larger, more heterogeneous alcoholic population showed definite depressive features. It is probable that the majority of alcoholics initially presenting for treatment will appear depressed, even though most alcoholics do not have affective disorder or a persistent state of depressed mood.

It is generally agreed that alcoholics who initially appear depressed deserve a second and third look, allowing the passage of sufficient observational time before a commitment is made to a course of antidepressant medication. There are four well recognized subgroups within the larger population of patients who present as acutely depressed alcoholics. The first are those patients who have, as has been verified in experimental intoxication studies (Tamerin & Mendelson, 1969) developed a "depressive syndrome of chronic intoxication" (Mayfield & Montgomery, 1972), which may be quite severe and may quite faithfully mimic retarded depression. This depressive syndrome disappears promptly and completely with cessation of drinking.

The second subgroup are those patients whose depression is the result of a situational crisis that brought them to treatment (Liskow, Mayfield, & Thiele, 1982). This reactive/situational depression is a response to the accumulation of drink-related problems and usually to some event which acutely disturbs the adjustment that had been made to accommodate the pathological drinking.

The third subgroup are those patients suffering with a "characterological depression"—a longstanding, relatively persistent depression that has an existence independent of drinking behavior or major life events (Akiskal, Rosenthal, Haykal, Lemmi, Rosenthal, & Scott-Strauss, 1980; Weissman & Myers, 1980). The fourth group are those patients suffering with an episode of affective disorder.

The first three subgroups are not candidates for antidepressant drug treatment. The depressive syndrome of chronic intoxication resolves spontaneously with interruption of alcohol intake. The reactive depression tends to resolve over a week or two in response to reduction in the intensity of the crisis and organization of a plan or denial of a problem. The characterological depressions persist, as they have for most of the patient's life, and are not benefitted by pharmacotherapy.

Only those patients in the fourth subgroup, who are suffering from an episode of affective disorder, are candidates for treatment with anti-depressant drugs. Such patients, when well selected, respond to treat-ment with a variety of antidepressant drugs in the same manner as do nonalcoholic patients with affective disorder. To select these patients well, it is necessary to evaluate them from a stable baseline. This means resisting the impulse to diagnose and start treatment immediately, even though the patient may be acutely symptomatic. It is valuable for even the most experienced clinicians to remind themselves that they cannot read through intoxication, withdrawal, situational depression, and so forth, to unerringly differentiate primary affective disorder from these conditions.

If an incorrect diagnosis of primary affective disorder is made and antidepressant drug treatment is started prematurely, the inevitable symptomatic improvement will be incorrectly attributed to drug treat-ment, and the patient will be locked into a lengthy drug-maintenance regimen. The unnecessary drug treatment, with all the attendant incon-venience, expense, and risk, tends to delay or sidetrack more appropri-ate efforts at treatment.

The opposite error, incorrectly attributing affective disorder symp-toms to some other cause (i.e., the failure to identify affective disorder when it is present) is less likely. However, it is well to remind ourselves that affective disorder is not rare among alcoholics. It is quite treatable, and failure to recognize and appropriately treat an episode of affective disorder will make all other efforts at treatment ineffective.

Antidepressant drugs do have a place in the treatment of certain syndromes that appear among excessive drinkers. This is, however, a more modest and a more circumscribed role than the one promoted in the past. Clinicians have generally become more conservative in their use of these drugs and less enthusiastic about their value. Of consider-ably more interest recently has been the potential value of lithium in the management of alcoholism.

1.5. Lithium Treatment in Alcoholism

Lithium has been known as a treatment for mania (Cade, 1949) for several decades. It has been widely recognized as useful for mainte-nance/prophylactic treatment of recurrent affective disorder for over a decade (Baastrup & Schou, 1967). Patients with alcoholism and well defined episodes of mania or recurrent affective disorder have been treated and have benefitted, as have nonalcoholic patients. This use of lithium differs little from its general use and excites little interest, com-ment, or special study.

More uncertain and more intriguing, however, is the proposed use of lithium in patients with alcoholism alone. This use is based upon the assumption that alcoholism is a *forme fruste* (i.e., an entity that looks different but has the same origin as another) of affective disorder, and that modulation of some underlying process might be manifested in a corresponding moderation in the excessive drinking.

The first report on the use of lithium in alcoholics was that of Fries (1969). It was a clinical report, not too enthusiastic, and it received little attention. Flemenbaum (1973) advocated the use of lithium in alcoholism on theoretical grounds. Kline, Wren, Cooper, Varga, and Canal (1974) were the first to report a controlled study of the effects of lithium in chronic alcoholics. Their results and clinical impressions were very favorable and excited a good deal of interest in psychopharmacological and alcohol treatment circles. The main problem with their study was that the large loss of sample to follow-up was probably beyond acceptable limits. Reports of subsequent studies have ranged from moderately positive (Reynolds, Merry, & Coppen, 1979) to negative (Peck, Pond, Becker, & Lee, 1981; Pond, Becker, Vandervoort, Phillips, Bowler, & Peck, 1981), with a large loss of sample to follow-up continuing to be a problem. As is the rule for most new treatments, the initial enthusiasm has waned, and even now enough time has passed and enough preliminary study has been done to indicate that lithium will not be a bonanza to the treatment of alcoholism.

Definitive study of treatment of this sort in an illness of this type requires large samples and extended periods of study. A number of large, well designed studies are planned and a number are underway. Until the results of these studies are reported, it is difficult to assess the value of the treatment and impossible to establish guidelines for administration of lithium to alcoholics. In fact, doubt has been expressed whether the beneficial effect of lithium, if it exists at all, has anything to do with correction of mood swings in an underlying affective disorder (Mayfield, 1979). It may be that lithium maintenance alters the features of intoxication, thereby influencing the probability of the occurrence of drinking to excess. Lithium pretreatment has been demonstrated to mitigate the subjective intoxication produced by amphetamine (Flemenbaum, 1974; Van Kammen & Murphy, 1966). Such an effect has not been reported for alcohol, but Judd, Hubbard, Huey, Attewell, Janowsky, and Takahashi (1977) found that a confusional effect manifest at low levels of intoxication was reduced by lithium pretreatment. Mayfield (1968b) and Parker, Alkana, Birnbaum, Hartley, and Noble (1974) have demonstrated this cognitive and affective confusion to be a common finding at low levels of intoxication in alcoholics. It may be that lithium acts not by influencing mood but by mitigating the confusional effects of intoxication, thus sharpening the excessive drinker's awareness of what he is experiencing and enhancing both his control and his ability to stop drinking. Sinclair (1979) has demonstrated that rats drink less when pretreated with lithium, thus raising the possibility of mechanisms unrelated to mood or cognitive effects.

Almost 10 years after the first use of lithium in alcoholism, the drug is still of uncertain value. Lithium is not going to become a mainstay of alcoholism treatment. Whether it will be a valuable adjunct or of any value at all, and by what mechanism, are still aspects to be demonstrated.

1.6. Alcoholism and Affective Disorder

The evidence is firm in support of the conclusion that there is a positive association between affective disorder and alcoholism. As is often the case, however, what at first seemed so simple and so self-evident became, when studied, complex and full of apparent paradoxes. The relationship is not simply one mediated by the manifest mood disturbance and the easily observable psychopharmacological effects of alcohol. These apparent causal links may indeed be irrelevant or frankly misleading clues to the causal mechanism. There is reason to suspect that there is a common, underlying process that has neither been observed, identified, nor understood in regard to alcoholism and affective disorder. The identification of the common process in this coincidence of psychopathology would very likely be an important clue to the understanding of alcoholism and would probably shed light on the pathogenesis of affective disorder as well. It is because of this heuristic potential that the convergence of these two disorders is of such great research interest.

2. Other Drugs

Perhaps the most remarkable thing about the use of other recreational drugs is that, in contrast to alcohol, there is so little evidence of an association with affective disorder. The discrepancy between alcohol and other drugs in this respect is surprising because it is contrary to what would be predicted by most psychological theories of substance abuse. Both psychodynamic and behavioral theories hold that most other recreational drugs are used in a manner similar to alcohol, and at least some other drugs should have the same relationship to affective disorder. The literature on substance abuse and affective disorder hardly supports this assumption. The literature, however, is admittedly scanty; reports of actual studies are even more sparse; they do not extend far back; and there is little to suggest that the literature is becoming more abundant.

What is the explanation for this surprising difference between alcohol and all other substances of abuse? This difference may represent a real, qualitative dissimilarity between alcohol and the other drugs of abuse or it may merely represent an actuarial phenomenon. It would be surprising if there weren't some differences between the use of alcohol and other drugs, solely on an actuarial basis, in view of the risk of exposure. It is well known that there are many more users and abusers of alcohol than there are for any other drug or, indeed, for all other drugs of abuse combined. What is not so commonly appreciated is how much greater is the total exposure to alcohol in all cultures, and over a so much greater span of human existence.

Alcohol had established its place in all but the most primitive societies by the dawn of recorded history. The problem of drunkenness appears in the Bible with the Noah story. The making of beverages appears in some of the earliest Sumerian writings, as do admonitions about drunkenness in early Egyptian hieroglyphics (Budge, 1926; Rosenberg & Keller, 1971). While there is reason to believe that opiates may have been known in antiquity (Kramer, 1972), and cannabis probably had some use as early as the middle ages (World Health Organization, 1971), alcohol has a historical/cultural place not to be challenged by any of the Johnnie-come-lately substances that at times seem to dominate the awareness of contemporary society.

The pervasiveness of alcohol in modern Western society may also be hard to appreciate. It is in our cuisine, in our religious ceremonies, in our social rituals, in our commerce, and in all manner of interpersonal encounters. No other drug can begin to compare with alcohol at present, in the past, or probably in the future. It is this very pervasiveness that dulls our awareness of alcohol so that newer drugs may be portrayed as the "new danger" or the "major threat" to public or private well-being. In fact, society is usually receiving a light dusting with the new drugs while continuing to be marinated in alcohol.

For this reason, a behavior or a diagnosis is much more likely to be tested vis-à-vis alcohol use than with respect to any other group of drugs. This is apt to be even more pronounced in the case of family sutdies where the drug abuse of the mother, father, or sisters of an index case is more apt to be influenced by availability or custom than is the case for alcohol abuse.

"Seek and ye shall find" is as much a truism in psychiatric research as anywhere else. The failure to report an association between affective disorder and other drug abuse may be merely the artifact of failure to look for the phenomenon. It is tempting to assume that if there were a phenomenon that were at all robust, it would have intruded itself upon the awareness of clinicians and researchers, prompting them to pursue appropriate studies. To a certain extent this must be true, but there are too many examples of failure to study the obvious (e.g., increased drinking in mania) to take it for granted. Nonalcoholic substance abuse in affective disorder has not been much studied. There seems to be little enthusiasm for study; perhaps for good reason, perhaps not. The nagging question remains: Is there lack of study because the phenomenon does not exist, or is the phenomenon unknown for lack of study?

If the findings are not accounted for by artifactual or actuarial factors, then an important qualitative difference between alcohol and other drugs may exist. Identifying this difference and then understanding it could have enormous heuristic potential for explaining the mechanisms involved in the etiology of alcoholism.

There are an amazing number, and an increasing number, of chemicals that have been used as recreational drugs. These substances fall into a myriad of chemical, pharmacological, and botanical categories and have been used and abused in a variety of ways. Fortunately, nonalcoholic substances of abuse fall into two groups—opiates and all other substances—with respect to study *vis-à-vis* affective disorder. The only drug group with a substantial body of studies reported in this area is the opiate group.

2.1. Opitaes

Opiate abusers have been more thoroughly studied than other substance abusers for several reasons. They are better subjects because, aside from alcoholics, they are the only abusers to be found in any numbers who are dedicated to only one drug. Polydrug abuse has increasingly been the pattern of use in all abusers, including opiate addicts, but this has bedeviled researchers much less for opiates than has been the case with sedative-hypnotics, stimulants, hallucinogens, and so forth. Opiate abusers are also more accessible insofar as they are more commonly a "captive audience" in specialized treatment programs. There has been something of a tradition of research in opiate drugs dating back to pre-World War II years under the leadership of the Lexington Federal Narcotics Hospital.

The coincidence or concomitance of opiate abuse and affective disorder can be studied by either looking for opiate abuse among patients with affective disorder or by seeking affective disorder among opiate abusers. When affective disorder populations have been scrutinized for opiate abuse, few abusers have been found. The studies of Cassidy *et al.* (1957) and Mayfield and Coleman (1968) are typical, with no opiate abusers found in substantial samples of manic-depressive patients. There is a small body of work relating opiate abuse to diagnostic considerations. The first attempt to classify opiate addicts in terms of psychopathological entities was reported by Kolb (1925). He categorized the patients in an idiosyncratic schema not entirely translatable into current diagnostic systems, but he did not identify affective disorder, nor did he delineate a group that seemed to be a counterpart of affective disorder as we presently apply the term. Gerard and Kornetsky (1955) also did not identify any subjects with affective disorder among their adolescent delinquent addicts. Vaillant (1966) followed up 100 New York addicts indexed 12 years before at the Lexington Narcotics Hospital. He found no subjects with clearly diagnosable affective disorder. Personality disorder and sociopathy were the two leading diagnostic categories. He noted that the subjects "rarely received a depressive diagnosis and symptoms of depression were not prominent in their mental status." Only three subjects had psychiatric hospitalization for depression. He found no mania at all. In referring to studies at Lexington at three different points in history (the mid 1930s, early 1950s, and mid 1960s), he noted that personality disorder and sociopathic personality were the leading categories, while affective disorder was never identified.

In recent years most of the scrutiny of the opiate abuse population has been focused on methadone maintenance patients. They have been in existence only in recent years, but they are obviously the most available of any drug-abusing subjects. They are a group of legitimate and especial interest, but they are probably not entirely representative of the full range of opiate abusers to be found in their natural habitat.

Recent studies have also been more rigorous of design and more descriptively reliable—consistent with this trend in psychiatric research generally. Along with this trend more depression is being identified and more attention is being directed at the question of the presence of affective disorder.

Weissman, Slobetz, Prusoff, Mezritz, and Howard (1976) found 30% of methadone maintenance patients depressed on the Raskin Depression Scale. They felt these patients were "clinically depressed," but they were quite uncertain about what kind of depression—primary or secondary, affective disorder, or other. Woody, O'Brien, and McLellan (1979) concluded that 50% of a small group of methadone maintenance addicts had a current or past depressive illness by Research Diagnostic Criteria (RDC) derived from the Schedule for Affective Disorders and Schizophrenia Lifetime Version (SADS-L) interview and the Beck Depression Inventory (BDI). Rounsaville, Weissman, Crits-Christoph, Wilber, and Kleber (1982a) also derived RDC from the SADS-L and the BDI in newly admitted methadone maintenance addicts whom they reexamined in 6 months. They found that "relatively few met RDC criteria for depression despite a high number of complaints of dysphoria." They considered these subjects in terms of primary/secondary schema and concluded that they were predominantly secondary depression with a low level of symptoms. They also noted a great deal of change, mostly improvement, over the 6-month interval.

Woody, O'Brien, and Rickels (1975) conducted a placebo and antidepressant drug (doxepin) double-blind study with 35 mildly depressed methadone maintenance patients. They found greater improvement in the doxepin group on a number of depression factors, but the study did not attempt to make detailed diagnostic distinctions, and a high dropout rate made interpretation of results difficult. Dorus and Senay (1980) and Rounsaville *et al.* (1982a) found considerable improvement in depressive symptoms in methadone maintenance addicts followed over a similar period without any antidepressant treatment.

The question of the occurrence of mania or bipolar disorder has been very lightly touched on in reports of opiate abusers, and the reports that are available are mixed. Rounsaville, Weissman, Kleber, and Wilber (1982b) felt that both mania and schizophrenia were diagnosed no more frequently in their opiate abusers than might be expected in the general population. Rounsaville *et al.* (1982a) said that "only a small minority met criteria for manic or hypomanic disorders." On the other hand, Khantzian and Treece (1979), in commenting on preliminary findings in an ongoing cooperative study, noted the finding of an unusually high frequency of hypomania in the subjects' histories. It has been noted, however, that reliable accounts of past affective states are notoriously difficult to obtain from longstanding drug abusers (Rounsaville *et al.*, 1982a).

The relationship between opiate abuse and affective disorder has certainly not been exhaustively studied. Enough creditable work has been reported, however, to make it seem unlikely that there is an undiscovered high positive correlation; that is, an association comparable to that found with alcohol abuse.

2.2. Nonopiates

Studies correlating psychiatric diagnoses with drugs of abuse other than alcohol or opiates are few and far between, and even more inconclusive than reports of studies on opiate abusers.

Nace, Meyers, O'Brien, Ream, and Mintz (1977) interviewed a sample of 200 Vietnam veterans 2 years after their return and found a high incidence of depression in mood and a high association of both drug and alcohol abuse with depression. Dorus and Senay (1980) found that polydrug abusers scored high on the Beck and Hamilton depression scales initially, but underwent considerable improvement when retested 4 and 8 months later. These researchers concluded that this was "secondary" depression.

Those studies that have been directed at establishing diagnostic correlations have not given strong support for coincidence of affective disorder and substance abuse. McLellan and Druley (1977) did find an interesting differential association of diagnosis with drug type. They systematically interviewed, using a questionnaire designed to detect substance abuse, patients admitted to psychiatric wards with non-drug abuse diagnoses. They found about 50% with serious alcohol or drug problems that had not been detected at initial work-up. Comparison was made of the patient's type of drug abuse and his psychiatric diagnosis, made independently. They found barbiturate abuse had a high association with depressive diagnoses and low association with schizophrenic diagnoses. Alcohol and heroin abuse had about the same high percentage of depressive diagnosis as the nondrug population. These contrasted with amphetamine and hallucinogen drug abuse, which was highly correlated with a paranoid schizophrenic diagnosis and a low incidence of depressive diagnoses. It was not possible to determine from the study how common was primary affective disorder in the various groups. In a somewhat similar study Hall, Stickney, Gardner, Perl, and LeCann (1979) made a thorough inquiry into the drug use of patients admitted to a research ward with a policy of excluding patients with previous drug abuse. They found, admission policy notwithstanding, that 58% of the patients were drug abusers. Depression and manic-depressive disease were underrepresented among this very mixed group of abusers.

McLellan, Woody, and O'Brien (1979) took advantage of the opportunity to study a group of drug abusers longitudinally. These patients had initially been hospitalized for drug abuse of various types and subsequently had been rehospitalized a number of times at the same facility over a 6-year period. The patients fell into three groups: stimulant, opiate, and depressant drug abusers. McLellan and his colleagues found that the stimulant abusers had a high incidence of evolving to a schizophrenic diagnosis, that the depressant group (including alcohol) evolved into depressive diagnoses, and that the opiate abusers did not evolve into anything. the relationship of affective disorder to this large group of miscellaneous recreational drugs has been understudied. It is becoming more difficult to carry out clean studies in this area, since polydrug abuse is increasingly the pattern of use. Especially confounding is the increasing combination of alcohol and polydrug abuse.

3. Summary

We have seen an increase in reports of studies relating alcoholism and substance abuse as a result of the recent emphasis on diagnostic rigor, the renewal of interest in correlates of psychopathology, and increasing interest in affective disorder. There is every indication that these trends will continue and thus contribute to understanding in this area.

A number of very consistent findings strongly suggest an association between alcoholism and affective disorder. Both disorders share suicide as a common cause of death. There is a remarkable coincidence of alcoholism and bipolar affective disorder. An extraordinary incidence of alcoholism has been noted in families of patients with affective disorder, and an extraordinary incidence of affective disorder has been reported in families of patients with alcoholism. The reasons for this association remain elusive. Results of experimental study do not support a mechanism mediated by the effect of mood on drinking behavior. An understanding of the mechanisms that account for this association should provide valuable clues to the etiology of alcoholism and to certain aspects of affective disorder as well. In fact, inquiry into this area might prove to be the most useful strategy for the study of the pathogenesis of alcoholism. Research has thus far failed, however, to find a relationship between nonalcohol substances of abuse and affective disorder.

4. References

Akiskal, H. S., Rosenthal, T. L., Haykal, R. F., Lemmi, H., Rosenthal, R. H., & Scott-Strauss, A. Characterological depressions. *Archives of General Psychiatry*, 1980, 37, 777–783.

Allman, L. R., Raylor, H. A., & Nathan, P. E. Group drinking during stress: Effects on drinking behavior, affect, psychopathology. *American Journal of Psychiatry*, 1972, *129*, 669–678.

Amark, C. A study in alcoholism. *Acta Psychiatrica Scandinavia*, 1951, *70 (Suppl.)*, 283.

Angst, J. Genetic aspects of depression. In P. Kielholz (Ed.), *Depressive illness*. Berne: Hans Huber, 1972.

Baastrup, P. C., & Schou, M. Lithium as a prophylactic agent against recurrent depressions and manic depressive psychosis. *Archives of General Psychiatry*, 1967, *16*, 162–172.

Behar, D., & Winokur, G. Research in alcoholism and depression: A two-way street under construction. In R. W. Pickens & L. L. Heston (Eds.), *Psychiatric factors in drug abuse*. New York: Grune & Stratton, 1979.

Budge, E. W. *Dwellers on the Nile*. New York: Arno Press, 1926.

Cade, J. F. J. Lithium salts in the treatment of psychotic excitement. *Medical Journal of Australia*, 1949, *2*, 349–352.

Campanella, G., & Fossi, G. Considerazioni sui rapport: Fra alcoolismo e manifestazioni depressive. *Rassegna Di Studi Psichiatrici*, 1963, *52*, 617–632.

Cassidy, W. L., Flanagan, N. B., Spellman, M., & Cohen, M. E. Clinical observations in manic-depressive disease. *Journal of the American Medical Association*, 1957, *164*, 1535–1546.

Ditman, K. Review and evaluation of current drug therapies in alcoholism. *Psychosomatic Medicine*, 1966, *28*, 667–677.

Dorus, W., & Senay, E. C. Depression, demographic dimensions, and drug abuse. *American Journal of Psychiatry*, 1980, *137*, 699–704.

Dunner, D. L., Hensel, B. M., & Fieve, R. R. Bipolar illness: Factors in drinking behavior. *American Journal of Psychiatry*, 1979, *136*, 583–585.

East, W. N. On attempted suicide with an analysis of 1,000 consecutive cases. *Journal of Mental Sciences*, 1913, *59*, 428–478.

Flemenbaum, A. Affective disorders & "chemical dependence": Lithium for alcohol and drug addiction? *Diseases of the Nervous System*, 1973, *35*, 281–284.

Flemenbaum, A. Does lithium block the effects of amphetamine?: A report of three cases. *American Journal of Psychiatry*, 1974, *131*, 820–821.

Freed, E. X. Alcohol abuse by manic patients. *Psychological Reports*, 1969, *25*, 280.

Freed, E. X. Alcoholism and manic-depressive disorders. *Quarterly Journal of Studies on Alcohol*, 1970, *31*, 62–89.

Fries, H. Experience with lithium carbonate treatment at a psychiatric department in the period 1964–1967. *Acta Psychiatrica Scandinavia*, 1969, *207*, 44–48.

Gerard, D. L., & Kornetsky, C. Adolescent opiate addiction: A study of control and addict subjects. *Psychiatric Quarterly*, 1955, *29*, 457–486.

Goldman, M. S. To drink or not to drink: An experimental analysis of group drinking decisions by four alcoholics. *American Journal of Psychiatry*, 1974, *131*, 1123–1130.

Hall, R. C. W., Stickney, S. K., Gardner, E. R., Perl, M., & LeCann, A. F. Relationship of psychiatric illness to drug abuse. *Journal of Psychedelic Drugs*, 1979, *11*, 337–342.

Hamm, J. E., Major, L. F., & Brown, G. L. The quantitative measurement of depression and anxiety in male alcoholics. *American Journal of Psychiatry*, 1979, *136*, 580–582.

Judd, L. L., Hubbard, B., Huey, L. Y., Attewell, P. A., Janowsky, D. S., & Takahashi, K. I. Lithium carbonate and ethanol induced "highs" in normal subjects. *Archives of General Psychiatry*, 1977, *34*, 463–467.

Keeler, M. H., Taylor, C. I., & Miller, W. C. Are all recently detoxified alcoholics depressed? *American Journal of Psychiatry*, 1979, *136*, 586–588.

Khantzian, E. J., & Treece, C. J. Heroin addiction: The diagnostic dilemma for psychiatry. In R. W. Pickens & L. L. Heston (Eds.), *Psychiatric factors in drug abuse.* New York: Grune & Stratton, 1979.

Kline, N. S., Wren, J. C., Cooper, T. B., Varga, E., & Canal, O. Evaluation of lithium therapy in chronic and periodic alcoholism. *The American Journal of the Medical Sciences,* 1974, *268,* 15–22.

Kolb, L. Types and characteristics of drug addicts. *Mental Hygiene,* 1925, *9,* 300–313.

Kramer, J. C. A brief history of heroin addiction in America. In D. E. Smith & G. R. Gay (Eds.), *Heroin in perspective.* Englewood Cliffs, N.J.: Prentice-Hall, 1972.

Lemere, F. What happens to alcoholics? *American Journal of Psychiatry,* 1953, *109,* 674–676.

Liskow, B., Mayfield, D. & Thiele, J. Alcohol and affective disorder: Assessment and treatment. *Journal of Clinical Psychiatry,* 1982, *43,* 144–147.

Mayfield, D. G. Psychopharmacology of alcohol: I. Affective change with intoxication, drinking behavior and affective state. *Journal of Nervous and Mental Disease,* 1968, *146,* 314–321. (a)

Mayfield, D. G. Psychopharmacology of alcohol: II. Affective tolerance in alcohol intoxication. *Journal of Nervous and Mental Disease,* 1968, *146,* 322–327. (b)

Mayfield, D. G. Alcohol and affect: Experimental studies. In D. W. Goodwin & C. K. Erickson (Eds.), *Alcoholism and affective disorders.* New York: Spectrum, 1979.

Mayfield, D., & Allen, D. Alcohol and affect: A psychopharmacological study. *American Journal of Psychiatry,* 1967, *123,* 1346–1351.

Mayfield, D. G., & Coleman, L. L. Alcohol use and affective disorder. *Diseases of the Nervous System,* 1968, *29,* 467–474.

Mayfield, D. G., & Montgomery, D. Alcoholism, alcohol intoxication, and suicide attempts. *Archives of General Psychiatry,* 1972, *27,* 349–353.

McGuire, M. T., Mendelson, J. H., & Stein, S. Comparative psychosocial studies of alcoholic and non-alcoholic subjects undergoing experimentally induced ethanol intoxication. *Psychosomatic Medicine,* 1966, *28,* 13–26.

McLellan, A. T., & Druley, K. A. Non-random relation between drugs of abuse and psychiatric diagnosis. *Journal of Psychiatric Research,* 1977, *13,* 179–184.

McLellan, A. T., Woody, G. E. & O'Brien, C. P. Development of psychiatric illness in drug abusers. *New England Journal of Medicine,* 1979, *301,* 1310–1314.

Mendelson, J. H. (Ed.). Experimentally induced chronic intoxication and withdrawal in alcoholics. *Quarterly Journal of Studies on Alcohol,* 1964, Suppl. 2.

Morrison, J. B. The family histories of manic-depressive patients with and without alcoholism. *Journal of Nervous and Mental Disease,* 1975, *160,* 227–230.

Nace, E. P., Meyers, A. L., O'Brien, C. P., Ream, N., & Mintz, J. Depression in veterans two years after Viet Nam. *American Journal of Psychiatry,* 1977, *134,* 167–170.

Norwig, J., & Borge, N. A follow-up of 221 alcohol addicts in Denmark. *Quarterly Journal of Studies on Alcohol,* 1956, *17,* 633–640.

Parker, E. S., Alkana, R. L., Birnbaum, I. M., Hartley, J. T., & Noble, E. P. Alcohol and the disruption of cognitive processes. *Archives of General Psychiatry,* 1974, *31,* 824–828.

Parker, J. B., Meiller, R. M., & Andrews, G. W. Major psychiatric disorders masquerading as alcoholism. *Southern Medical Journal,* 1960, *53,* 560–564.

Pauleikoff, B. Uber die Seltenheit von Alkohol abusus bei zyklothym Depressiven. *Nervenartz,* 1953, *24,* 445–448.

Peck, C. C., Pond, S. M., Becker, C. E., & Lee, K. An evaluation of the effects of lithium in the treatment of chronic alcoholism: II. Assessment of the two-period crossover design. *Alcoholism: Clinical and Experimental Research,* 1981, *5,* 252–255.

Pitts, F. N., & Winokur, G. Affective disorder: VII. Alcoholism and affective disorder. *Journal of Psychiatric Research,* 1966, *4,* 37–50.

Pond, S. M., Becker, C. E., Vandervoort, R., Phillips, M., Bowler, R. N., & Peck, C. C. An evaluation of the effects of lithium in the treatment of chronic alcoholism: I. Clinical results. *Alcoholism: Clinical and Experimental Research*, 1981, 5, 247–251.

Pottenger, M., McKernon, J., Patrie, L. E., Weissman, M. M., Ruben, H. L., & Newberry, P. The frequency and persistence of depressive symptoms in the alcohol abuser. *Journal of Nervous and Mental Disease*, 1978, 166, 562–570.

Reich, L. H., Davies, R. K., & Himmelhoch, J. M. Excessive alcohol use in manic-depressive illness. *American Journal of Psychiatry*, 1974, 131, 83–86.

Reynolds, C. M., Merry, J., & Coppen, A. Prophylactic treatment of alcoholism by lithium carbonate: An initial report. In D. W. Goodwin & C. K. Erickson (Eds.), *Alcoholism and affective disorders*. New York: Spectrum, 1979.

Robins, E., Murphy, G. E., Wilkinson, R. H., Gassner, S., & Kayes, J. Some clinical considerations in the prevention of suicide based on a study of 134 successful suicides. *American Journal of Public Health*, 1959, 49, 888–899.

Rosenberg, S. S., & Keller, M. (Eds.). *Alcohol and health* (DHEW Publication Number (HSM) 72-9099). Washington, D.C.: U.S. Government Printing Office, 1971.

Rounsaville, B. J., Weissman, M. M., Crits-Christoph, K., Wilber, C., & Kleber, H. Diagnosis and symptoms of depression in opiate addicts. *Archives of General Psychiatry*, 1982, 39, 151–156. (a)

Rounsaville, B. J., Weissman, M. M., Kleber, H., & Wilber, C. Heterogeneity of psychiatric diagnosis in treated opiate addicts. *Archives of General Psychiatry*, 1982, 39, 161–166. (b)

Shaw, J. A., Donley, P., Morgan, D. W., & Robinson, J. A. Treatment of depression in alcoholics. *American Journal of Psychiatry*, 1975, 132, 641–644.

Sherfey, M. J. Psychopathology and character structure in chronic alcoholism. In O. Diethelm (Ed.), *Etiology of chronic alcoholism*. Springfield, Ill.: Charles C Thomas, 1955.

Sinclair, J. D. Ethanol intake and lithium in rats. In D. W. Goodwin & C. K. Erickson (Eds.), *Alcoholism and affective disorders*. New York: Spectrum, 1979.

Smith, R. C., Parker, E. S., & Noble, E. P. Alcohol and affect in a dyadic social interaction. *Psychosomatic Medicine*, 1975, 37, 25–40.

Sullivan, W. C. Alcoholism and suicidal impulses. *Quarterly Journal of Inebriety*, 1900, 22, 17–29.

Tamerin, J. S., & Mendelson, J. H. The psychodynamics of chronic inebriation: Observations of alcoholics during the process of drinking in an experimental group setting. *American Journal of Psychiatry*, 1969, 125, 886–899.

Tyndel, M. Psychiatric study of one thousand alcoholic patients. *Canadian Psychiatric Association Journal*, 1974, 19, 21–24.

Vaillant, G. E. A 12-year follow-up of New York narcotic addicts. *Archives of General Psychiatry*, 1966, 15, 599–609.

Van Kammen, D. P., & Murphy, D. Blockading of d-and L-amphetamine-induced activation and euphoria by lithium ion. *Comprehensive Psychiatry*, 1966, 7, 197–206.

Viamontes, J. A. Review of drug effectiveness in the treatment of alcoholism. *American Journal of Psychiatry*, 1972, 128, 100–121.

Warren, G. H., & Raynes, A. E. Mood changes during three conditions of alcohol intake. *Quarterly Journal of Studies on Alcohol*, 1972, 33, 979–989.

Weissman, M. M., & Myers, J. K. Clinical depression in alcoholism. *American Journal of Psychiatry*, 1980, 137, 372–373.

Weissman, M. M., Slobetz, F., Prusoff, B., Mezritz, M., & Howard, P. Clinical depression among narcotic addicts maintained on methadone in the community. *American Journal of Psychiatry*, 1976, 133, 1434–1438.

Williams, A. F. Social drinking, anxiety, and depression. *Journal of Personality and Social Psychology*, 1966, 3, 689–693.

Winokur, G. Alcoholism and depression in the same family. In D. W. Goodwin & C. K. Erickson (Eds.), *Alcoholism and affective disorders*. New York: Spectrum, 1979.

Winokur, G., Rimmer, J., & Reich, T. Alcoholism IV: Is there more than one type of alcoholism? *British Journal of Psychiatry*, 1971, *118*, 525–531.

Winokur, G., Reich, T., Rimmer, J., & Pitts, F. N. Alcoholism: III. Diagnosis and familial psychiatric illness in 259 alcoholic probands. *Archives of General Psychiatry*, 1970, *23*, 104–111.

Woody, G. E., O'Brien, C. P., & McLellan, A. T. Depression in narcotic addicts: Possible causes and treatment. In R. W. Pickens & L. L. Heston (Eds.), *Psychiatric factors in drug abuse*. New vork: Grune & Stratton, 1979.

Woody, G. E., O'Brien, C. P., & Rickels, K. Depression and anxiety in heroin addicts: A placebo-controlled study of doxepin in combination with methadone. *American Journal of Psychiatry*, 1975, *132*, 447–450.

World Health Organization. *The use of cannabis*, (Technical Report Series No. 478). Author, 1971.

5

Substance Abuse in Psychopathic States and Sociopathic Individuals

DAVID C. KAY

1. Introduction

In the minds of most people, and even of some professionals, substance abusers are automatically "sociopaths," and therefore unable to profit from therapy. However, some members of this population are capable of much personal change when the process of therapy operates within certain principles. This chapter will describe some characteristics of sociopathic individuals that might guide attempts to help them change. It is also important to differentiate individuals with long-term antisocial problems that antedate drug abuse (sociopaths) from individuals who experience mood and behavior disorders after introduction to chronic psychoactive drug use. This chapter will differentiate each syndrome, and its relationship to substance abuse.

Since its development by Maxwell Jones and its popularization by Synanon, the therapeutic community has been a staple in the treatment of substance abusers. Yet the essential characteristics of such a community for therapy of antisocial behaviors are still being discovered. Herein are some characteristics that we have found helpful in eliciting socialization.

DAVID C. KAY • Departments of Psychiatry and Pharmacology, Baylor College of Medicine, Houston, Texas 77030.

1. Substance Abuse as Antisocial Behavior

2.1. Definition of Substance Abuse and Antisocial Behavior

2.1.1. Substance Abuse

Although it might seem too elementary to define substance abuse in this chapter, the scope and type of interest by professionals and society in this behavior have most often been reflected by the use of specific labels. Such labels have started as neutral descriptors, only to end as pejorative terms reflecting society's rejection of the individuals involved. Thus, *drug addiction* has been superseded by *drug dependence,* and now by *substance abuse.* To paraphrase Wikler's (1971) definition of drug dependence, *substance abuse* could be defined as *habitual nonmedical substance-seeking and substance-taking behavior resistant to extinction or suppression by its adverse social or pharmacological consequences.* This definition includes the sniffing of solvents such as toluene, ether, or gasoline, as well as the abuse of alcohol, caffeine, and nicotine (by DSM criteria). It also logically could include the abuse of cigarettes, food, and money, although many smokers, hearty eaters, and extravagant spenders might object.

The DSM-III classification of substance use disorders (American Psychiatric Association, 1980) includes two forms. *Substance abuse* is differentiated from nonpathological use by a pattern of pathological use, impairment in social or occupational functioning due to substance use, and a minimal duration of one month. *Substance dependence* is identified by tolerance or withdrawal, with signs of abuse also required for alcohol and cannabis. Although the DSM-III classification may have some heuristic value, using CNS modification (e.g., tolerance, withdrawal) as the indicator of more serious abuse furthers a common misperception. Most, if not all, psychoactive drugs, including neuroleptics and thymolytics, produce CNS modification that becomes evident when they are abruptly stopped after chronic use, but compulsive harmful use of these latter drugs is uncommon. Also, stimulants such as cocaine, which show minimal abstinence symptoms upon withdrawal, demonstrate maximal compulsive drug-seeking behavior in most animal species, including man.

2.1.2. Antisocial Behavior

In contrast with the specificity of the DSM-III definition of antisocial personality disorder, a wide range of antisocial behaviors are seen in substance abusers and other psychiatric patients. In this chapter, anti-social behavior includes all violations of socially approved behaviors, even if they are not codified into laws and regulations (i.e., crimes). Individuals who break big rules invariably violate several small rules along the way, as a reflection of an egocentric disregard of rules.

2.2. Relationship of Substance Abuse to Antisocial Behavior

Human experience with substance abuse dates from antiquity, especially involving alcohol and cannabis. Scientific investigation is relatively recent, with a duration of about 100 years, and most validated information has been discovered only in the last 40–50 years.

The relationships among alcoholism, opiate abuse, and antisocial behavior was early recognized in scientific studies of addiction. Kolb (1925b) found that 39% of the 230 opiate addicts whom he examined had either a history of spree drinking (20.5%) or of continuous excessive drinking (18.7%). Besides the spree drinkers (inebriate personality), Kolb (1925a) and Pescor (1939) also described thrill-seekers (psychopathic diathesis) and habitual criminals (psychopathic personality) as types of addicts. Felix (1944) later included all these groups within a psychopathic personality category.

It has been a common finding to note an elevated *Pd* (psychopathic deviate) scale on the Minnesota Multiphasic Personality Inventory (MMPI) in alcoholics and opiate addicts, even though many different personality patterns are clinically apparent. Astin (1959) described five nonorthogonal factors in the *Pd* scale in opiate addicts: Self-esteem, Hypersensitivity, Social Maladaption, Emotional Deprivation, and Impulse Control. Monroe (Monroe, Miller, & Lyle, 1964) modified and expanded such an analysis, and derived six major orthogonal factors: Intrapunitiveness, Denial of Shyness, Hypersensitivity, Impulse Control, Emotional Deprivation, and Social Maladaption.

Hill, Haertzen, and Davis (1962) analyzed the MMPI patterns of 571 alcoholics, opiate addicts, and criminals. Their composite profiles were markedly similar. Only the differences in the *D* (depression) scale appeared practically significant: Alcoholics and addicts were higher than criminals in this scale. Factor analyses extracted three well-defined factors: (1) an "undifferentiated psychopath," with a single elevation of the *Pd* scale; (2) a bipolar factor, with a "primary psychopath" (elevations of *Pd* and *Ma* [mania] scales), and a "depressed neurotic psychopath" (elevations of *D*, *Hy* [hysteria] and *Pt* [psychasthenia] scales); and (3) a third factor that differentiated neurotic from schizoid patterns.

Components of this MMPI profile in alcoholics, opiate addicts, and criminals (Hill *et al.*, 1962) have also been reported by other authors in related populations (Black & Heald, 1975; Frame & Osmond, 1956; Hewitt, 1943; Rosen, 1960). Haertzen (1978) found that opiate addicts have an elevated *Pd* scale regardless of demographic (age, sex, ethnicity) or institutional status (prisoner, probationer, civil commitment, volunteer). Abusers of stimulants, sedative-hypnotics, or hallucinogens also have similar MMPI scale elevations (Cox & Smart, 1972; Penk & Robinowitz, 1976; Schoolar, White, & Cohen, 1972; Smart & Jones, 1970).

2.3. Correlation of Substance Abuse and Sociopathy

Barton (1982) found that 68% of a sample of 5,300 local jail inmates had a history of illicit drug use, and that 44% had used such drugs in the month prior to jail; among the convicted inmates, 21% reported being under the influence of such drugs during the offense for which they were convicted. Of those who had used drugs, 24% had been enrolled at one time in a drug treatment program, but only 6% were enrolled when jailed. Greene (1981) found that a significant number of 1,544 members of drug and alcohol treatment centers were involved in crimes during 1975–1976. Alcohol was more associated with crimes against persons, or with victimless crimes, while use of other drugs was more associated with property crimes.

Although such correlations have been common, differentiations can be made. Vaillant (1982) found in a prospective study of 400 inner city men aged 47 that sociopathy and alcoholism appeared to be independent disorders. Adult individuals diagnosed as sociopathic by the Robins criteria (Robins, 1966) appeared more socially isolated and lacking in a sense of self than did adult alcoholics. Best childhood predictors of adult alcoholism were parental ethnicity and number of family members with alcoholism; for adult sociopathy, best predictors were boyhood ego strength (high), multiproblem family membership, and infant restlessness. Wilber, Rounsaville, Weissman, and Kleber (1982) have been able to divide 363 opiate addicts into three groups: (1) an initial childhood trauma (e.g., parental death; child abuse) group (31%) had more severe psychopathology and poorer social functioning; (2) an initial delinquency group (24%) had heavier opiate use, more arrests for violent crimes, and more diagnoses of antisocial personality disorder; and (3) an initial drug use group (45%) had lowest rates of arrests and psychiatric disorders, and most adequate social and occupational functioning.

3. Sociopathic Personality

3.1. Definition of Sociopathic Personality

According to the DSM-III (American Psychiatric Association, 1980), the antisocial personality disorder is characterized by "a current age over 17; an onset before age 15 of (3 or more) truancy, school misbehavior, delinquency, running away from home, persistent lying, repeated casual sexual intercourse, repeated substance abuse, thefts, vandalism, school underachievement, chronic rule violations, initiation of fights; evidence after 18 of (4 or more) inability to sustain consistent work behavior, lack of ability to parent responsibly, repeated illegal behavior, marital failure, physical aggressiveness, financial failure, impulsivity, repeated lying, or recklessness; no spontaneous break in antisocial behavior over 5 years after the age of 15; and absence of mental retardation, schizophrenia or mania."

Hare and Cox (1978) found that 40% of 145 white, male, Canadian, imprisoned criminals met these criteria for antisocial personality disorder, while 76% met the criteria of two behaviors prior to age 15 and three behaviors after age 18 (Hare, 1980). The DSM-III diagnosis clearly entails difficulties in socializing since childhood, with involvement in multiple areas of life. When a person's problems with socializing are more limited in scope or time, the diagnosis is less certain. Most (75–90%) individuals involved in substance abuse have less severe social pathology than that seen in individuals who qualify for the DSM-III diagnosis of antisocial personality disorder.

3.2. History of the Concept of Sociopathic Personality

The type of antisocial behavior now labeled as sociopathic has been recognized for several centuries. It was considered a type of mental disorder by the time of Pinel (1809), who used the label of *manie sans delire* to describe aimless antisocial behavior (Maughs, 1941; Pichot, 1978). J. C. Pritchard (1835) coined the term *moral insanity* for such antisocial behavior, and Esquirol (1838) included this behavior among his *monomanias.*

Morel (1857), Magnan (Magnan & Legrain, 1895), and Lombroso (1911) espoused the idea of *moral degeneracy:* that certain individuals have a hereditary predisposition to moral pathology. Koch (1891) developed the term *psychopathic inferiority*, and Kraepelin (1915) the terms *psychopathic state* and *psychopathic personality.* However, these and later

German authorities (Schneider, 1950) applied such terms to many personality disorders besides those expressed in antisocial behavior. English and American authorities have generally used the term *psychopathic* (or *sociopathic*) *personality* in the sense of Pritchard (a disorder associated with antisocial behavior), and we shall use that term in such a fashion throughout this discussion. *Sociopathic personality disturbance: antisocial reaction* was introduced as a modern equivalent to Pritchard in the first edition of the APA Diagnostic and Statistical Manual (American Psychiatric Association, 1952), and *antisocial personality* in the DSM-II (American Psychiatric Association, 1968).

Cleckley (1941) has provided the most extensive clinical profile of the sociopath. He considers 16 features to be characteristic: (1) superficial charm and good "intelligence"; (2) absence of delusions and other signs of irrational thinking; (3) absence of "nervousness" or psychoneurotic manifestations; (4) unreliability; (5) untruthfulness and insincerity; (6) lack of remorse or shame; (7) inadequately motivated antisocial behavior; (8) poor judgment and failure to learn from experience; (9) pathologic egocentricity and incapacity for love; (10) general poverty in major affective relations; (11) specific loss of insight; (12) unresponsiveness in general interpersonal relations; (13) fantastic and uninviting behavior with drink and sometimes without; (14) threats of suicide, rarely carried out; (15) sex life impersonal, trivial, and poorly integrated; and (16) failure to follow any life plan. As can be seen by studying this list, Cleckley's definition rests on several clinical judgments, rather than just observable phenomena. However, it appears to be internally consistent and clinically reproducible. He views the sociopath as having a disorder equivalent to psychosis, but with a "mask of sanity."

Investigators at Washington University (Robins, 1966) also have defined the sociopathic personality, with the exclusion of schizophrenia, chronic brain syndrome, and mental retardation. Their behavioral criteria for sociopathic personality include: (1) chronic failure to conform with social norms; (2) failure to maintain close personal relationships; (3) a poor work record; (4) engaging in illegal activities; (5) problems maintaining support; (6) sudden changes in plans; and (7) a low frustration tolerance. Except for the last two, these criteria primarily are historical, and should reflect the most persistent patterns of sociopathic behavior.

The Washington University criteria have also been dominant in the current APA definition of *antisocial personality disorder* (American Psychiatric Association, 1980). Such objective criteria should at least ensure definition of a consistent group of individuals for research and clinical studies. A major difficulty, however, which would stem from this dominant historical approach, is the resultant inability to discover significant therapeutic shifts because the history cannot be changed.

There has been a recent revival in interest in sociopathy and psychopathy, as evidenced in several books (Hare, 1970; Hare & Schalling, 1978; Reid, 1978a; Smith, 1978; Yochelson & Samenow, 1976, 1977). Discussion of psychopathy has even penetrated into the general press (Harrington, 1971).

3.3. Current Research on Sociopathy

3.3.1. Biological correlates of Sociopathic Personality.

Hare (Hare, 1970; Hare & Schalling, 1978) has summarized several studies that support the hypothesis that biological defects underlie the phenomena of sociopathic behavior. EEG bursts of positive temporal spikes (6–8, and 14–16 cycles per second) have been associated with conscious, impulsive, aggressive, and remorseless behavior (Hughes, 1965); this has often been seen in sociopathic individuals (Kurland, Yeager, & Arthur, 1963). Sociopathic individuals show a pattern of autonomic response characteristic of drowsy, cortically underaroused subjects (Forssman & Frey, 1953; Hare, 1968; Stern & McDonald, 1965). Sociopaths (Gellhorn, 1957) are more likely to have only a brief hypotensive response to mecholyl (N response), although some have a large and prolonged hypotensive response (E response). During stress, N responders tend to become aggressive and E responders anxious (Fine & Sweeney, 1968; Kaplan, 1960). Quay (1965) and Zuckerman (1978) view the sociopath's impulsivity, need for excitement, and intolerance of boredom as the result of a pathological need for stimulation. Petrie (1967) found sociopathic individuals to be stimulus reducers (underestimators) in their modulation of sensation, and thus more upset by isolation than by pain.

Trasler (1978) views sociopaths as defective in the ability to learn in a passive avoidance conditioning mode; they fail to respond to signals of impending punishment (conditional aversive stimuli), while having no difficulty with conditioned appetitive stimuli. Mednick and Christiansen (1977) present experimental and epidemiological evidence to support the concept that delinquency (antisocial or asocial behavior) is associated with autonomic nervous system hyporeactiveness and delayed electrodermal recovery.

3.3.2. Psychological Measurement of Sociopathic Personality

The most prevalent measure of sociopathy in the United States is the Psychopathic Deviate (*Pd*) scale of the MMPI. This scale consists of 50 items that were initially derived (McKinley & Hathaway, 1944) from a largely juvenile population undergoing psychiatric evaluation for delinquent behavior (noncapital offenses). Probably as a reflection of the original derivation of its items, the *Pd* scale tends to be more sensitive to low levels of sociopathy (a short history of antisocial behavior), and not so sensitive to changes (such as drug effects) in individuals with a high level of sociopathy (i.e., a long history of antisocial behavior). It includes many expressions of angry rebellion, especially directed toward the family. The Mania (*Ma*) scale of the MMPI has also been used as a measure of sociopathy, usually in association with the *Pd* scale. This scale consists of 46 items that were developed to estimate the "overactivity, emotional excitement and flight of ideas" of hypomania (Dahlstrom & Welch, 1960).

Some newer MMPI scales have been developed that are better candidates for measuring degree of sociopathy in older delinquents. The Antisocial (*Ant*) scale was developed by Haertzen (Haertzen, Hill, & Monroe, 1968) from the Social Maladaption scale (Astin, 1959; Monroe *et al.*, 1964). It differentiates degree of criminality and level of addiction in substance abusers. In both its trait and state format, this scale differentiated addicts and alcoholics from non-drug-using controls, whereas the *Pd* scale did not (Haertzen, Martin, Hewett, & Sandquist, 1978).

When seeking an MMPI scale that would differentiate prisoner populations on the basis of Cleckley-like clinical dimensions, Spielberger (Spielberger, Kling, & O'Hagan, 1978) developed a Sociopathy (*Spy*) scale. The major characteristic of this 20-item scale is toughness. Individuals rating high on this scale deny any physical weakness, deny shyness, and glorify aggressiveness.

A Psychopathic (*Pyp*) scale has been developed for the Addiction Research Center Inventory (ARCI). This 71-item scale (Haertzen, 1974) was revised from one developed earlier by Haertzen and Panton (1967) in 785 subjects in order to distinguish sociopathic (criminal, addict, alcoholic) groups from nonsociopathic (normal, mentally ill) groups. In its modified form (Haertzen, 1974), it has high internal reliability (KR-20 = 0.75) and is useful as a measure of individual differences. Criminals and addicts are the most deviant and best differentiated from normals and mentally ill samples, with alcoholics intermediate. The ARCI itself is a 550-item questionnaire that contains many items that are presumptive indicators of sociopathic deviancy. It has been developed to demonstrate specific patterns of drug effects, as well as characteristics of personality and clinical diagnoses (Haertzen, 1974; Hill, Haertzen, Wolbach, & Minter, 1963). The California Psychological Inventory (CPI; Gough, 1957) also has scales that tend to reflect sociopathic tendencies: the Socialization (So) scale and the Responsibility (Re) scale.

3.3.3. Sociological Measurement of Sociopathic Individuals

Robins (1966) studied the outcome 30 years later of 524 child guidance clinic patients and 100 controls. Of the patients, 406 had been referred for antisocial behavior and 118 for other reasons. Using criteria developed at the Washington University department of psychiatry, they considered (but did not mandate) the diagnosis of sociopathic personality in these adults only if the individual demonstrated violations of societal goals in at least 5 of the 19 life areas described. All but one of these criteria (truancy) referred to behavior after age 18. It is important to note that they did not include in their criteria the 1952 DSM idea that these persons "profit neither from experience nor punishment, and maintain no real loyalties to any person, group or code" (American Psychiatric Association, 1952).

Robins and associates located 90% of the 624 subjects (and records on 98%); 94 of these adult subjects were diagnosed as having a sociopathic personality. A majority of these 94 individuals had histories of poor work functioning (85%), financial dependency (79%), repeated arrests (75%), poor marital stability (74%; 81% of those married), heavy drinking (72%), school problems and truancy (71%), impulsive behavior (67%), sexual promiscuity or perversion (64%), recklessness as an adolescent (62%), vagrancy (60%), physical aggression (58%), and social isolation (56%). Poor marital history, impulsiveness, vagrancy, and the use of aliases distinguished the adult subjects with sociopathic personality from those with anxiety neurosis, hysteria, schizophrenia, or alcoholism. The group diagnosed as alcoholics (29 men) were very similar to the sociopaths: 70% of the alcoholics had adult antisocial symptoms in at least five life areas, although these problems were not as severe as those of the sociopathic group. It must be recognized that, for these authors, these two diagnoses were exclusive: If an individual had enough evidence to be diagnosed sociopathic, his alcohol problems would be described as part of his sociopathy rather than being the basis of an additional diagnosis of alcoholism.

Antisocial behavior as a child was significantly associated with sociopathic personality as an adult. However, 406 children had antisocial behavior referrals and only 94 adults were diagnosed as sociopathic. This is a reflection of the stricter criteria for the adult diagnosis, as well as a reflection of differing outcomes of those with antisocial childhood problems. Antisocial symptoms in children were powerful predictors of sociopathy, hysteria, and alcoholism. They also predicted the level of antisocial behavior in adults who were psychotic, or who had less anti-social behavior than that necessary to be diagnosed as sociopathic. So-ciopathic personality could be predicted about equally by three child-hood measures: (1) the variety (number of areas) of antisocial behavior, (2) the number of episodes of antisocial behavior, or (3) the seriousness of the antisocial behavior (court involvement). Many measures of pover-ty or deprivation were not predictors of sociopathic disease independent of the child's antisocial behavior (Robins, 1966).

Hewett and Martin (1980) have also found evidence for a long histo-ry of social problems in alcoholics and opiate addicts, in contrast to control subjects with little drug use. They developed a Personal History Questionnaire (PHQ) with scales to assess alcohol and drug abuse, adult sociopathy, adult nonsociopathic difficulties, and developmental (child-hood and adolescent) difficulties. They found that both groups of sub-stance abusers scored greater than controls in PHQ measures of devel-opmental and adult *nonsociopathic* difficulties: Substance abusers had more adult depression, sleep disorder, daytime restlessness, and per-ceptual dysfunction. They also had more childhood and adolescent learning problems, sleep disorders, and daytime restlessness.

Childhood antisocial behavior was greater than controls in sub-stance abusers. As adults, they had more antisocial behavior, criminal behavior, work instability, legal difficulties (both misdemeanors and felonies), and problems with alcohol and other drugs. Incidentally, these substance abusers also had significantly higher scores on several standard MMPI scales (*Pd, Sc, Ma, Pt,* and *D*), although the Antisocial scale differentiated them best (Hewett & Martin, 1980).

3.3.4. Self-Derogation as an Antecedent to Antisocial Behavior

Kaplan (Kaplan & Pokorny, 1969) has developed a measure of self-rejection that involves: (1) a wish to have more respect for self, (2) a general dissatisfaction with self, (3) a feeling of not having much for which to be proud, (4) a feeling of being a failure, (5) the lack of a positive attitude toward self, (6) at times, thinking of self as no good at all, and (7) feeling useless at times. Kaplan theorized that individuals move into social deviancy to reduce their self-derogation. Initial high scores on this self-derogation scale have been associated with later social deviancy and reduction in self-derogation scores (Kaplan, 1975, 1978). Such self-rejection is related to prior devaluation by the individual's membership group, as well as to his degree of defenselessness (Kaplan, 1976). Thus, self-derogation could be considered compatible with a temporary dysphoric mood prior to antisocial behavior, even in adults.

4. Psychopathic States

4.1. Definition of Psychopathic States

One of the more recent explanations of some antisocial behaviors in individuals who do not qualify as having an antisocial personality disorder is that they are experiencing a psychopathic state. *Psychopathic states* may be defined as *unpleasant feeling states of variable character and duration, associated with an urge toward impulsive antisocial behavior.* They appear to be biological as well as psychological states, and may be induced by chronic intake of psychoactive drugs of abuse. An alternate way to view antisocial personality disorders (including sociopathic personality) is that they are severe, chronic psychopathic states.

4.2. History of the Concept of Psychopathic States

4.2.1. Biological Phenomena of Psychopathic States

Characteristic abstinence patterns are seen upon withdrawal of opioids (Martin & Jasinski, 1969; Martin, Jasinski, Haertzen, Kay, Jones, Mansky, & Carpenter, 1973) or sedative-hypnotics (Fraser, Isbell, Eisenman, Wikler, & Pescor, 1954; Isbell, Altschul, Kornetsky, Eisenman, Flanary, & Fraser, 1950; Isbell, Fraser, Wikler, Belleville, & Eisenman, 1955). Up to the first 6–9 weeks after withdrawal of morphine or other opioids, humans experience a relative hypertension, tachycardia, hyperthermia, mydriasis, hyperpnea, anorexia, restlessness, and weight loss; nausea, emesis, diarrhea, muscle cramps, sweating, rhinorrhea, and insomnia are also part of this acute abstinence pattern. The MMPI increases in Hypochondriasis (*Hs*), Schizophrenia (*Sc*), and Hysteria (*Hy*) scales. The ARCI shows an increase in the *PCAG* (pentobarbital-chlorpromazine- alcohol group), *weak, tired, social withdrawal,* and *chronic opiate* (negative feeling state) scales, with a decrease in the *MBG* (morphine- benzedrine group), *efficiency* and *competitive* (positive feeling state) scales and an increase in *alcohol withdrawal* and *opiate withdrawal* scales (Martin *et al.*, 1973).

About 9 weeks after withdrawal of an opioid, a different pattern begins to emerge. In humans this is characterized by relative hypotension, bradycardia, hypothermia, miosis, tachypnea (Martin & Jasinski, 1969; Martin *et al.*, 1973), decreased respiratory center sensitivity to carbon dioxide (Martin, Jasinski, Sapira, Flanary, Kelly, Thompson, & Logan, 1968), increased delta sleep and REM sleep (Kay, 1975), increased cold pressor response (Himmelsbach, 1941), and increased norepinephrine secretion (Eisenman, Sloan, Martin, Jasinski, & Brooks, 1969). In laboratory rats, there is seen increased metabolic rate and increased fluid intake (Martin, Wikler, Eades, & Pescor, 1963), with "wet-dog shakes" and increased sleep (Khazan & Colasanti, 1972) and increased aggression (Gianutsos, Hynes, Puri, Drawbaugh, & Lal, 1974). In laboratory dogs, there is seen mydriasis, hypopnea, and hyperalgesia (Martin, Eades, Thompson, Thompson, & Flanary, 1974).

The existence of such protracted abstinence patterns was first established in rats (Martin *et al.*, 1963), and then in humans (Martin & Jasinski, 1969) and dogs (Martin *et al.*, 1974). There is evidence that the abstinence syndrome lasts more than 4 months in humans (Martin *et al.*, 1973).

Withdrawal of barbiturates and of alcohol is also known to be associated with a characteristic abstinence pattern. Upon the initial withdrawal of alcohol, tremulousness, nausea, perspiration, and insomnia are followed by marked tremor, weakness, vomiting, diarrhea, hyperreflexia, fever, and hypertension. The most severe abstinence symptoms are associated with convulsions ("rum fits") and delirium (Isbell *et al.*, 1955), as well as abnormalities (spikes and slow-wave bursts) in the EEG. Upon initial withdrawal of barbiturate (Fraser *et al.*, 1954; Isbell *et al.*, 1950), weakness, tremor, great anxiety, anorexia, nausea and vomiting, rapid weight loss, tachycardia, tachypnea, fever, hypertension, postural hypotension, *grand mal* convulsions, and psychosis develop. The withdrawal psychosis is characterized by anxiety, agitation, insomnia, confusion, disorientation (time and place), delusions, and auditory and visual hallucinations. Anxiety temporarily decreases after each convulsion. No gross abnormalities could be detected 2–3 months after withdrawal of alcohol or barbiturates, although the existence of a (subtle) protracted abstinence syndrome was not postulated at the time of those studies.

A search for protracted abstinence patterns after sedative-hypnotics, including alcohol, or other drugs of abuse, has not been accomplished, although several physiological measures are known to be persistently abnormal after drug withdrawal in alcoholics.

4.2.2. Psychological Characteristics of Psychopathic States

The phenomena of protracted abstinence, as detailed above, have been defined primarily in the biological sphere, with some indication that psychological status (feeling state) and behavior are also involved. In evolving a hypothesis about the pharmacological factors in drug abuse, Martin (Martin, Haertzen, & Hewett, 1978; Martin, Hewett, Baker, & Haertzen, 1977) proposed that the state of protracted abstinence might well be a model for feeling states in individuals *before* as well as *after* chronic psychoactive drug intake.

Because such psychological states appeared to be dominated by a particular type of unpleasant feeling state ("hypophoria") that was reversible by psychoactive drugs, Martin hypothesized that drug abusers (including both opiate addicts and alcoholics) were afflicted by an increased need-state (Martin *et al.*, 1977). This increased need-state naturally would lead to the several behaviors labeled as "psychopathic" or "sociopathic."

Martin designed a maturation (*Mat*) scale with five subscales to include pertinent dimensions of such pathological states: Impulsivity, Egocentricity, Need, Hypophoria, and Sociopathy. His total *Mat* scale is actually scored positively for immaturity. The use of the maturation concept incorporates the idea that these behaviors and feelings are evidence of immaturity, which diminishes with age and further development as such needs diminish. Martin (Martin *et al.*, 1977) tested this instrument and others in three populations: 53 exalcoholics, 24 imprisoned (and presumably drug-free) opiate addicts, and 54 controls (persons who used very little of any psychoactive drug). All *Mat* subscales (except Egocentricity) and the total *Mat* score clearly differentiated controls from both groups of substance abusers, and most subscales differentiated addicts from alcoholics. In the MMPI, the *Pd*, *Ma*, and *D* scales also differentiated controls from substance abusers, as did plasma levels of testosterone and LH.

As part of this endeavor to define the current feeling states of substance-free alcoholics and addicts, Haertzen (Haertzen *et al.*, 1978) developed the Social Experience Questionnaire (SOEX). This questionnaire was constructed by systematic revision of several MMPI, CPI, and ARCI psychopathy scales, so that only comparable current feelings, thoughts, motives, and actions were assessed. This instrument was then tested also in the 53 alcoholics, 28 opiate addicts, and 54 low-drug-using controls described above. It was found to significantly differentiate controls from substance abusers in 58% of its 560 items. Thus, the current status of substance abusers' life experience and condition, including mood (current contextless feelings), is different from that of nonabusers.

Haertzen conceived the SOEX as measuring the current *state of psychopathy*, and thus *psychopathic state* (Haertzen *et al.*, 1978; Haertzen, Martin, Ross, & Neidert, 1980). He further refined the concept by developing six rational psychopathic state scales. Extensive scales were first derived that included Martin's *Mat* items, and which had been judged to measure: (1) the search for highs, (2) impulsivity, (3) egocentricity, (4) increased needs, (5) hypophoria, and (6) sociopathic attitudes. After these had been developed and then tested in the three populations, a shorter (90 item) version (the Psychopathic State Inventory; PSI) was developed with items that best correlated with the parent scales, and best differentiated substance abusers from controls (Haertzen *et al.*, 1980). It should be noted that the PSI scales are not exactly the same as the *Mat* subscales developed with the same names by Martin (Martin *et al.*, 1977). The PSI scales have had the benefit of both rational development (from a large pool of SOEX items) and of empirical testing prior to definition.

4.3. Current Research on Psychopathic States

4.3.1. The Delineation of Hypophoria

A crucial element of this discussion of psychopathic states is Martin's concept of hypophoria. Martin defined this negative feeling state as comparable to depression, but without feelings of unworthiness, sleep disorders, anorexia, decreased libido, or inability to experience joy (Martin *et al.*, 1977, 1978). The Hypophoria scale in the *Mat* "includes items relating to a general negative perception of life; a poor self-image; feelings of being disrespected, disapproved of and unappreciated; feelings of inefficiency or ineptness; and withdrawal from competition, worry and anger." Martin conceptualized this feeling state as a reaction to the social frustration of the increased need-state of sociopathic individuals. Although he defined hypophoria as a feeling state, Martin included elements of attitudes (poor self-image, unpopular, inefficient, withdrawal) that are quite complex.

Hewett (Hewett & Martin, 1980) cited the work of Winokur (Cadoret & Winokur, 1974; Pitts & Winokur, 1966; Winokur, 1974; Winokur, Cadoret, Baker, & Dorzab, 1975; Winokur, Cadoret, Dorzab, & Baker,

1971; Woodruff, Guze, Clayton, & Carr, 1973) in support of a "depressive spectrum disease" that appears more related to alcoholism than it is to primary depression. She found evidence in the personal history of substance abusers for persistent low moods without severe depression; these started at an early age before the onset of drug abuse. She concluded that these data are evidence for the existence of an "affective-antisocial spectrum disorder characterized by early onset, low moods, and alcohol abuse, drug abuse and/or antisocial behaviors."

Haertzen (Haertzen *et al.*, 1980) utilized the concept of hypophoria to encompass "negative feeling states such as depression" and developed the Hypophoria scale in the PSI with items that were correlated with low mood items. These items do not include much about secondary "feelings," such as judgments about how others react.

In considering hypophoria, Cowan and Kay (Cowan, Kay, Neidert, Ross, & Belmore, 1980) distinguished four elements in this concept: lack of confidence, unpopularity, anergia, and joylessness. After they segregated SOEX items that appeared to express these four elements, factor analysis demonstrated only two components: Defeated (lack of confidence, unpopularity, and anergia), and Joyless. The Defeated scale differentiated both substance-abuser groups from controls, but not from each other; the Joyless scale differentiated control and alcoholic groups from addicts, but not from each other.

In considering sociopathic individuals (not just substance abusers), Reid (1978b) conceptualizes the psychopath ("aneothopath") as suffering a type of mood disorder ("sadness"), but considers it primary rather than reactive to social constrictions, and to be present in individuals with several varieties of antisocial behavior. He sees that the "nothing which . . . lies at the center of the psychopath is a lack of energy, or living force," with defensive avoidance of such an endogenous depression.

4.3.2. Use of the Psychopathic State Inventory (PSI)

We have been using the PSI (Haertzen *et al.*, 1980) to assess the degree of initial pathology in substance abusers. In Table 1, the mean, standard deviation, and range of values for each subscale and the total PSI scale are presented for 370 dependent polydrug abusers admitted to a substance abuse treatment unit. As can be seen in comparison to the populations tested by Haertzen, described earlier, our population shows significant elevation over the low-drug users on all scales. Mean elevations on the drug-craving (High), Egocentric, and Hypophoria scales tend to exceed those of imprisoned, nondependent ("clean") opioid

Table 1
Psychopathic State Inventory in Dependent Polydrug Abusers[a]

PSI scale	370 Dependent abusers on admission			Low-drug user (mean)	Nondependent alcohol abusers	Nondependent opiate abusers
	Mean	Standard deviation	Range			
High	6.75	3.60	0–15	0.61	3.01	5.07
Impulsive	7.35	3.04	0–14	1.80	5.19	8.07
Egocentric	8.12	2.51	1–14	2.52	6.02	6.86
Need	7.62	2.33	1–14	2.37	5.77	7.86
Hypophoria	9.65	3.49	0–15	1.32	4.55	7.96
Sociopathy	6.81	3.32	0–15	1.28	3.85	8.18
Total	46.2	13.3	7–81	9.89	28.38	44.00

[a]Average responses to the PSI by 370 polydrug abusers admitted for inpatient treatment, as compared to the means of populations of 54 low- (or non-) drug users, 53 nondependent ex-alcoholics, and 28 nondependent (imprisoned) opiate abusers (Haertzen *et al.*, 1980).

addicts. However, imprisoned addicts tend to score higher than our polydrug abusers in the Impulsive and Sociopathy scale scores. We would expect that most of these PSI scores would be high upon an individual's admission to a substance abuse inpatient treatment program, and that later scores would decrease to the degree that his drug-induced psychopathic state responded to treatment or time.

We also have been monitoring the progress of patients in our program by repeated PSI evaluations. As illustrated in Figure 1, the PSI reflects an initial pattern of denial in a stimulant abuser (A.B.), with a subsequent progression in therapy to experiencing his psychopathic state (i.e., less denial), and then gradual resolution. There is also an indication in this patient of recurrent monthly aggravations of unpleasant mood, which suggests that a psychopathic state might have a periodicity.

In Figure 2, the PSI reflects the mood changes in a more seriously disturbed polydrug abuser (C.D.). He gradually improved long after admission, shortly before assuming the role of an officer in the patients' therapeutic community, and then reverted to his prior dysphoric state after a familial disturbance, but prior to stepping down as a community officer. He secretly resumed episodic drug use a few days after his PSI scores had reverted to a high level. It can also be seen that his PSI scores after passes lagged behind his before-pass PSI scores in the hospital, both as he improved and as he worsened.

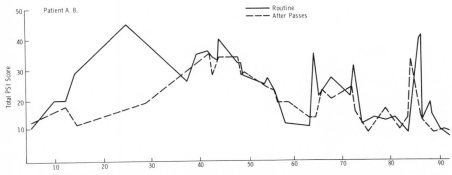

Figure 1. The progression of total PSI scores during treatment of patient A.B. Maximum potential total PSI score is 90 on any occasion. Time is marked every 10 days. PSI was tested on admission, episodically during hospital course, and regularly before and after any pass. PSI scores after passes are indicated separately by the dashed line. This patient shows initial denial, and then increased scores while working on his problems.

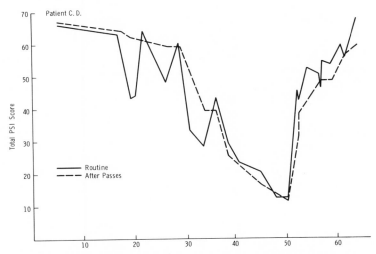

Figure 2. The progression of total PSI scores during treatment of patient C.D. Maximum potential total PSI score is 90 on any occasion. Time is marked every 10 days. PSI was tested on admission, episodically during hospital course, and regularly before and after any pass. PSI scores after passes are indicated separately by the dashed line. This patient shows temporary improvement (or increased denial) while holding a position of responsibility.

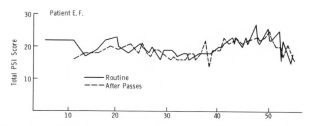

Figure 3. The progression of total PSI scores during treatment of patient E.F. Maximum potential total PSI score is 90 on any occasion. Time is marked every 10 days. PSI was tested on admission, episodically during hospital course, and regularly before and after any pass. PSI scores after passes are indicated separately by the dashed line. This patient shows a very guarded response until late in his stay.

In Figure 3, one of the major problems in recurrent testing can be demonstrated: Some patients will resist transmitting information unless active feedback is maintained. Some patients on our unit "forgot" to complete the PSI, or some items on it, until they lost a pass as a result. This patient (E.F.), as did others, then made exact or slightly modified copies of an earlier response that they had formulated; after staff feedback, E.F. began reflecting his mood variations in latter reports, as can be seen in Figure 3. We have had to be alert to obvious behavioral and mood changes that can be seen by observers, but are not reflected in the PSI. Such clinical validation, as well as supervised group answers to an oral presentation (to obviate reading problems and the aggravation of having to alternately read the questions and then locate the appropriate answer blanks) have been helpful in establishing the reliability of such a screening device.

4.3.3. Other Evidence of Increased Psychopathology in Substance Abusers

Kleber and associates (Rounsaville, Weissman, Kleber, & Wilber, 1981; Wilber, Rounsaville, Weissman, Critz-Christoph, & Kleber, 1981) found that 69% of 638 opiate addicts currently met criteria for a psychiatric disorder in addition to their substance use disorder. Depression (39%), alcoholism (14%), phobias (8%), antisocial personality (26%), labile personality (18%), and cyclothymic personality (4%) were much more prevalent in these opiate abusers than in a comparable community sample (Rounsaville *et al.*, 1981). In 157 opiate addicts, over half had elevated Beck Depession Inventory scores on admission to treatment, and one-fifth met Research Diagnostic Criteria for a major depressive episode (Wilber *et al.,* 1981). Most depressive symptomatology improved spontaneously within 6 months, but 25–30% of the group had persistent mild to moderate depression, associated with poorer functioning. Addicts with a major depressive disorder on admission had increased relapse to drug use and poorer social functioning 6 months later. Ward and Hemsley (1982) found elevated perceptual and cognitive dysfunction in 38 polydrug abusers on admission to treatment, with subsequent decrease in the 24 patients still present at 4 weeks and the 15 patients still present at 8 weeks. These abusers also had persistently elevated Psychoticism and Neuroticism scores on the Eysenck Personality Questionnaire. Graf, Baer, and Comstock (1977) found elevated MMPI profiles consistent with psychosis on admission of 66 non-narcotic drug abusers, with resolution into a sociopathic profile in 42 patients remaining 2 weeks after admission. Although heroin addicts may be mildly depressed on the Beck Depression Inventory, their degree of suicidal intent was more correlated with their degree of hopelessness than with their depression (Emery, Steer, & Beck, 1981).

5. Discussion

5.1. Psychopathology in Sociopathy and Substance Abuse

5.1.1. Sociopathic Personality as a Psychiatric Disease

One of the most consistent and reproducible dimensions in defining a sociopathic personality is a long and repetitive history of social problems (sociopathy, in the sense of *social pathology*). This phenomenon has led to the Washington University and American Psychiatric Association criteria for diagnosis of antisocial personality disorder. A history of antisocial problems is also included to varying degrees in psychological tests of psychopathy, as discussed above.

The clearest evidence for the existence of sociopathic personality as a psychiatric disorder is presented by Robins (1966), who argues that a common set of symptoms appearing in persons with a similar age of onset, a similar family history of the disorder, and a similar course of the disorder, constitutes a psychiatric disease. She observed that sociopathic personality:

> occurs in children whose fathers have a high incidence of the disease and whose siblings and offspring also appear to have an elevated incidence. The

symptoms follow a predictable course, beginning early in childhood with
illegal behavior and school discipline problems and continuing into adulthood as illegal behavior, marital instability, social isolation, poor work history, and excessive drinking.

To the degree that sociopathic personality is an expression of a
biological disease, then biochemical or pharmacological treatment might
become possible, and would extend the current limits of psychological
or social treatment.

5.1.2. Acute or Chronic Depression in Substance Abusers

Many psychiatric observers have remarked on the relationship of
substance abuse to depressive symptoms or mood disorders. Rado explained the impulse to use drugs on the presence of a drug-induced
"narcotic riddance" of unpleasant feelings, which riddance contributes
to a "narcotic superpleasure" (Rado, 1964). As Kleber and his associates
have noted recently (Rounsaville *et al.*, 1981; Wilber *et al.*, 1981), depression and other psychiatric disorders are common in substance abusers.

It is noteworthy that most of the symptoms of depression are modified by psychoactive drugs: A *sad mood* can be eased (temporarily) by the
euphoria produced by opioids, sedative-hypnotics, stimulants, and hallucinogens, and it can be increased during chronic use of several psychoactive drugs; a *thought disorder* (hallucinations, delusions) can be produced by stimulants and hallucinogens and by withdrawal of sedative-
hypnotics, and partially blocked by opioids; *psychomotor retardation* can be
produced by acute and chronic opioids, chronic sedative-hypnotics, and
acute and chronic hallucinogens; *increased psychomotor activity* can be seen
during acute use of opioid, stimulant, or hallucinogenic drugs, or during
withdrawal from opioids, stimulants, sedative-hypnotics, or hallucinogens; *weight (and appetite) loss can occur during stimulant use or withdrawal
from opioids or sedative-hypnotics; weight gain* can occur during chronic
opioid or sedative-hypnotic use (unless food competes with drug for the
abuser's money); *sleep loss* can occur with acute or chronic stimulant use,
acute opioid use or withdrawal, or sedative-hypnotic withdrawal; *excessive sleep* can be produced by acute or chronic sedative-hypnotic use, or
withdrawal from stimulants; *decreased sexual activity* can occur with acute
opioid use, or chronic use of any psychoactive drug; and, *increased sexual
activity* has been reported (but not documented) during acute stimulant,
hallucinogen, and sedative-hypnotic use (possibly from disinhibition),
and during withdrawal from opioids. Therefore, monitoring the presence
or progress of depressive symptoms or signs is difficult in an individual

who is repeatedly modifying his emotional and mood state by use of psychoactive drugs (or abstinence from them).

5.1.3. Hypophoria in Substance Abusers

As noted earlier, Martin has postulated the presence of an altered mood state distinct from depression in substance abusers. He used the term *hypophoria* to refer to a complex negative feeling state without feelings of worthlessness, sleep disorder, anorexia, decreased libido, or anhedonia (Martin *et al.*, 1977). This may be conceptualized (but not by Martin) as the absence of pleasure rather than the presence of displeasure, and it can be counteracted (temporarily) by psychoactive drugs of abuse. Other theorists have postulated comparable concepts in *boredom*, *anomie*, or *ennui*, as associated with general social problems.

Martin (1977) sees hypophoria as a feeling state opposite to the euphoric symptoms expressed in the *MBG* scale of the ARCI (Haertzen, 1974), and hypothesizes that hypophoria is one of several complex but unique feeling states that have specific neuronal and neurochemical substrates.

5.1.4. Acute and Chronic Psychopathic States in Substance Abusers

Several benefits would result from the concept of a definable psychopathic state: First, the use of such a "tonic" state measure could provide a much better estimate of the significant and variable shifts in mood in individuals with psychopathic (sociopathic) characteristics, rather than tabulating historical events or waiting for episodic ("phasic") phenomena such as drug taking.

Second, antisocial behavior could be differentiated according to the presence of a concurrent aversive mood state, and the concept of chronic psychopathic state could be used to investigate shifts of feeling state in individuals who are resistant to modification of their behavior (and thus show little obvious behavioral effect initially).

Third, treatment programs for alcoholism, opiate addiction, and other psychopathic disorders would benefit from the use of measures of psychopathic state to assess the degree and duration of improvement from various treatments. This could be accomplished without requiring drug intake (usually a measure of failure) to be the measure of success.

Fourth, such psychological measures of psychopathic state could provide validation for biological and behavioral markers of minor types of psychopathy, and thus lead to better understanding of the neuropathology of psychopathic states.

5.2. Principles of Value in the Treatment of Psychopathic States

5.2.1. Behavioral Inhibition as a Precondition for Therapy

The theory has been proposed that individuals with antisocial behavior problems do not respond adequately to negative reinforcement
because of difficulty in inhibiting their behavior (Hare, 1970; Hare &
Schalling, 1978; Mednick & Christiansen, 1977). This leads to the concept that adequate therapy for sociopathic (psychopathic) individuals
must begin with behavioral and/or chemical inhibition of their antisocial
behavior. In the Houston VAMC substance abuse program, this has led
to: (1) the use of daily observed urine collection (and random testing for
common drugs of abuse) to inhibit drug usage, as well as (2) confrontation by members of the therapeutic community to inhibit rule-ignoring
or violations, and (3) confrontation in therapy groups to inhibit common
nonproductive (transactional analysis) "games." The provision of
lithium or of neuroleptics has also benefited some individuals in their
attempts to inhibit their behavior, although these medications are not
popular in this population.

5.2.2. Apparent Limitations in the Utility of Verbal Psychotherapy

It is clear that ordinary psychotherapeutic contracts cannot be ar-
ranged with sociopathic individuals because of their mistrust and manipulation of words. It would be appropriate to demand specific
behaviors before members of a therapeutic community qualify for group
psychotherapy, or for individual psychotherapy. In the Houston VAMC
substance abuse program, only ward officers qualify for individual therapy, on the basis that they are experiencing maximum inhibition of
antisocial behavior (and therefore, maximum stress). Because a population of substance abusers learns more by doing than by saying, variants
of psychodrama and other "action" therapies deserve testing. Because
the abuse of psychoactive drugs is associated with the active avoidance
of suffering, learning an increased tolerance of suffering would appear
to be a logical focus of therapy with substance abusers.

5.2.3. Money Abuse as a Basic Behavioral Disorder

When the annual income of substance abusers is carefully estimated, it quickly becomes evident that many of these individuals could
qualify as very respectable citizens on the basis of income. It therefore is
pertinent to explore how and why such individuals waste their re-

sources, and to view substance abuse as one form of money abuse. In the treatment of substance abusers, who usually need money to buy a car and to rent an apartment, it is common to find that they have great difficulty in saving the money that they go out to earn. Cab fares, expensive meals and dates, gifts to family and friends, and similar habits of the "good life" rapidly deplete the resources garnered by these individuals as ordinary wage-earners. When they do have a significant sum of money on their person, a resurgence of drug craving can destabilize and overwhelm their new and carefully nurtured life pattern of abstinence. Extinction of such money-induced craving should be tried, to increase the short-term success of graduates of substance abuse programs.

5.2.4. Early Predictors of Success or Failure in Substance Abusers

Two types of prognostic predictors could be of value in substance abusers: (1) An initial evaluation of patients, so that individuals likely to benefit from a specific program might be differentiated from those unlikely to benefit; and (2) a concurrent and repetitive evaluation of patient progress, to detect both improvement and deterioration. Most tests, such as the *Ant* (Haertzen *et al.*, 1968) and *Spy* (Spielberger *et al.*, 1978) have been developed to meet the first requirement. The PSI (Haertzen *et al.*, 1980) can also be used for initial evaluation of individuals, on the basis that high PSI scores would have a poorer prognosis. The PSI also can be used to monitor the progress of patients in a program, as illustrated earlier.

Other than measures of mood or emotional state, certain measures of behavior hold promise as predictors of improvement or deterioration in a program. If a patient's concern for himself and the program is evidenced by "good" behaviors, his stopping of such behaviors would have face validity as an early indicator of difficulty, even before drug use or other misbehaviors emerge.

6. Summary

Substance abuse is a modern term designed to subsume older problems, such as alcoholism and drug abuse, as well as more modern concerns, such as inhalant abuse, nicotine abuse, and caffeine abuse. However, because sugar, chocolate, and money are all substances, and can be abused, then substance abuse could well include these problems, too. Such an expanded definition might bring needed attention to these problems, but it would also increase the scope and decrease the focus of any substance abuse treatment program. Even if "substance abuse" is now a less negative term than "drug dependence" or "drug addiction," it logically can be expected to become a perjorative term when individuals identified with it make others angry, as drug abusers so often do. On a pragmatic level, "substance abuse" is still used to refer primarily to opioid, sedative-hypnotic (including alcohol), stimulant, or hallucinogen drug abuse.

Sociopathy is also a modern term that applies to a very old concept: the psychopathic (or antisocial) personality. Substance abuse has been identified with sociopathy for many years, although some recent evidence supports the concept that, in some ways, they might be independent processes (associated with different antecedent events). Substance abuse has also been associated for years with depression and mood disorders. Some newer approaches have invoked the concepts of hypophoria and psychopathic state in the understanding of the pathology of substance abuse. *Hypophoria* can be considered the absence of a positive mood (euphoria) rather than the presence of a negative mood (dysphoria). *Psychopathic state* incorporates the idea of a variable mood or feeling state (including hypophoria or dysphoria) that may be acute, or chronic (in which case it is equivalent to psychopathic personality). It often is associated with chronic psychoactive drug use ("secondary," or drug-induced, psychopathic state), which has a better prognosis than does a chronic psychopathic state associated either with a major mood disorder or persistent antisocial behavior.

The treatment of substance abusers is vitally related to the degree of external control needed to help them inhibit some of their behaviors. It is obvious that the most severely disturbed individuals will need to find their treatment while imprisoned, or under strict limitations. One hope of neuropsychiatrists or psychopharmacologists is that medications will be developed that can help such individuals recover their self-control enough to learn (or resume) a more fulfilling life. A promising approach to the treatment of substance abusers is to repeatedly monitor the degree of their psychopathic state, so that the maintenance of any beneficial effects can be measured, and any adverse responses can be quickly identified and modified.

7. References

American Psychiatric Association. *Diagnostic and Statistical Manual: Mental Disorders* (DSM-I). Washington, D.C.: A.P.A., 1952.
American Psychiatric Association. *Diagnostic and Statistical Manual: Mental Disorders* (DSM-II). Washington, D.C.: A.P.A., 1968.

American Psychiatric Association: *Diagnostic and Statistical Manual: Mental Disorders* (DSM-III). Washington, D.C.: A.P.A., 1980.

Astin, A. A factor study of the MMPI psychopathic deviate scale. *Journal of Consulting Psychologists*, 1959, 23, 550–554.

Barton, W. I. Drug histories and criminality of inmates of local jails in the United States (1978): Implications for treatment and rehabilitation of the drug abuser in a jail setting. *International Journal of Addictions*, 1982, 17, 417–444.

Black, F. W., & Heald, A. MMPI characteristics of alcohol and illicit drug abusers enrolled in a rehabilitation program. *Journal of Clinical Psychology*, 1975, 31, 572–575.

Cadoret, R., & Winokur, G. Depression in alcoholism. *Annals of the New York Academy of Science*, 1974, 233, 34–39.

Cleckley H. *The mask of sanity* (1st ed.). St. Louis: Mosby, 1941.

Cowan, J. D., Kay, D. C., Neidert, G. L., Ross, F. E., & Belmore, S. Defeated and joyless: potential measures of change in drug abuser characteristics. *Journal of Nervous and Mental Disease*, 1980, 168, 391–399.

Cox, C., & Smart, R. G. Social and psychological aspects of speed use: A study of types of speed users in Toronto. *International Journal of Addictions*, 1972, 7, 201–217.

Dahlstrom, W. G., & Welch, G. S. *An MMPI handbook: A guide to use in clinical practice and research*. Minneapolis: University of Minnesota Press, 1960.

Eisenman, A. J., Sloan, J. W., Martin, W. R., Jasinski, D. R., & Brooks, J. W. Catecholamine and 17-hydroxycorticosteroid excretion during a cycle of morphine dependence in man. *Journal of Psychiatric Research*, 1969, 7, 19–28.

Emery, G. D., Steer, R. A., & Beck, A. T. Depression, hopelessness, and suicidal intent among heroin addicts. *International Journal of Addictions*, 1981, 16, 425–429.

Esquirol, E. *Des maladies mentales considerées sous les rapports medical, hygienique et medico-legal*. Paris: J. B. Baillière, 1838.

Felix, R. H. An appraisal of the personality types of the addict. *American Journal of Psychiatry*, 1944, 100, 462–467.

Fine, B. J., & Sweeney, D. R. Personality traits, and situational factors, and catecholamine excretion. *Journal of Experimental Research Personality*, 1968, 3, 15–27.

Forssman, H., & Frey, T. S. Electroencephalograms of boys with behavior disorders. *Acta Psychiatrica Neurologica Scandinavia*, 1953, 28, 61–73.

Frame, M. C., & Osmond, W. M. G. Alcoholism: psychopathic personality and psychopathic reaction type. *Medical Proceedings of Johannesburg*, 1956, 2, 257–261.

Fraser, H. F., Isbell, H., Eisenman, A. J., Wikler, A., & Pescor, F. T. Chronic barbiturate intoxication: Further studies. *AMA Archives of Internal Medicine*, 1954, 94, 34–41.

Gellhorn, E. *Autonomic imbalance and the hypothalamus: Implications for physiology, medicine, psychology, and neuropsychiatry*. Minneapolis, University of Minnesota Press, 1957.

Gianutsos, G., Hynes, M. D., Puri, S. K., Drawbaugh, R. B., & Lal, H. Effect of apomorphine and nigrostriatal lesions on aggression and striatal dopamine turnover during morphine withdrawal: Evidence for dopaminergic supersensitivity in protracted abstinence. *Psychopharmacologia*, 1974, 34, 37–44.

Gough, H. G. *California Psychological Inventory manual*. Palo Alto, Calif.: Consulting Psychologists Press, 1957.

Graf, K., Baer, P. E., & Comstock, B. S. MMPI changes in briefly hospitalized non-narcotic drug users. *Journal of Nervous and Mental Disease*, 1977, 165, 126–133.

Greene, B. T. An examination of the relationship between crime and substance use in a drug/alcohol treatment population. *International Journal of Addictions*, 1981, 16, 627–645.

Haertzen, C. A. *An overview of Addiction Research Center Inventory scales (ARCI): An appendix and manual of scales*. (DHEW Publication No. (ADM)74–92) Rockville, Md.: U.S. Government Printing Office, 1974.

Haertzen, C. A. Clinical psychological studies. In W. R. Martin & H. Isbell (Eds.), *Drug addiction and the U.S. Public Health Service*. Washington, D.C.: U.S.Government Printing Office, 1978.

Haertzen, C. A., & Panton, J. H. Development of a "psychopathic" scale for the Addiction Research Center Inventory. *International Journal of Addictions*, 1967, *2*, 115–127.

Haertzen, C. A., Hill, H. E., & Monroe, J. J. MMPI scales for differentiating and predicting relapse in alcoholics, opiate addicts, and criminals. *International Journal of Addictions*, 1968, *3*, 91–106.

Haertzen, C. A., Martin, W. R., Hewett, B. B., & Sandquist, V. Measurement of psychopathy as a state. *Journal of Psychology*, 1978, *100*, 201–214.

Haertzen, C. A., Martin, W. R., Ross, F. E., & Neidert, G. L. Psychopathic State Inventory (PSI): Development of a short test for measuring psychopathic states. *International Journal of Addictions*, 1980, *15*, 137–146.

Hare, R. D. Psychopathy, autonomic functioning, and the orienting response. *Journal of Abnormal Psychology*, 1968, Monograph Supp. 73 (No. 3. Part 2), 1–24.

Hare, R. D. *Psychopathy: Theory and research*. New York: Wiley, 1970.

Hare, R. D. A research scale for the assessment of psychopathy in criminal populations. *Personality and Individual Differences*, 1980, *1*, 111–119.

Hare, R. D., & Cox, D. N. Clinical and empirical conceptions of psychopathy. In R. D. Hare & D. Schalling (Eds.), *Psychopathic Behaviour: Approaches to research*. Chichester, England: Wiley, 1978.

Hare, R. D., & Schalling, D. *Psychopathic behavior: Approaches to research*. New York: Wiley, 1978.

Harrington, A. The coming of the psychopath. *Playboy*, 1971, 18(12), 203–335.

Hewett, B. B., & Martin, W. R. Psychometric comparisons of sociopathic and psychopathological behaviors of alcoholics and drug abusers versus a low drug control population. *International Journal of Addictions*, 1980, *15*, 77–105.

Hewitt, C. C. A personality study of alcohol addiction. *Quarterly Journal of Studies on Alcohol*, 1943, *4*, 368–386.

Hill, H. E., Haertzen, C. A., & Davis, H. An MMPI factor analytic study of alcoholics, narcotic addicts and criminals. *Quarterly Journal of Studies on Alcohol*, 1962, *23*, 411–431.

Hill, H. E., Haertzen, C. A., Wolbach, A. B., & Minter, E. J. The Addiction Research Center Inventory: Standardization of scales which evaluate subjective effects of morphine, amphetamine, pentobarbital, alcohol, LSD-25, pyrahexyl and chlorpromazine. *Psychopharmacologia*, 1963, *4*, 167–183.

Himmelsbach, C. K. Studies on the relation of drug addiction to the autonomic nervous system: Results of cold pressor tests. *Journal of Pharmacology and Experimental Therapeutics*, 1941, *73*, 91–98.

Hughes, J. R. A review of the positive spike phenomenon. In W. Wilson (Ed.), *Applications of electroencephalography in psychiatry*. Durham, N. C.: Duke University Press, 1965.

Isbell, H., Altschul, S., Kornetsky, C. H., Eisenman, A. J., Flanary, H. G., & Fraser, H. F. Chronic barbiturate intoxication: An experimental study. *Archives of Neurology and Psychiatry*, 1950, *64*, 1–28.

Isbell, H., Fraser, H. F., Wikler, A., Belleville, R. E., & Eisenman, A. F. An experimental study of the etiology of "rum fits" and "delirium tremens." *Quarterly Journal of Studies on Alcohol*, 1955, *16*, 1–33.

Kaplan, H. B. Increase in self-rejection as an antecedent of deviant responses. *Journal of Youth and Adolescence*, 1975, *4*, 281–292.

Kaplan, H. B. Antecedents of negative self attitudes: Membership group devaluation and defenselessness. *Social Psychiatry*, 1976, *11*, 15–25.

Kaplan, H. B. Deviant behavior and self-enhancement in adolescence. *Journal of Youth and Adolescence,* 1978, 7, 253–277.

Kaplan, H. B., & Pokorny, A. D. Self-derogation and psychosocial adjustment. *Journal of Nervous and Mental Disease,* 1969, 149, 421–434.

Kaplan, S. D. A visual analog of the Funkenstein test. *Archives of General Psychiatry,* 1960, 3, 383–388.

Kay, D. C. Human sleep and EEG through a cycle of methadone dependence. *Electroencephalography and Clinical Neurophysiology,* 1975, 38, 35–43.

Khazan, N., & Colasanti, B. Protracted rebound in rapid eye movement sleep time and EEG voltage output in morphine-dependent rats upon withdrawal. *Journal of Pharmacology and Experimental Therapeutics,* 1972, 183, 23–30.

Koch, J. L. A. *Die Psychopathischen Minderwertigkeiten, Erste Abteilung. Einleitung—Die Angeborenen Andauernden Psychopathischen Minderwertigkeiten.* Ravensburg: Otto Maier, 1891.

Kolb, L. Drug addiction in its relation to crime. *Mental Hygiene* 1925, 9, 74–89. (a)

Kolb, L. Types and characteristics of drug addicts. *Ment. Hygiene* 1925, 9, 300–313. (b)

Kraepelin, E. *Psychiatrie, Ein Lehrbuch fur Studierende und Aerzte. Siebente, vielfach umgearbeitete Auflage. II. Band. Klinische Psychiatrie.* Leipzig: J. A. Barth, 1915.

Kurland, H. D., Yeager, C. T., & Arthur, R. J. Psychophysiologic aspects of severe behavior disorders. *Archives of General Psychiatry,* 1963, 8, 599–604.

Lombroso, C. *Crime, its causes and remedies.* (H. P. Horton, trans.). Boston: Little, Brown, 1911.

Magnan & Legrain, *Les degeneres (Etat mental et syndromes episodiques).* Paris: Rueff et Cie, 1895.

Martin, W. R. *Drugs and drug addiction.* Paper presented to the 39th annual scientific meeting of the Committee on Problems of Drug Dependence, Cambridge, Massachusetts, July 6–9, 1977.

Martin, W. R., & Jasinski, D. R. Physiological parameters of morphine dependence in man: Tolerance, early abstinence, protracted abstinence. *Journal of Psychiatric Research,* 1969, 7, 9–17.

Martin, W. R., Wikler, A., Eades, C. G., & Pescor, F. T. Tolerance to and physical dependence on morphine in rats. *Psychopharmacologia,* 1963, 4, 247–260.

Martin, W. R., Jasinski, D. R., Sapira, J. D., Flanary, J. G., Kelly, O. A., Thompson, A. K., & Logan, C. R. The respiratory effects of morphine during a cycle of dependence. *Journal of Pharmacology and Experimental Therapeutics,* 1968, 162, 182–189.

Martin, W. R., Jasinski, D. R., Haertzen, C. A., Kay, D. C., Jones, B. E., Mansky, P. A., & Carpenter, R. W. Methadone: A reevaluation. *Archives of General Psychiatry,* 1973, 28, 286–295.

Martin, W. R., Eades, C. G., Thompson, W. O., Thompson, J. A., & Flanary, H. G. Morphine physical dependence in the dog. *Journal of Pharmacology and Experimental Therapeutics* 1974, 189, 759–771.

Martin, W. R., Hewett, B. B., Baker, A. J., & Haertzen, C. A. Aspects of the psychopathology and pathophysiology of addiction. *Drug and Alcohol Dependence,* 1977, 2, 185–202.

Martin, W. R., Haertzen, C. A., & Hewett, B. B. Psychopathology and pathophysiology of narcotic addicts, alcoholics, and drug abusers. In M. A. Lipton, A. DiMascio, & K. F. Killam (Eds.), *Psychopharmacology: A generation of progress.* New York: Raven Press, 1978.

Maughs, S. A concept of psychopathy and psychopathic personality: Its evaluation and historical development. *Journal of Clinical Psychopathology,* 1941, 2, 465–499.

McKinley, J. C., & Hathaway, S. R. The MMPI: V. Hysteria, hypomania and psychopathic deviate. *Journal of Applied Psychology*, 1944, *28*, 153–174.

Mednick, S., & Christiansen, K. O. *Biosocial bases of criminal behavior*. New York: Gardner Press, 1977.

Monroe, J. J., Miller, J. S., & Lyle, W. H. Extension of psychopathic deviancy scales for the screening of addict patients. *Educational and Psychological Measurement*, 1964, *24*, 47–56.

Morel, B. A. *Traite des degenerescences physiques intellectualles et morales de l'espêce humaine et des causes qui produisent ces varietés maladives*. Paris: J. B. Baillière, 1857.

Penk, W. E., & Robinowitz, R. Personality differences of volunteer and nonvolunteer heroin and nonheroin drug users. *Journal of Abnormal Psychology* 1976, *85*, 91–100.

Pescor, M. J. The Kolb classification of drug addicts. *Public Health Reports Supplement*, 1939, *155*, 1–10.

Petrie, A. *Individuality in pain and suffering*, Chicago: University of Chicago Press, 1967.

Pichot, P. Psychopathic behavior: A historical overview, In R. D. Hare & D. Schalling (Eds.), *Psychopathic behavior: Approaches to research*. New York: Wiley, 1978.

Pinel, P. *Traite medico-philosophique sur l'alienation mentale*. Paris: J. Ant. Brosson, 1809.

Pitts, F. N., & Winokur, G. Affective disorder—VII: Alcoholism and affective disorder. *Journal of Psychiatric Research*, 1966, *4*, 37–50.

Pritchard, J. C. *A treatise on insanity and other disorders affecting the mind*. London: Sherwood, Gilbert and Piper, 1835.

Quay, H. C. Psychopathic personality as pathological stimulation seeking. *American Journal of Psychiatry*, 1965, *122*, 180–183.

Rado, S. Hedonic self-regulation of the organism. In R. G. Heath (Ed.), *The role of pleasure in behavior*. New York: Harper & Row, 1964.

Reid, W. H. *The psychopath: A comprehensive study of antisocial disorders and behaviors*. New York: Brunner/Mazel, 1978. (a)

Reid, W. H. The sadness of the psychopath. In W. H Reid (Ed.), *The psychopath: A comprehensive study of antisocial disorders and behaviors*. New York: Brunner/Mazel, 1978, 7–21. (b)

Robins, L. N. *Deviant children grown up: A sociological and psychiatric study of sociopathic personality*. Baltimore: Williams and Wilkins, 1966.

Rosen, A. C. A comparative study of alcoholic and psychiatric patients with the MMPI. *Quarterly Journal of Studies on Alcohol*, 1960, *21*, 253–266.

Rounsaville, B. J., Weissman, M., Kleber, H., & Wilber, C. *Psychiatric disorders in opiate addicts*. Paper presented at the 134th annual meeting of the American Psychiatric Association, New Orleans, May 9–15, 1981.

Schneider, K. *Die Psychopathische Personlichkeiten, 9*, Aufl. Wein: Franz Deuticke, 1950.

Schoolar, J. C., White, E. H., & Cohen, C. P. Drug abusers and their patient counterparts: A comparison of personality dimensions. *Journal of Consulting and Clinical Psychology*, 1972, *39*, 9–14.

Smart, R. G., & Jones, D. Illicit LSD users: Their personality characteristics and psychopathology. *Journal of Abnormal Psychology*, 1970, *75*, 286–292.

Smith, R. J. *The psychopath in society*, New York: Academic Press, 1978.

Spielberger, C. D., Kling, J. K., & O'Hagen, S. E. J. Dimensions of psychopathic personality: Antisocial behavior and anxiety. In R. D. Hare & D. Schalling (Eds.), *Psychopathic behavior: Approaches to research*. New York: Wiley, 1978.

Stern, J. A., & McDonald, D. G. Physiological correlates of mental disease. In P. R. Farnsworth (Ed.), *Annual Review of Psychology*. Palo Alto, Calif.: Annual Reviews, 1965.

Trasler, G. Relations between psychopathy and persistent criminality: Methodological and theoretical issues. In R. D. Hare & D. Schalling (Eds.), *Psychopathic behaviour: Approaches to research.* New York: Wiley, 1978.

Vaillant, G. E. *Is alcoholism the cart to sociopathy or the horse?* Paper presented at the 135th annual meeting of the American Psychiatric Association, Toronto, May 15–21, 1982.

Ward, E. S., & Hemsley, D. R. The stability of personality measures in drug abusers during withdrawal. *International Journal of Addictions,* 1982, *17,* 575–583.

Wikler, A. Present status of the concept of drug dependence. *Psychological Medicine,* 1971, *1,* 377–380.

Wilber, C., Rounsaville, B. J., Weissman, M. M., & Kleber, H. D. *The importance of events preceding drug addiction.* Paper presented at the 135th annual meeting of the American Psychiatric Association, Toronto, May 15–21, 1982.

Wilber, C., Rounsaville, B., Weissman, M., Critz-Christoph, K., & Kleber, H. *Course of depression in opiate addicts.* Paper presented at the 134th annual meeting of the American Psychiatric Association, New Orleans, May 9–15, 1981.

Winokur, G. The division of depressive illness into depression spectrum disease and pure depressive disease. *International Pharmacopsychiatry,* 1974, *9,* 5–13.

Winokur, G., Cadoret, R., Baker, M., & Dorzab, J. Depression spectrum disease versus pure depressive disease: Some further data. *British Journal of Psychiatry,* 1975, *127,* 75–77.

Winokur, G., Cadoret, R., Dorzab, J., & Baker, M. Depressive disease: A genetic study. *Archives of General Psychiatry,* 1971, *24,* 135–144.

Woodruff, R. A., Guze, S. B., Clayton, P. J., & Carr, D. Alcoholism and depression. *Archives of General Psychiatry,* 1973, *28,* 97–100.

Yochelson, S., & Samenow, S. E. *The criminal personality. Volume I: A profile for change.* New York: J. Aronson, 1976.

Yochelson, S., & Samenow, S. E. *The criminal personality. Volume II: The change process.* New York: J. Aronson, 1977.

Zuckerman, M. Sensation seeking and psychopathy. In R. D. Hare and D. Schalling (Eds.), *Psychopathic behaviour: Approaches to research.* New York: Wiley, 1978.

6

Substance Abuse in Psychiatric Patients

ETIOLOGICAL, DEVELOPMENTAL, AND TREATMENT CONSIDERATIONS

Arthur I. Alterman

1. General Introduction

Alcoholism and drug abuse are known to occur in combination with a number of other disorders. Perusal of the recent Diagnostic and Statistical Manual (American Psychiatric Association, 1980) indicates that substance abuse problems may be among those exhibited in a number of psychiatric disorders, as in many of the personality and anxiety disorders and in the affective disorders. Many of the chapters in this book confirm this fact. Additionally, research by this writer (Alterman, Erdlen, McLellan, & Mann, 1980; Alterman, Erdlen, & Murphy, 1981) and others (McLellan, Druley, & Carson, 1978) have revealed that problems of alcoholism and drug abuse are a frequent complicating condition in schizophrenia. In this world of specialization, the existence of alcoholism or drug abuse as secondary diagnoses for persons with other primary diagnoses, does, indeed, represent a complicating circumstance. As Alterman's research on a VA psychiatric inpatient population indicated (Alterman, Erdlen, McLellan, & Mann, 1980; Alterman et al., 1981), general psychiatric staff feel, and are generally, ill equipped to treat psychiatric patients with concomitant alcohol or drug abuse prob-

Arthur I. Alterman • Research Department, Philadelphia VA Medical Center, Philadelphia, Pennsylvania 15206.

lems. Although informal programs certainly exist for the treatment of conjoint serious psychiatric and addiction problems, there are almost no programs designed specifically for this purpose. Among the possible reasons for this situation has been the traditional segregation of alcoholism/drug abuse treatment functions from the mainstream of psychiatry, a situation that has been dominant until recently. As a result, staff treating schizophrenic patients have felt that they must focus their attention on this disorder and that substance abuse represents an alien complication. At the same time, alcoholism and drug abuse treatment programs have, until recently, emphasized direct, interpersonal approaches that were relatively bereft of psychiatric and other professional considerations. However, the recent inroads made in the use of structured psychiatric interviewing procedures and more formal diagnostic criteria appear to be one influence that is increasing awareness and lowering thresholds of staff to the presence of multiple, concomitant psychiatric disorders in the same patient, whether the problem of present focus is schizophrenia, depression, or alcoholism. Also, researchers investigating serious psychiatric disorders, such as schizophrenia, have as a rule excluded subjects with complicating conditions, such as alcoholism and drug abuse, from their studies, since these conditions have been seen as potentially confounding and invalidating the interpretation of their findings. With some exceptions (Powell, Penick, Othmer, Bingham, & Rice, 1982), and for many of the reasons already cited, alcoholism researchers have paid only limited attention to other psychiatric disorders in their patients. Thus, although the occurrence of conjoint psychiatric and addiction problems are apparently a quite common and significant phenomenon in the mental patient, this condition has received very little of the attention of the clinician or researcher.

2. Diagnosis/Identification

There are basically no major, formal obstacles in identifying patients or individuals who qualify for diagnoses of both schizophrenia and substance abuse. Although some discussion continues to be carried on by workers in each problem area concerning the precision and utility of various diagnostic criteria, there exists general consensus about the diagnosis of schizophrenia and substance abuse (i.e., alcoholism or drug abuse). On the other hand, numerous writers have stressed the frequent failure to make this diagnosis on the part of many of those professionals who come into contact with secondary substance abuse. These writers have suggested remedies, including the use, for example, of brief self-report checklists or laboratory assays (Page, 1979). Thus, it is generally believed that secondary substance abuse problems are underidentified and that the conjoint occurrence of substance abuse and schizophrenia is by no means an exception. It should be noted also that some of the more formal diagnostic criteria used for research purposes may place limits upon the diagnosis of schizophrenia, given the presence of a pre-existing substance abuse disorder. Thus, the Feighner criteria (Feighner, Robins, Guze, Woodruff, Winokur, & Munoz, 1972) eliminate schizophrenia as a possible diagnosis if the individual were addicted to drugs or alcohol prior to the onset of schizophrenic symptoms. A similar strategy has been adopted in the Research Diagnostic Criteria (RDC) (Spitzer, Endicott, & Robins, 1978).

3. Incidence and Prevalence

As the foregoing implies, there is only limited information currently available concerning the co-occurrence of substance abuse disorders with schizophrenia, so that any effort to provide a precise estimate of the incidence and prevalence of these problems would be premature. Nevertheless, there is some information available, and this will be presented in the immediately following sections. Since the literature on conjoint alcoholism/psychiatric disorders and that for drug abuse/psychiatric disorders have received virtually independent treatment, these two topics will be treated separately.

3.1. Conjoint Alcoholism/Psychiatric Problems

At the outset, it should be made clear for this and the following drug abuse section that the focus of this chapter is on persons who are being treated in psychiatric, and not in substance abuse, units; that is, the current focal problem is considered to be schizophrenia rather than substance abuse. Others have described the reverse situation; that is, the development of schizophrenia in individuals who were initially alcoholics (Freed, 1975; Gottheil & Waxman, 1982) or the development of other serious psychiatric symptomatology in drug (primarily stimulant) abusers (McLellan, Woody, & O'Brien, 1979). We recognize that the distinction we are presently making, with respect to schizophrenia being the *primary* disorder, may be artificial. In the long run, this remains a question for investigation.

3.1.1. Schizophrenia and Alcoholism

Alcohol abuse is generally recognized as being a common problem among psychiatric patients receiving treatment in psychiatric hospitals. Several studies have, indeed, noted that persons diagnosed as alcoholics represent a significant proportion of the admissions of mental hospitals. For example, Crowley and his colleagues (Crowley, Chesluk, Dilts, & Hart, 1974) reported that alcoholics contributed to one-fourth of the admissions to a psychiatric hospital, but there was no indication that these patients had psychiatric diagnoses other than alcoholism or that they continued to drink while in the hospital. Whittier and Korenji (1961) found that 20% and 23%, respectively, of first admissions over 60 years old to a mental hospital were the result of alcoholism. Aside from chronic brain syndrome, these patients did not appear to carry any other psychiatric diagnoses, however, and it was not clarified whether they continued to drink while under treatment. Another study conducted by Stenback, Achte, and Rimon (1965) in Finland indicated that nearly a third of mental hospital patients committing suicide were alcoholics. Again, there was no indication that the alcoholics carried other psychiatric diagnoses or that acute intoxication had contributed to the suicides.

A small body of studies has also accumulated that makes reference to the coexistence of alcoholism in patients with psychoses or other psychiatric disorders. Malzberg (1955), for example, reported that intemperance was relatively high among first admissions for "dementia praecox" to New York State mental institutions. Parker (Parker, Meiller, & Andrews, 1960) noted that 22% of the schizophrenic and 33% of the manic-depressive patients in a VA hospital were also diagnosed as alcoholics. Nichols (Nichols, Pike, Richter, & Sculthorpe, 1961) discussed the merits of foster home placement of 22 "psychotic" patients with a history of problem drinking; Scott (1966) described a group therapy approach used in treating 24 alcoholic schizophrenic outpatients; and Panepinto and his colleagues (Panepinto, Higgins, Keane-Dawes, & Smith, 1970) described using supportive drug therapy with 60 schizophrenic alcoholics who constituted 18% of the alcoholics undergoing treatment. Several other papers have reported on the treatment of single cases of apparently alcoholic, schizophrenic individuals (Markham, 1957; Winship, 1957). More recently, McLellan (McLellan et al., 1978) reported that about 15% of the schizophrenic patients in a VA psychiatric facility had a prior history of alcohol abuse.

In investigating the multiple readmission patients in a VA psychiatric facility, Pokorny (1965) found that about 15% of the hospital population had a diagnosis of alcoholism, and that nearly half of a group of patients with six or more readmissions had received a diagnosis of alcoholism at some point in time. Eighty-eight percent of the patients in this latter group had primary diagnoses other than alcoholism. Thirty-one percent had been diagnosed as schizophrenic (over half, paranoid), 26% were neurotic, and 24% had personality disorders. Several more recent investigations by Alterman and his colleagues (Alterman, Erdlen, McLellan, & Mann, 1980; Alterman et al., 1981) have provided additional evidence for the existence of conjoint schizophrenia and alcoholism in about 10–15% of the psychiatric population. In one study in a VA inpatient psychiatric facility, about 15% of the schizophrenic population were found to have secondary diagnoses of alcoholism. Almost two-thirds of these alcoholic schizophrenics were diagnosed as paranoid schizophrenic (Alterman, Erdlen, & McLellan, 1980). A second study conducted about a year later in the same institution (Alterman et al., 1981) found that 131 out of the 979 (13.4%) psychiatric (nonsubstance abuse) patients carried either secondary or tertiary diagnoses of alcoholism. About one-third of these patients (active plus nonactive drinkers) had diagnoses of paranoid schizophrenia. Psychiatric patients with conjoint diagnoses of alcoholism were also more likely than their nonalcoholic cohorts to be black (31% versus 20%).

It should be mentioned that the diagnoses described in all of the studies reported were essentially derived from patient charts or from unstructured psychiatric interviews employing DSM-II criteria. The reliability and validity of such procedures has frequently been questioned. Insofar as such criticisms are justifiable, the validity of the data we have thus far presented can be questioned. Therefore Alterman and his colleagues (Alterman, Ayres, & Williford, 1984) conducted an investigation in which they attempted to ascertain the extent to which patients' hospital-based diagnoses of schizophrenia and alcoholism could be confirmed when more reliable diagnostic procedures were employed. Sixteen inpatients at a VA psychiatric facility with a diagnosis of schizophrenia and a history of alcohol abuse noted in their charts were therefore administered the Renard Diagnostic Interview (RDI). This is a structured diagnostic interview (Helzer, Robins, Croughan, & Welner, 1981) that has been shown to have satisfactory reliability when used in conjunction with any of the major systems of diagnostic criteria. All 16 patients were found to qualify for a diagnosis of alcoholism using either the RDC (Spitzer et al., 1978), Feighner (Feighner et al., 1972) or DSM-III (American Psychiatric Association, 1980) criteria, while 11 of the 16 patients satisfied the criteria for a lifetime diagnosis of schizophrenia on these criteria. Thus, although not all of the VA psychiatric inpatients with diagnoses of both schizophrenia and alcoholism qualified for both diagnoses, when more rigorous, standardized interviewing procedures and criteria were employed, 69% did meet the criteria for both diagnoses. These results accordingly increased the confidence of Alterman and his colleagues in the validity of co-occurring schizophrenia and alcoholism.

It seems clear, then, that schizophrenics with complicating alcoholism represent a significant proportion of the patient population in psychiatric treatment facilities. Although the evidence indicates that they by no means represent a majority of schizophrenics, or of any other psychiatric disorder, they nevertheless constitute a population of sufficient number to pose a significant treatment problem for our society. At the same time, we wish to emphasize that the available information on the relative incidence of such problems is entirely derived from patient populations. The author is unaware of any data concerning the conjoint presence of schizophrenia and alcoholism in nonpatient populations. However, since the more severe cases are likely to be found in institutions, it is anticipated that the proportion of persons with conjoint diagnoses ultimately found in prospective epidemiological studies will not exceed the proportions identified in patient populations.

3.1.2. Schizophrenia and Drug Abuse

Just as for alcohol abuse, knowledge has existed for some time that a large proportion of the admissions to psychiatric facilities for general psychiatric problems have a history or present a problem of substance abuse. For example, Cohen and Klein (1970) reported, on the basis of a records review, that 5% of admissions to their facility under 25 years of age were illicit-drug users. Although some of the drug users were diagnosed as schizophrenic, they were more likely to have a character disorder, in contrast to non-drug-using psychiatric controls. Similarly, (Crowley et al., 1974) found that long-term drug abuse (exclusive of alcohol) contributed to the need for admission in about 10% of the cases they surveyed, and that drug users were less likely than the psychiatric patients to be schizophrenic and more likely to be sociopathic.

Findings of several other studies suggest a somewhat greater prevalence of drug abusers among psychiatric patients. Rockwell and Ostwald (1968) found, both from reviewing charts of consecutive admissions and from the analysis of urine samples, that approximately 15% of their psychiatric patients were amphetamine users. Similar results were obtained by Robinson and Wolkind (1980) using gas chromatography techniques, but Blumberg and his colleagues (Blumberg, Cohen, Heaton, & Klein, 1971), in a rather extensive laboratory-based study, found evidence of drug use in 60% of psychiatric inpatients tested, and Razani (Razani, Farina, & Stern, 1975) found that 52% of psychiatric inpatients were illicitly using drugs. Fischer, Halikas, Baker, and Smith (1975) and McLellan and his colleagues (McLellan et al., 1978) discovered from patient interviews that nearly one-third of their psychiatric population admitted to prior drug use problems (excluding alcohol). In contrast to other findings, McLellan (McLellan et al., 1978) and Hekimian and Gershon (1968) reported that a relatively large proportion of their drug users were schizophrenic. The former investigators found that 64% of the amphetamine users and the hallucinogen users were diagnosed as schizophrenic, as contrasted with only 41% of the heroin users and 22% of the barbiturate users. Hekimian and Gershon (1968) indicated that 75% of the marijuana users, 50% of the amphetamine users, and only 12% of the heroin users were diagnosed as schizophrenic. The results of this latter study are supported by recent research by McLellan and his associates (McLellan et al., 1979), which revealed that stimulant abusers were much more likely to develop psychotic symptomatology over time, including schizophrenia, than were opiate abusers (McLellan, Childress, & Woody, Chapter 7). A recently completed study by Alterman and his colleagues (Alterman, Erdlen, LaPorte, & Erdlen, 1982) provides additional information on the incidence of drug abuse in inpatient psychiatric patients. This investigation found that 18% of the non-substance-abuse patients of a VA psychiatric facility had a history or diagnoses of substance abuse. Of these patients, 57% carried a primary diagnosis of paranoid schizophrenia, as contrasted with 39% of a non-substance-abusing psychiatric comparison group. Substance-abusing psychiatric patients were more likely to be black (61% versus 31%) and were significantly younger (30 versus 48 years old) than non-substance-abusing psychiatric patients. The major substance of abuse was found to be marijuana. However, inasmuch as the research data were based on information obtained from ward nursing staffs, as contrasted with laboratory assays, the authors felt that the incidence of substance abuse was underestimated in the population studied. Thus, the limited information obtained to date provides clear indication that a fairly substantial number of general psychiatric admissions have a history/diagnosis of drug abuse, although precise incidence estimates are not currently available. It is also known that large numbers of substance-abusing psychiatric inpatients, including those with alcoholism problems, continue to consume their substance of choice even while undergoing treatment in a controlled institutional environment. This problem and its consequences will be treated in a subsequent section of this chapter.

4. Etiological and Developmental Considerations

Given the joint occurrence of schizophrenia and substance abuse disorders, a number of questions arise concerning the etiology, developmental course, and possible interactions between the disorders. At this point in time the available evidence is terribly meagre and is limited mainly to conjoint schizophrenia and alcoholism. Family studies constitute a line of investigation employed to search for genetic linkages between schizophrenia and alcoholism. In one such study, Rimmer and Jacobsen (1977) examined the biological families of Danish schizophrenic adoptees and found a low prevalence of alcohol problem histories. Studies of the family psychiatric histories of identified alcoholic patients have also revealed a relatively low incidence of schizophrenia in the relatives of alcoholics (Cloninger, Reich, & Wetzel, 1979; Winokur, Reich, Rimmer, & Pitts, 1970). Thus, family studies have failed to reveal any underlying genetic linkages between the disorders of schizophrenia and alcoholism. Evidence on the incidence of conjoint occurrence of these two disorders bears, at least indirectly, on the question of their relationship. That is, the research of Alterman (Alterman, Erdlen, McLellan, & Mann, 1980; Alterman et al., 1981) and others (McLellan et al., 1978), which have found about a 15% incidence of alcoholism in schizophrenic patients, also suggests that there are no unique, compelling interactions between schizophrenia and alcoholism, since estimates of alcoholism in the general population also approximate that 15% figure (Cahalan, Cisin, & Crossley, 1969).

In a recent investigation, Alterman (Alterman et al., 1984) identified eleven alcoholic schizophrenic patients by means of the RDI. Schizophrenic symptoms were reported as preceding alcoholism symptoms in 6 of the 11 patients (S-A), whereas alcoholism had an earlier onset in 5 of these patients (A- -S). Onset of schizophrenia in the former group occurred in childhood (average age = 9.2) and preceded alcoholism by an average of 12 years (age = 21.3), while onset of alcoholism occurred at age 19 for the A- -S group followed by schizophrenia at age 26. Thus, although alcoholism developed at about age 20 for both subgroups, onset of schizophrenia was quite early for one and later for the other. The subgroups did not differ in the number or pattern of their schizophrenic symptoms, nor did the schizophrenic symptomatology of alcoholic schizophrenics as a group differ from that of non-substance-abusing schizophrenic patients. Whether the two alcoholic schizophrenic subgroups can be shown to differ from each other in other ways, or whether alcoholic schizophrenics can be differentiated from non-substance-abusing schizophrenics, remains to be determined. An important question, therefore, is whether markers can be established that would differentiate schizophrenics with substance abuse problems from those not experiencing problems of substance abuse. A search for such markers might take a number of directions; for example, biochemical (MaO), behavioral (developmental problems), neurophysiological (evoked potentials), cognitive (cognitive asymmetry), and neurological (CAT scans) characteristics of the two groups might be examined for differences.

5. Treatment Considerations

A major difficulty in treating schizophrenic patients with complicating substance abuse problems is that these individuals may continue to abuse substances even while undergoing treatment. The consequences of such behavior are to a large extent unknown. The limited available evidence will first be reviewed with respect to alcohol abuse and then for drug abuse.

5.1. Alcohol Abuse in Psychiatric Inpatients.

Pokorny (1965) was probably the first investigator to point to the problem of alcohol abuse in psychiatric patients. In the process of commenting on the presence of alcohol abuse in a VA hospital's psychiatric population, he alluded to the occurrence of such drinking in the hospital's schizophrenic patients. Several papers by Alterman and his colleagues (Alterman, Erdlen, McLellan, & Mann, 1980; Alterman et al., 1981) appear to be the first systematic description of the occurrence of alcohol abuse in alcoholic schizophrenic inpatients and the problems associated with treating and managing these patients. This research, based on evaluations of nursing staff, revealed that about 15% of the schizophrenic population of a VA psychiatric facility had received secondary diagnoses of alcoholism and that about half of these individuals were continuing to abuse alcohol while undergoing treatment. These problem drinkers were significantly younger and more likely to be black, and had many more previous admissions than nondrinking alcoholic schizophrenic patients. Also, a larger proportion of the alcohol abusers had diagnoses of paranoid schizophrenia (77% versus 50%). Staff reported a number of negative treatment and management consequences associated with the illicit alcohol abuse of these patients, including mood and behavior changes, hallucinations, blackouts, and the shakes.

Other commonly occurring correlates of alcohol abuse were negative attitudes toward treatment, need for increased supervision, unruly behavior, poor money management, missing roll calls, not completing assignments, and problems with the family. It is also noteworthy that significantly more (49%) of these alcohol abusers than the nondrinking alcoholic schizophrenics were suffering from serious medical conditions contradictive of alcohol use (seizures, diabetes, hypertension, peripheral neuropathy, ulcers). Additionally, almost all of these patients were being medicated with one or a combination of the major tranquilizers, all of which had potential adverse cross-reactivities with alcohol (Vogel, Chapter 9). A subsequent review of the 24-hour nursing reports and the inpatient charts of known alcoholic schizophrenic patients (Alterman & Erdlen, 1983) revealed a number of extremely serious medical and psychiatric consequences of alcohol/drug abuse. For example, one 46-year-old paranoid schizophrenic patient repeatedly became violent toward fellow patients after imbibing alcohol on the grounds, while a 28-year-old patient with a diagnosis of catatonic schizophrenia returned from weekend passes, during which he used LSD, experiencing auditory hallucinations and symptoms of catatonia.

Abusing alcoholic schizophrenic patients were also found to have significantly more readmissions to the hospital than nonabusing alcoholic schizophrenic patients; almost half of them had six or more previous hospital admissions. Staff felt that many of these patients were not motivated for treatment and that their primary interest was in satisfying their craving for alcohol. There existed no clear-cut policy and much confusion about how these patients might best be treated and managed. Therefore, the preliminary evidence thus far acquired suggests that schizophrenic individuals with complicating alcoholism represent a difficult and virtually disregarded treatment problem.

5.2. Drug Abuse in Psychiatric Inpatients

The occurrence of inpatient drug abuse has been described by several researchers who made use of gas chromatography methods to detect derivatives of prior drug use in the patient's urine. Robinson and Wolkind (1980) analyzed urine samples of a group of inpatients for amphetamines, and obtained positive screens in 21 of 35 cases (60%) in which use was suspected; only 1 of 20 drug screens based on spot checks was positive. The 22 positive cases represented 15% of the psychiatric population treated that year. The authors discussed the part played by illicit amphetamine use in the overall clinical picture and described three cases of exacerbation of schizophrenia apparently attributable to covert amphetamine use. Blumberg (Blumberg *et al.*, 1971) evaluated 332 young psychiatric patients by means of repeated urinary analyses over periods ranging from one to 72 weeks ($M = 27$ weeks) and found that at least 60% ($n = 195$) of their patients had used abusable drugs such as barbiturates, amphetamines, or narcotics at least once. Barbiturates were most commonly detected. Marijuana, LSD, and glue use were not investigated because satisfactory assay methods were not available when that research was conducted. The authors discussed the potentially adverse interactions of abusable drugs with commonly prescribed psychotropic medications, and possible contamination of standard laboratory data by covert drug use. In a related study, Ranzani (Razani *et al.*, 1975) described the use of urinary analyses for detection of amphetamines, barbiturates, and narcotics in 44 consecutive psychiatric admissions under 41 years of age. Findings of covert drug abuse within the hospital setting were obtained for 23 patients, or 52%, of the population studied. In contrast to Blumberg *et al.*'s earlier findings, amphetamines, and not barbiturates, were the most commonly detected drugs.

A study by Alterman (Alterman *et al.*, 1982), however, suggested that, without the use of laboratory findings, staff is less aware of the presence and psychiatric effects of patients' drug abuse than they are of their alcohol abuse in the same setting. Marked mood changes were the only psychiatric symptom attributed by staff to drug abuse. The only other drug-related behavior perceived by staff as causing problems seemed to be mainly related to the drug procurement process (i.e., bootlegging/smuggling, secretiveness/cliquishness, and thefts). Drug abusing psychiatric patients were also more likely to have a greater number of previous admissions to the same facility, which were of shorter duration than those of a comparison group of non-substance-abusing psychiatric patients.

5.3. Treatment Implications

Research to date has therefore shown that a fairly substantial number of general psychiatric admissions have a history of substance abuse, and that drug use is a common occurrence within the treatment setting. Yet, questions concerning the implications of this behavior for patient treatment and management have been addressed in only the most preliminary fashion. Additionally, there is a conflict, and only limited current knowledge, concerning the extent to which drug abusers are likely to have schizophrenic, sociopathic, or other psychiatric diagnoses. Little is known about the relative amount of psychopathology or the comparative treatment course of abusers and nonabusers. The effects and contribution of covert drug abuse to psychiatric symptomatology, its probably adverse interaction with psychotropic medications, and its confounding effects on standard laboratory studies have largely been items of discussion, rather than of investigation.

Thus, there is as yet very little research on the treatment outcome of either alcohol- or drug-abusing schizophrenic patients. As mentioned previously, staff in a VA psychiatric inpatient unit seemed to be uncertain about how to treat schizophrenic patients with problems of alcoholism. Yet, in a study done about a decade ago, the authors (Panepinto et al., 1970) reported that alcoholics who were also schizophrenic were able to remain in therapy longer and maintain a better relationship with their counsellor in a nonintensive, supportive drug therapy program than alcoholics who had diagnoses of sociopathy. A study by Hall and his colleagues (Hall, Popkin, DeVaul, & Stickney, 1977), which examined the effects of illicit drug use by psychiatric outpatients on their diagnoses and treatment course, appears to be the most systematic effort to date to address such questions. These authors found that drug users were much more likely to be incorrectly diagnosed as schizophrenic, that they were more likely to miss therapy sessions and to discontinue therapy, that they induced more discomfort in their therapists, and that they were much less likely to benefit from treatment than non-drug-using psychiatric outpatients. These results undoubtedly point to some of the issues that will have to be addressed by prospective researchers.

6. Discussion

We have attempted in this chapter to provide some fundamental information on the incidence, etiology and development, and treatment problems of psychiatric, mainly schizophrenic, patients who are also alcohol or drug abusers. Although individuals with these conjoint psychiatric problems have been the object of only a limited amount of research, some information is available. This indicates that alcohol and drug abuse each occur in at least 10%–15% of the schizophrenic population, and our best judgment is that these figures may represent an underestimate. There is little evidence of an underlying genetic or etiological linkage between the two types of psychiatric disorders. At the same time, there undoubtedly exist functional relationships between them. A self-medication hypothesis has often been given explanatory status for the drinking behavior of the alcoholic who is attempting to suppress underlying psychopathology, and McLellan has discussed a similar possibility (McLellan *et al.*, Chapter 7) for certain drug abusers. Unfortunately, a similar hypothesis can be proposed to explain substance abuse behavior of schizophrenic and other psychiatric patients. Thus, such hypotheses would have to be articulated in much more detail before they can be of much explanatory value.

Further efforts to understand the etiology, developmental course, and interaction between schizophrenia and substance-abusing and non-substance-abusing schizophrenic individuals can be undertaken at this point in time in terms of a number of characteristics, such as their family psychiatric histories, nature of psychiatric symptomatology, cognitive functioning, and neurological and biochemical responsivity. The information obtained may contribute significantly to our understanding of both schizophrenia and alcoholism, and may also provide clues as to the approaches that would prove most effective in treating these individuals.

At this point in time, there is little known about how patients with conjoint diagnoses respond to treatment, although the limited available evidence suggests that they fare more poorly than do patients without substance abuse complications. But the critical question is how, specifically, do these patients differ? Since substance-abusing schizophrenic patients constitute a substantial proportion of the schizophrenic population, they by no means represent an insignificant problem. The problems of actively abusing schizophrenic patients and the corollary effects of such abuse on patient evaluation, physical condition, and susceptibility and reaction to the conventional psychotropic medications are in particular need of attention. There is clearly a need for systematic, prospective research to ascertain and delineate the long-term treatment course and outcome of these individuals, and for specialized programs or treatment strategies designed to evaluate and treat their multiple problems (see Chapter 13 by Harrison, Martin, Tuason, and Hoffman in this volume).

7. Summary

The existence of substance abuse problems in psychiatric patients has received very little attention to the present time for a variety of reasons. Yet there are no major obstacles to the identification of substance abuse problems in psychiatric patients. The advent of more structured psychiatric interviewing procedures and formal diagnostic criteria may help in this connection, although it should be noted that both the Renard Diagnostic and Feighner Criteria rule out a diagnosis of schizophrenia if there is any evidence of pre-existing substance abuse.

There is only limited available information concerning the occurrence of conjoint psychiatric and substance abuse problems. The findings to date indicate that approximately 15% of schizophrenic inpatients also have an alcoholism problem. The findings concerning conjoint occurrence of drug abuse and psychiatric problems are more variable, depending in part on whether the data were derived from laboratory assays or human observation; but in general, these findings indicate higher proportions of psychiatric patients with drug abuse problems than with alcohol problems. A preliminary diagnostic study has supported the validity of the conjoint schizophrenia and alcoholism diagnosis in about two of three cases. A similar investigation has thus far not been conducted with drug-abusing schizophrenic patients.

The relatively low incidence of conjoint schizophrenia and alcoholism suggests that there is no unique, dynamic interaction between the two disorders. The results of studies of the family psychiatric histories of both schizophrenic and alcoholic individuals leads to a similar conclusion. One preliminary study found that alcoholism preceded schizophrenia in half of the cases, but in the remaining cases schizophrenia had a very early onset. The significance of these findings is not clear at this time.

Although there may be no unique relationship between conjoint psychiatric and substance abuse disorders, there are undoubtedly a number of practical implications of this relationship. Many of the patients with substance abuse disorders are active abusers. Accordingly, the presence of a substance abuse disorder complicates and probably makes the treatment of the psychiatric patient more difficult. Questions concerning possible adverse drug interactions and the patient's response to prescribed medication, problems of proper diagnosis, the physical problems of substance abusers, and how these individuals respond to conventional treatment are among those that have not been investigated.

8. References

Alterman, A. I., & Erdlen, D. Illicit substance abuse in hospitalized psychiatric patients. *Journal of Psychiatric Treatment and Evaluation*, 1983, 5, 377–380.
Alterman, A. I., Ayres, F., & Williford, W. Diagnostic validation of conjoint schizophrenia and alcoholism. *Journal of Clinical Psychiatry*, 1984, 45, 300–303.
Alterman, A. I., Erdlen, F. R., & McLellan, A. T. Problem drinking in a psychiatric

hospital: Alcoholic schizophrenics. In E. Gottheil, A. T. McLellan, & K. A. Druley (Eds.), *Substance abuse and psychiatric illness*. New York: Pergamon Press, 1980.

Alterman, A. I., Erdlen, F., & Murphy, E. Alcohol abuse in the psychiatric hospital population. *Addictive Behaviors*, 1981, *6*, 69–73.

Alterman, A. I., Erdlen, D. L., LaPorte, D. J., & Erdlen, F. R. Effects of illicit drug use in an inpatient psychiatric population. *Addictive Behaviors*, 1982, *7*, 231–242.

Alterman, A. I., Erdlen, F. R., McLellan, A. T., & Mann, S. C. Problem drinking in hospitalized patients. *Addictive Behaviors*, 1980, *5*, 273–276.

American Psychiatric Association. *American Psychiatric Association Diagnostic and Statistical Manual of Mental Disorders* (DSM-III). Washinton, D.C.: APA, 1980.

Blumberg, A. G., Cohen, M., Heaton, A. M., & Klein, D. F. Covert drug abuse among voluntary hospitalized psychiatric patients. *Journal of the American Medical Association*, 1971, *217*, 1659–1661.

Cahalan, D., Cisin, I. & Crossley, H. *American drinking practices*. (Monograph No. 6). New Brunswick, N.J.: Rutgers Center of Alcohol Studies, 1969.

Cloninger, C. R., Reich, T. & Wetzel, R. Alcoholism and affective disorders: Familial associations and genetic models. In D. W. Goodwin & C. M. Erickson (Eds.), *Alcoholism and affective disorders*. New York: Spectrum Publications, 1979.

Cohen, M., & Klein, D. F. Drug abuse in a young psychiatric population. *American Journal of Orthopsychiatry*, 1970, *40*, 448–455.

Crowley, T. J., Chesluk, D., Dilts, S., & Hart, R. Drug and alcohol abuse among psychiatric admissions. *Archives of General Psychiatry*, 1974, *30*, 13–20.

Feighner, J., Robins, E., Guze, S., Woodruff, R., Winokur, G., & Munoz, R. Diagnostic criteria for use in psychiatric patients. *Archives of General Psychiatry*, 1972, *26*, 57–63.

Fischer, D. E., Halikas, H. A., Baker, J. W., & Smith, J. B. Frequency and patterns of drug abuse in psychiatric patients. *Diseases of the Nervous System*, 1975, *36*, 550–553.

Freed, E. Alcoholism and schizophrenia: The search for perspectives. *Quarterly Journal of Studies on Alcohol*, 1975, *36*, 853–881.

Gottheil, E., & Waxman, H. M. Alcoholism and schizophrenia. In M. Pattison & E. Kaufman (Eds.), *Encyclopedic handbook of alcoholism*. New York: Gordon Press, 1982.

Hall, R. C. W., Popkin, M. K., DeVaul, R., & Stickney, S. K. The effect of unrecognized drug abuse on diagnosis and therapeutic outcome. *American Journal of Drug and Alcohol Abuse*, 1977, *4*, 455–465.

Hekimian, L. J., & Gershon, S. Characteristics of drug abusers admitted to a psychiatric hospital. *Journal of the American Medical Association*, 1968, *205*, 75–80.

Helzer, J. E., Robins, L. N., Croughan, J. L., & Welner, A. Renard diagnosis interview. *Archives of General Psychiatry*, 1981, *38*, 393–398.

Malzberg, B. Use of alcohol among white and negro mental patients: Comparative statistics of first admissions to New York State Hospitals for mental disease. *Quarterly Journal of Studies on Alcohol*, 1955, *16*, 668–674.

Markham, J. Casework treatment of an alcoholic woman with severe underlying pathology. *Quarterly Journal of Studies on Alcohol*, 1957, *18*, 475–491.

McLellan, A. T., Druley, K. A., & Carson, J. E. Evaluation of substance abuse problems in a psychiatric hospital. *Journal of Clinical Psychiatry*, 1978, *39*, 425–226. 429–230.

McLellan, A. T., Woody, G. E., & O'Brien, C. P. Development of psychiatric disorders in drug abusers. *New England Journal of Medicine*, 1979, *301*, 1310–1314.

Nichols, S., Pike, A. W., Richter, M. H., & Sculthorpe, W. B. Foster home placement of psychotic patients with histories of problem drinking. *Quarterly Journal of Studies on Alcohol*, 1961, *22*, 298–311.

Page, J. B. Identifying drinking problems in VA hospital patients. Quarterly Journal of Studies on Alcohol, 1979, 40, 447–456.

Panepinto, W. C., Higgins, M. J., Keane-Dawes, W. Y., & Smith, D. Underlying psychiatric diagnosis as an indicator of participation in alcoholism therapy. Quarterly Journal of Studies on Alcohol, 1970, 31, 950–956.

Parker, J. B., Meiller, R. M., & Andrews, G. W. Major psychiatric disorders masquerading as alcoholism. Southern Medical Journal, 1960, 53, 560–564.

Pokorny, A. D. The multiple readmission psychiatric patient. Psychiatric Quarterly, 1965, 39, 70–78.

Powell, B. J., Penick, E. C., Othmer, E., Bingham, S. F., & Rice, A. S. Prevalance of psychiatric syndromes among male alcoholics. Journal of Clinical Psychiatry, 1982, 10, 404–407.

Razani, J., Farina, F. A., & Stern, R. Covert drug abuse among patients hospitalized in psychiatric ward of a university hospital. International Journal of Addiction, 1975, 10, 693–698.

Rimmer, J., & Jacobsen, B. Alcoholism and schizophrenics and their relatives. Quarterly Journal of Studies on Alcohol, 1977, 38, 1781–1784.

Robinson, A. E., & Wolkind, S. N. Amphetamine abuse among psychiatric inpatients: The use of gas chromatography. British Journal of Psychiatry, 1980, 116, 643–644.

Rockwell, D. A., & Ostwald, P. Amphetamine use and abuse in psychiatric patients. Archives of General Psychiatry, 1968, 18, 612–618.

Scott, E. M. Group therapy for schizophrenic alcoholics in a state-operated out-patient clinic: With hypnosis as an integrated adjunct. International Journal of Clinical and Experimental Hypnosis, 1966, 14, 232–242.

Spitzer, R. L., Endicott, J., & Robins, E. Research diagnostic criteria. Archives of General Psychiatry, 1978, 35, 773–782.

Stenback, A., Achte, K. A., & Rimon, R. H. Physical disease, — hypochondria and alcohol addiction in suicides committed by mental hospital patients. British Journal of Psychiatry, 1965, 111, 933–937.

Whittier, J. R., & Korenji, C. Selected characteristics of aged patients: A study of mental hospital admissions. Comprehensive Psychiatry, 1961, 2, 113–120.

Winokur, G., Reich, T., Rimmer, J., & Pitts, F. Alcoholism III: Diagnosis and familial psychiatric illness in 259 alcoholic probands. Archives of General Psychiatry, 1970, 23, 104–111.

Winship, G. M. Disulfiram as an aid to psychotherapy in the case of an impulsive drinker. Quarterly Journal of Studies on Alcohol, 1957, 18, 666–672.

7

Drug Abuse and Psychiatric Disorders

ROLE OF DRUG CHOICE

A. THOMAS MCLELLAN, ANNA ROSE CHILDRESS, AND GEORGE E. WOODY

1. Introduction

Perhaps the most widely studied issue within the field of addiction has been the relationship of psychopathology to drug use. Because of the compulsive, antisocial, and self-destructive character of addiction, drug use has been considered a psychiatric illness in itself (DSM-III). However, there has also been a traditional belief that drug abuse is intimately related to other psychiatric disorders, although there has been considerable controversy regarding the nature of these relationships.

In fact, several diagnostic and psychological testing studies have reported evidence of personality disorders (Herl, 1976; Hill, Haertzen, & David, 1962), depression (Dorus & Senay, 1980; Pittel, 1971; Woody & Blaine, 1979), anxiety (Heller & Mordkoff, 1972; Zuckerman, Sola, & Masterson, 1975), and even psychosis (Cox & Smart, 1972; Hekimian & Gershon, 1968), among segments of the abusing population. In contrast to these findings, several other studies have failed to show evidence of significant psychiatric problems other than neuroses and sociopathy

A. THOMAS MCLELLAN, ANNA ROSE CHILDRESS, AND GEORGE E. WOODY • Drug Dependence Unit, Philadelphia VA Medical Center, University and Woodland Avenues, Philadelphia, Pennsylvania 19104.

(Berzins, Ross, & English, 1974; Crowley, Chesluk, Dilts, & Hart, 1974; Gendreau & Gendreau, 1970). Further, there has been no demonstration of a common pattern or cluster of personality traits among drug abusers that can effectively discriminate them from other abnormal populations (Gendreau & Gendreau, 1970; Nyswander, 1967).

In an attempt to resolve the confusing and contradictory findings in this area, and to obtain more precise information regarding the possible relationships between drug use and psychopathology, our research group has performed a series of investigations concentrating upon the relationship between the use of particular types of street drugs and the occurrence of specific psychological symptomatology. The results of these studies indicate the presence of significant psychopathology within the drug abusing population and also show that specific patterns of drug use are differentially (and perhaps causally) associated with acute and chronic expressions of particular psychiatric disorders. Part I of this chapter will summarize our work in this area, while Part II will discuss the implications of this research for more appropriate treatments of these disorders.

2. Part I—Background

2.1. Initial Work

Our initial work in this area resulted from confidential interviews with a large group of psychiatric inpatients ($N = 136$). These patients revealed significant alcohol and/or drug abuse problems that they *had not previously reported* to the treatment staff at the time of hospital admission (McLellan & Druley, 1977). Thus, this group of primary psychiatric patients had concurrent, secondary substance abuse problems but had been admitted and given psychiatric diagnoses without knowledge of their substance abuse problems.

A comparison of this group with a non-drug-abusing group ($N = 143$) of primary psychiatric patients who were selected randomly and interviewed at the same time, revealed what most prior diagnostic studies had shown: that there were no significant ($p > .10$) differences in type or frequency of psychiatric disorders between the two samples. Thus, the presence of a substance abuse problem alone did not differentiate that group psychiatrically from the nonabusing patients. However, when we divided the substance-abusing sample into groups based upon their preferred drugs, two clear findings emerged. First, the majority of substance-abusing subjects reported regular use (at least three times per week) of more than one drug, but virtually all of these subjects used combinations of drugs having *similar psychological effects*. Thus, concurrent abuse of alcohol, barbiturates (secobarbitol, pentobarbitol), and benzodiazepines (Valium, Serax, Librium) was common, as was the concurrent abuse of amphetamine, methylphenidate (Ritalin), and hallucinogens. However, there was very little evidence of regular conjoint use of barbiturates and amphetamine, or benzodiazepines and hallucinogens. Second, there was a significant relation between use of specific drugs and the psychiatric diagnosis received. This relationship is illustrated in Table 1, which presents the distribution of diagnoses for the nonabusing population (last line), the total abusing population (first line), and each of the primary drug use groups from the total abusing population. As can be seen, regular abuse of amphetamine or hallucinogens was associated with a high proportion of schizophrenia (specifically paranoid form) diagnoses and a low incidence of depression. Barbiturate abuse was associated with a high proportion of depression diagnoses, and a low proportion of schizophrenia. Although alcohol and heroin groups showed a high proportion of depression diagnoses, these two groups were not significantly different from the nonabusing population.

Given the interesting, albeit correlational, results of this first study investigating drug abuse patterns in a primarily psychiatric population, it was logical to examine the same issues through an investigation of the psychological profiles of a primary substance abuse population (McLel-

Table 1
Distribution of Psychiatric Diagnoses in a Male Veteran Inpatient Population

	Paranoid schiz.	Undiff. schiz.	Depression	Other	Total
	N (%)	N (%)	N (%)	N (%)	N (%)
Drug-problem population (Total)	32 (24)	32 (24)	38 (27)	34 (25)	136 (100)
Alcohol	9 (21)	11 (24)	15 (34)	9 (21)	44 (100)
Amphetamine	11 (44)	5 (20)	2 (8)	7 (28)	25 (100)
Barbiturate	0 (0)	5 (22)	12 (52)	6 (26)	23 (100)
Hallucinogens	9 (41)	5 (23)	1 (5)	7 (33)	22 (100)
Heroin	3 (14)	6 (27)	8 (36)	5 (23)	22 (100)
Nondrug population	34 (24)	43 (30)	49 (28)	26 (18)	143 (100)

lan, MacGahan, & Druley, 1980). In this later study, we analyzed the Minnesota Multiphasic Personality Inventories (MMPI) of a sample of 158 consecutive drug abuse admissions, divided into drug preference groups, to determine if specific patterns of drug use were correlated with particular patterns of psychopathology.

A group of 33 primary stimulant users (amphetamine, hallucinogens), 110 opiate users (heroin, methadone), and 15 mixed drug users (stimulant and opiate) were identified on the basis of self-reported drug use confirmed by urinalysis. Results of the supervised, postdetoxification MMPI tests are presented in Table 2. As can be seen, the tests revealed generally lower scores in the opiate group (indicating lower levels of psychopathology), with the exception of elevated symptoms of depression (D) and psychopathy (Pd). The stimulant and the mixed drug use groups had higher levels of psychological symptoms generally, and specific elevations in the symptom areas of schizophrenia (Sc), psychasthenia (Pt), and paranoia (Pa). Thus, as in the first study, the majority of subjects reported regular use of drugs having psychophysiologically similar effects, and the pattern of use was differentially associated with the type of psychopathology.

One finding, which was seen here but not in the first study, was significant symptoms of depression among the opiate users. The rela-

Table 2
Mean MMPI T-Values for 158 Male, Veteran Drug Abuse Patients

	Psychostimulant	Mixed	Opiate	Sig.[a]
N	15	33	110	
MMPI SCALE				
L	52	48	49	
F	69[b]	70	63	+
K	48	47	49	
Hypochondriasis	62	63	64	
Depression	70	72	74	
Hysteria	63	64	64	
Psychopathic deviate	75	78	74	
Male–female	62	63	61	
Paranoia	72	67	61	*
Psychasthenia	75	73	66	+
Schizophrenia	82	78	67	*
Mania	73	74	72	
Social introversion	56	56	53	

[a] $+ = p < .05$; $* = p < .01$.
[b] Underlined pairs do not differ $p > .05$.

tion between opiate use and depression has been studied and reviewed by Woody and Blaine (1979) and Woody, O'Brien, and Rickels (1975) in several articles. These investigators concluded that depression is significantly higher in opiate addicts than in nonabusing subjects matched for age and socioeconomic status. They noted, however, that much of the depression was situational or reactive, and that the intensity of symptoms appeared to diminish significantly with methadone stabilization. Several other authors have noted the rapid reduction in depression following methadone stabilization and have suggested its role as an antidepressant (Gold, Donabedian, & Dillard, 1977; Kauffman, 1974; Senay, Dorus, & Meyer, 1982), although the appropriate long-term studies of this issue have not been done.

2.2. Six-Year Outcome Study

Although we had seen evidence of depression in our opiate addicts and elevated psychiatric symptomatology associated with nonopiate drug use in several studies, the majority of the data had been simply correlational, offering no clear indications regarding the nature of the relationships. In the course of this investigation, we identified a population of chronic drug abusers who had been initially treated at our inpatient facility in 1972, and who had been readmitted at least once each year for the following 6 years (McLellan, Woody, & O'Brien, 1979).

The readmission records of these patients provided information on intake status, psychiatric assessments, psychological testing, and within-treatment progress over the course of the 6-year period. This offered an opportunity to examine the longitudinal relationships between patterns of extended substance abuse and the development of psychiatric disorders.

This sample consisted of 51 male veterans who had been admitted to inpatient drug abuse treatment at the Coatesville VA Medical Center during 1971–1972, and subsequently readmitted at that facility a minimum of six times between 1972 and 1978. This sample was selected retrospectively during 1978, and represented *all* of the patients who met the frequency of readmission criterion described above.

These patients were divided into three groups based upon their primary drug preferences in 1971–1972. Patients in the Stimulant group ($N = 11$) had reported primary use of hallucinogens, amphetamines, and/or inhalants. Patients in the Depressant group ($N = 14$) had initially reported primary use of barbiturates, benzodiazepines, and/or sedative hypnotics. Opiate patients ($N = 26$) had reported primary use of narcotics such as heroin, methadone, and/or synthetic opiates.

The major purpose of this research was to examine the course of change in drug problems and psychiatric status within these three groups over the course of their 6-year treatment history.

2.2.1. Admission Status 1972

Table 3 compares the admission characteristics of the three groups at the time of their initial treatment at the Coatesville VA Medical Center in 1972, and summarizes the characteristics of the remainder of the 1972 drug patient admissions. As can be seen from the data in this table, the three study groups were generally similar in terms of overall status, and surprisingly comparable to the remaining 1972 population. Besides the obvious differences in patterns of drug usage, the only significant differences between the groups were in racial composition ($\chi^2 = 6.38$, $df\ 2$, $p < .05$) and in years of regular (at least three times per week) drug use ($F = 3.24$, $df\ 2.48$, $p < .05$). The remaining variables of age, education, number of previous drug treatments, and work history were not found to differ significantly between the three study groups or between the pooled data of the study groups and the remaining population.

It is important to note that while virtually all patients reported use of multiple drugs in 1972 (modal number $= 2$), only two (both in the Stimulant group) reported *regular* concurrent use of chemical agents having dissimilar psychological effects. These patients reported regular use of stimulants with depressants or with opiates. The remainder of the subjects chose drugs having similar psychological effects (e.g., benzodiazepines and barbiturates; cocaine and amphetamine), and as indicated previously, we have found this to be the typical pattern of abuse in the large majority of our clients. In this study, only alcohol and marijuana were regularly used by a majority of patients in all groups.

2.2.2. Treatment Progress 1972–1978

These clients were specifically selected because they had been in treatment at the Coatesville VA Medical Center at least once each year from 1972 to 1978. Table 4 summarizes the course of their treatment experiences at Coatesville and elsewhere during that period. As can be seen, these patients as a group averaged more than eight treatment attempts during this 7-year period, while maintaining a total of less than 11 months of abstinence and only 2.4 months of continuous abstinence during that period.

The data on duration of treatment, frequency of treatment attempts, and months of abstinence provide a generally representative picture of

Table 3

Three Groups of Male Veteran Drug Abuse Inpatients, Admission Status 1971–1972

	Group I Psychostimulant	Group II Psychodepressant	Group III Narcotics	Remaining 1972 population	Between groups sig.[a]
N	11	14	26	122	
Mean age	21	21.5	24	23	+
Race					
Black	31%	44%	51%	54%	
White	69%	56%	49%	46%	
Mean years of education	10.8	11.3	10.5	11.2	
High school diploma	62%	66%	59%	66%	
Major drugs					
1st	Psychedelics 79%	Barbiturates 89%	Heroin 100%	Heroin 73%	
2nd	Amphetamine 64%	Benzodiazepines 59%	Methadone 90%	Amphetamines 10%	
3rd	Inhalants 19%	Sedative hypnotics 21%	Synthetic opiates 76%	Barbiturates 6%	
Mean years of regular drug use	2.6	3.2	3.4	2.6	
Mean number of previous drug treatments	2	1.8	2.1	2	
Worked in past year	36%	41%	35%	39%	
Secondary psychiatric diagnoses	9%	13%	13%	11%	

[a] + = $p < .05$.

Table 4
Treatment History from 1972–1978 for Three Groups of Drug Abusers

	Group I Psychostimulants	Group II Psychodepressants	Group III Narcotics	Total	Sig.[a]
N	11	14	26	51	
Treatments	7.8	8.2	8.6	8.3	
Different programs	4.4	3.8	2.4	3.3	+
Months in treatment	7.8	8.0	11.3	10.2	+
Months abstinent	12.0	10.9	10.6	10.8	
Months continuous abstinence	2.8	2.9	1.9	2.3	

[a] + = $p < .05$.

the patients' status during that period. All attempts at drug-free treatment must be considered unsuccessful by any criterion for all patients. Not only were attempts at abstinence unsuccessful; the readmission data for these patients across the 6-year period of study showed evidence of progressive deterioration in the abuse problem, as well as the social, criminal, and self-support concomitants of the addiction.

With regard to within-treatment results, the data are equally disappointing. Forty-seven percent of the treatment attempts were prematurely terminated by these patients, while an additional 28% of the treatment attempts were terminated prematurely by program staff for disciplinary reasons. Only 25% of treatments resulted in satisfactory or favorable completions, and no patient had completed more than 35% of his total treatment attempts.

2.2.3. 1972 Comparisons

2.2.3.1. Psychological Testing. Two standard psychological tests were administered by the drug rehabilitation program in 1972, 10–14 days following completion of detoxification. The Minnesota Multiphasic Personality Inventory (MMPI) was included to evaluate personality and affective disorders, and the Shipley Institute of Living Scale (SILS) was included as an index of intelligence and cognitive impairment. Table 5 presents the mean scores for each of the three groups on those tests. Data from the SILS on both the derived WAIS IQ (Simes, 1958) and the CQ (Cognitive Quotient, to measure abstracting ability) scales indicated no significant differences between the three study groups ($F = 1.71$, $df = 2, 48$, $p > .10$). Since this test was administered irregularly to patients who entered the rehabilitation program in 1972, it was not possible to compare these patients with the remaining 1972 population. In absolute terms, the mean derived IQ score for the pooled subjects in 1972 was 102, while the CQ score averaged 92.3. Both are approximately equal to those of normal populations.

Data from the 1972 MMPI profiles were analyzed for each of the three groups and indicated no significant overall differences between them ($F = 1.16$, df 2, 28, $p > .10$). An additional repeated measures analysis of variance on the 1972 MMPI data was computed between the pooled scores for all the study subjects, and the scores for the remaining 1972 drug abuse clients ($N = 122$). Results of this analysis suggested a marginal difference ($F = 2.64$, df 1, 171, $p > .10$) between these two groups. Subsequent specific comparisons (paired t-test) on each of the individual scale scores indicated significantly ($p < .05$) higher scores on

Table 5

Psychological Testing Results 1972–1978, of Three Groups of Drug Abusers[a]

	Group I Psychostimulants			Group II Psychodepressants			Group III Narcotics		Sig. between groups 1978
	1972	1978		1972	1978		1972	1978	
N	11	11		14	14		26	26	
Mean IQ	101	103		102	94	*	104	102	+
Mean C.Q.	93	94		93	81	*	92	91	+
MMPI									
L	52	54		54	51		56	55	
F	64	96	*	58	78	*	62	66	*
K	56	58		53	58		51	56	
1. Hypochondriasis	61	67	+	61	72	+	59	61	*
2. Depression	55	58		60	94	*	60	66	*
3. Hysteria	60	71		54	58		54	54	+
4. Psychopathic deviate	70	76	+	68	74	+	72	78	
5. Male–female	64	70		60	58		68	66	
6. Paranoia	63	84	*	57	55		59	61	*
7. Psychasthenia	56	57		58	62		58	62	
8. Schizophrenia	66	98	*	60	63		64	61	+
9. Mania	64	87	*	63	66		65	59	*
10. Social introversion	48	55		55	59	+	56	58	*
Previous psychiatric treatment	4%	63%		0%	24%		4%	8%	*

[a] + = $p < .05$; * = $p < .01$.

the F (general pathology) and D (depression) scales for the study subjects, while the groups did not differ in the other scales.

Modal MMPI profiles for the study patients (and for the remaining
1972 population) were types 2-4 and 2-4-9, indicative of psychopathic
maladjustment, poor impulse control, but low levels of psychotic
symptoms.

2.2.3.2. Admission Interviews. All patients admitted to the Drug
Rehabilitation Program were interviewed by the staff physician and psychologist following detoxification. Results of these initial interviews suggested only low to moderate levels of psychiatric symptoms in all
groups. Only 12% of the patients were given either a primary or secondary psychiatric diagnosis other than drug abuse at the time of admission
in 1972. Comments from the psychologists and physicians who originally interviewed these patients tended to confirm the results of the psychological testing, indicating no incidence of psychosis or major affective disorder in any member of the three groups. Further, results of the
Mental Status Examination routinely performed on all patients entering
treatment showed little evidence of significant symptomatology in any
of the three groups, or in the remainder of the 1972 patient population.
Thus, as in the case of the psychological testing data, interviews revealed no evidence of meaningful differences in psychological status
between the three groups, or between the pooled data for the three
groups and the remainder of the 1972 drug patient population.

2.2.4. 1978 Comparisons

An examination of the 1978 admission data for the same patients
indicated significant and specific changes in psychological status among
the three groups. The findings for each of the three types of drug-
abusing groups will be described separately.

2.2.4.1. Stimulant Abusers. Although the patients in this group
could still be characterized as stimulant users generally, there was evidence of significant and pervasive change in their patterns of abuse by
1978. For example, the majority (79%) of these subjects reported regular
use of amphetamine or methylphenidate (Ritalin) injected intravenously, and little regular use of psychedelics or hallucinogens (with the exception of marijuana). In addition, 28% now reported irregular use of
chemicals with dissimilar psychological effects (i.e., barbiturates, benzodiazepines). A similar pattern of change in drug preference away from
psychedelics to amphetamine has been reported by Smith (1969) and
appears to reflect a legitimate shift in preference rather than a change in

availability. Only 23% of these patients reported any opiate use, and no period of physical addiction.

As can be seen in the comparison of 1972–1978 psychological testing for these subjects (Table 5), the MMPI results showed significant overall increases in psychological symptoms ($F = 12.12$, df 1,274, $p < .001$) over the years. Although no decrements in intellectual or conceptual functioning were evident, the MMPI data demonstrated pervasive differences, especially with respect to areas of general pathology (F), hysteria (Hy), Paranoia (Pa), Schizophrenia (Sc), and Mania (Ma). High scores on these scales suggest the presence of psychotic symptomatology, especially paranoid form. The modal MMPI profiles in this group were 2-8-9 (31%) and 4-8-9 (14%), as contrasted with the 2-4 and 2-4-9 profiles seen in 1972.

The results of the psychological testing (Table 6) verified the results of the psychological interviews performed by the staff physician and psychologist at each readmission. These interviews in 1977–1978 indicated the presence of psychopathology sufficient to warrant a primary or secondary psychiatric diagnosis in 71% of the subjects and referral to primary psychiatric treatment in 45% of the cases. The diagnosis of schizophrenia (Paranoid or Undifferentiated) was rendered in 63% of the cases, and 8% were characterized as having sociopathic personalities. The admission symptoms reported by these subjects were also interesting, with 72% of the subjects claiming visual or auditory hallucinations (during drug-free periods) and 18% expressing suicidal ideation. It is most important to stress that these symptoms were *not* associated with the acute stage of toxic amphetamine psychosis nor the brief period following physical withdrawal from amphetamine and other stimulants (Brown & Chaitkin, 1980; Ellinwood, 1979). Many of the symptoms reported were clearly psychotic in nature, such as auditory and (especially) visual hallucinations, as well as considerable paranoid ideation. By the time of their latest hospitalization, the symptoms shown by the patients were of prolonged duration, and were ameliorated only by standard neuroleptic therapy. Four of the stimulant-abusing patients were still to be found in inpatient psychiatric status a year after their 1978 admission.

2.2.4.2. Depressant Abusers. These patients remained primarily depressant users, although a significant proportion (34%) reported abuse of alcohol instead of, or in addition to, barbiturate and benzodiazepine abuse.

Again, like the stimulant users, the depressant-abusing patients evidenced significant changes in their psychological status by the 1978 admission. However, unlike the stimulant users, this group exhibited

Table 6
Admission Status by 1978

	Group I Psychostimulants	Group II Psychodepressants	Group III Narcotics
N	11	14	26
Major drugs			
1st	Amphetamine 71%	Barbiturates 82%	Methadone 86%
2nd	Cocaine 41%	Benzodiazepines 78%	Synthetic opiate 59%
3rd	Phencyclidine 20%	Alcohol 34%	Heroin 52%
Psychiatric diagnoses			
1st	Paranoid schiz. 36%	Dep. neurosis 34%	Psychopathic pers. 16%
2nd	Undiff. schizoph. 27%	Anxiety reaction 21%	Sociopathic pers. 12%
3rd	Sociopathic pers. 8%	Organic brain syn. 14%	Other
Admission symptoms			
Suicidal ideation	18%	77%	16%
Suicide attempts	9%	35%	8%
Hallucinations	72%	6%	8%
Memory change	9%	35%	4%
Problems concentrating	9%	35%	0%
Percentage referred for psychiatric treatment	45%	21%	8%

few psychotic symptoms. As can be seen in the Table 5 data, the most significant (paired t-tests) changes in the MMPI test results were found in the general pathology (F) and especially the depression (D) scales, although the entire group profiles were found to differ between 1972 and 1978 ($F = 3.99$, $df\ 1,349, p < .05$). Again, the results of psychological testing were also reflected by the diagnostic impressions of the admitting staff (Table 6). Anxiety and depressive disorders were diagnosed in 55% of these patients, and 21% were referred to primary inpatient psychiatric treatment. Especially significant is the fact that by 1978, fully 77% of this group reported suicidal ideations, and over 35% had actually attempted suicide. In addition, the incidence of drug overdoses was four times higher than in either of the other two groups. The symptoms of depression in these patients did not abate following the completion of physical detoxification. The treatment of choice for the majority of these individuals was symptom management with tri-cyclic antidepressants and referral to inpatient rehabilitation on an alcohol abuse treatment ward. This particular referral was chosen partly because of the high incidence of concurrent alcoholism in these patients, but primarily because the concomitant problems of confusion, anxiety, and depression are similar to those of the chronic alcoholic patient population.

Finally, the data from the Shipley Institute of Living Scale indicated a significant ($t = 6.34$, $df\ 13$, $p < .001$) reduction in the mean conceptual quotient scores of this group between 1972 and 1978. It is noteworthy that, while all patients in this group were within normal limits in 1972, and repeated administrations typically produce improved scores, fully 35% of this group showed worse results on the test and presumptive evidence of brain damage by 1978. Although the Shipley Institute of Living Scale is only a gross measure of cognitive impairment, the large decrements in test performance clearly suggested the presence of cognitive deficits. Previous authors (Greenblatt & Shades, 1974; Judd & Grant, 1975; Schuster & Wilson, 1972) have found evidence of cognitive impairment in a sample of younger polyabusers, the majority of whom used barbiturates. However, their findings revealed subtle cognitive deficits, and not the gross deficits reflected by the Shipley. It may be that the longer period of use coupled with the generally higher severity of abuse were responsible for the degree of impairment found in the current study. It is also possible that the cognitive effects of the barbiturates require longer amounts of time for recovery than other depressant or polydrug combinations. It is interesting that, despite earlier suggestive reports (Grant, Mohns, & Miller, 1978), we found no evidence of gross cognitive impairment in any of the psychostimulant users. Our finding does not preclude the possibility that stimulant users have subtle cognitive deficits not revealed in the SILS.

2.2.4.3. *Opiate Abusers.* The patients in this group remained primarily opiate users with only slight evidence of regular nonopiate drug use. However, there was a clear shift in primary drug preferences from heroin in 1972 to methadone (licit and illicit) and synthetic opiates by 1978. However, this shift may have been more a function of reduced quality and availability of heroin, than a change in drug preference. An additional problem, which was increasingly reported by this group, was alcohol abuse. While only one subject had changed his primary abuse problem to alcohol, most subjects had reported greater day-to-day use of alcohol than at the time of their 1972 admission.

Data from the 1978 SILS (Table 5) indicated no significant changes in group IQ or conceptual quotient (brain damage) scores, both of which were within normal population ranges. Similarly, results of personality testing with the MMPI showed no general between-years effect ($F = 1.06$, $df\ 1,649$, $p > .10$), and indicated only moderate levels of depression and low levels of psychotic symptomatology. Modal profiles for this group were comparable to those found in 1972; 2-4 (31%), 2-4-9 (11%), again suggesting little systematic change in psychological status over the 6-year period.

Summaries of readmission interviews tended to support the results of the psychological testing. Comments from ward physicians and psychologists indicated a general group trend toward moderate increases in symptoms of depression, but little suggestion of psychotic symptomatology (Table 6). By 1978, only two patients had been referred for primary psychiatric treatment, both due to prominent suicidal ideation and recent attempts.

2.2.5. Discussion

These data are subject to some limitations. They were generated from a relatively small, selected sample of patients, studied in retrospective fashion: Diagnostics were derived from admission interviews rather than interviews structured according to Research Diagnostic Criteria. The cognitive testing employed (SILS) targeted gross, unsubtle, cognitive deficits. Yet the overall findings are generally consistent and are not likely the product of biased collection methods. All data were collected during standard admission evaluations, 10 to 14 days *after* completing detoxification on a locked ward. Thus, the results cannot be attributed to acute or toxic drug reactions. Since the data were collected independently, not as part of an experimental study, the diagnoses and evaluations were made without possible bias resulting from awareness of our experimental hypothesis. Furthermore, the pattern of results seen cannot satisfactorily be explained as products of change in staff diagnostic methods or the generally damaging effects of high stress "street life." Though these *general* effects may have played some role, they cannot explain the *specific* pattern of results seen in the independent psychological testing—or the absence of increased symptoms in the opiate group.

No single theory can fully account for this pattern of results. However, we offer two possible explanations for the data collected thus far. With regard to nonopiate drug use, these substances may actually have or (more likely) may appear to have a short-term palliative effect on certain underlying psychological symptoms. However, they may exacerbate the underlying psychiatric illness in the long term. Indeed, even the stresses associated with drug-procuring may actually exacerbate the existing symptoms.

This "self-medication" view is a plausible explanation for the available data and for the tendency of chronic substance abusers to selectively use combinations of drugs with similar psychological effects. The reported association between psychodepressant use and depression is quite consistent with this view. This view would suggest that individuals with underlying symptoms of depression such as dysphoria, poor appetite, agitation, anxiety, somatic symptoms, and sleeplessness would be preferentially drawn to barbiturates, benzodiazepines, and sedative hypnotics. These drugs would have particular short-term appeal due to their soporific, anxiolytic, and sedative effects. However, after prolonged use, the effects of tolerance and rebound withdrawal could even exacerbate these symptoms.

Interestingly, we are aware of a small group of stimulant-abusing patients who also appear to be medicating symptoms of depression. These patients often appear to be suffering from more lethargic, retarded, anergic symptoms (especially loss of interest in previously stimulating events) than are other stimulant abusers. It may be that the sedative drugs are preferred by patients with more agitated forms of depression, while the stimulant drugs are selected for their energizing qualities by patients with retarded forms of depression.

The self-medication view would seem to be less plausible with respect to the use of hallucinogens and/or amphetaminelike compounds for psychotic symptomatology. On the surface, it would seem that these agents might be particularly disliked by individuals with hypomania or soft signs of psychosis, since they could exacerbate hallucinations and underlying feelings of suspiciousness and confusion. In fact, there are studies that indicate that diagnosed schizophrenics, even those in remission, are extremely sensitive to the psychotomimetic properties of amphetamine and the hallucinogens (Bell, 1973; Griffith, Cavanaugh, & Oates, 1968). However, it is at least conceivable that the acute effects of the amphetaminelike compounds, including the energizing qualities, feelings of power, conceptual clarity, and heightened awareness (Griffith, Fann, & Oates, 1972; Ellinwood, 1967; Pope, 1979) might be reinforcing to these individuals. It is also possible that patients who regularly use the amphetaminelike compounds may have the "paradoxical" calming reaction to these drugs so often found in hyperkinetic children and adolescents. There does appear to be anecdotal support for this possibility (Pope, 1979; Schecter, 1980), and it would be consistent with a self-medication view of this type of drug use.

A contrasting "causation" view of drug use would suggest that the chronic self-administration of nonopiate (sedatives or stimulants) drugs actually *produced* the observed psychiatric illness, regardless of underlying or pre-existing personality factors. According to this causation view, psychiatric symptoms could be produced by dysfunctions in the CNS monoamine system (Baldassarini, 1975; Schildkraut, 1965), in the brain reward and punishment systems (Stein, 1964), or in the physical structure of the limbic system (Duarte-Escalante & Ellinwood, 1972; Schuster, 1982).

This view is consistent with the acute toxic conditions produced in humans by these drugs. It is also consistent with animal data from laboratory studies in which chronic high doses of amphetamine were associated with prolonged psychosislike symptoms (Ellinwood & Duarte-Escalante, 1972; Schuster & Wilson, 1972), while chronic administration of barbiturates produced protracted anergia and a depressive-like state (Sharpless, 1965). However, many of these animal studies have used very high dosages and few of the symptoms seen in the laboratory animals have been reported in clinical studies. The amphetamine-induced psychosis is a case in point. Griffith *et al.* (1968) have demonstrated the amphetamine psychosis in "normal" subjects given very high doses of amphetamine. However, Bell and his colleagues, using lower dosages of amphetamine, were able to demonstrate this psychosis only in schizophrenics in remission, but not "normals" (Bell, 1973). Further, there are only a few clinical reports of protracted, long-term psychoses associated with psychostimulant use, and these come from post-World War II amphetamine "epidemics" in Japan (Tatetsu, 1972) and Sweden (Goldberg, 1968).

With regard to the possible production of depression in depressant users, there are some clinical data suggesting that severity of depression is increased by chronic administration of benzodiazepines. The data on brain damage associated with chronic benzodiazepine and especially barbiturate use are consistent with the "causation" model and are more compelling. Extensive work by Grant (Grant & Judd, 1976; Grant, Mohns, & Miller, 1978) in the field of cognitive function and drug abuse indicates a distinct and long-term relation between barbiturate use and cognitive impairment as measured by several rigorous neuropsychological instruments.

With regard to chronic opiate use, the lack of severe long-term psychiatric symptoms in our patients is intriguing. Given the severe stresses associated with opiate addiction and the documented incidence of depression shown in opiate-addicted patients at the time of admission to treatment and even shortly after detoxification (Senay et al., 1976) or methadone stabilization (Woody et al., 1975), it might be expected that longer term symptoms would be seen with more protracted use. However, the opiates may have a medicinal or even prophylactic effect against psychiatric illness. For example, opiates have been used in this country as an antipsychotic (Comfort, 1977; Carlsson & Simpson, 1963) and have been shown to release high levels of prolactin (generally considered a pharmacological marker of antipsychotic activity; Ellingboe, Mendelson, & Kuehnle, 1980). With regard to the possible prophylactic effects of the opiates, we have recently examined the diagnostic and psychological testing records of a sample of approximately 4,000 opiate addicts admitted to our inpatient facility. We have found that the incidence of schizophrenia is only 10% of that which would be expected in a "normal" population, despite the fact that demographic and socioeconomic data for the sample would predict a 25% *greater* than "normal" incidence of this condition (McLellan, Woody, & O'Brien, 1981).

In summary, the results of our studies investigating the relationships between drug abuse pattern and psychiatric illness has suggested four general conclusions. First, the majority of chronic substance abusers seem to select combinations of drugs with similar psychological effects for regular, concurrent use. So-called polypharmacy is seen in a minority of patients or on an aperiodic basis in others.

Second, acute heavy use of nonopiate drugs produce distinct, short-term, toxic psychiatric syndromes (psychostimulants—psychosis, mania; psychodepressants—depression, cognitive confusion). These syndromes generally remit in most patients shortly after detoxification.

Third, *chronic*, regular use of nonopiate drugs has been associated with long-term psychiatric illness (psychostimulants—psychosis, mania; psychodepressants—depression, cognitive impairment). These syndromes apparently *do not* remit even one year after detoxification.

Fourth, acute or chronic use of opiates is associated with short-term symptoms of minor depression (which typically improve rapidly). We have seen little evidence of significant long-term psychotic effects from opiate use.

In attempting to explain the results for the nonopiate users, we have offered two views. A "self-medication" view provides a particularly good explanation for the observed tendency of subjects to select drugs with psychologically similar effects. A "causation" view is most useful in explaining the development of new symptomatology over time. In discussing these explanations we have recognized that those views are *not* mutually exclusive and may very well be simultaneously operative. Patients with a given set of symptoms who choose a nonopiate drug for self-medication may indeed receive symptom relief from the *short-term* effects of the drug. Unfortunately, the *long-term*, chronic effect of these nonopiate drugs may actually exacerbate the original symptoms and even create symptoms or illness *de novo*. Patients with the same set of symptoms who choose opiates for self-medication may also experience the short-term symptom relief, but apparently without the long-term psychiatric consequences.

3. Part II—Treatment Implications

Clearly, the data relating acute and chronic use of particular drugs to acute and chronic patterns of psychiatric symptomatology is potentially important to our understanding of the etiologies of both drug abuse and psychiatric illness. However, it will be necessary to continue this line of research with more exacting, prospective studies if we are to advance our understanding of etiology. Meanwhile, we are left with the important day-to-day issue of how best to treat patients who *already* have engaged in chronic abuse of one or more chemical agents and may now be suffering the associated psychiatric symptoms. Of particular concern is the treatment of choice for a growing population of patients presenting with mixed patterns of substance abuse. By "mixed abuse" we mean the regular (at least three times per week), concurrent use of drugs from more than one class of chemicals: opiates, stimulants, depressants, or hallucinogens. During the course of pursuing the research summarized in Part I of this paper, we gained valuable information regarding the kinds of treatment problems that were potentially associated with the mixed abuse patient. In the remainder of this paper we will describe a research strategy, based on this early work, that attempted to determine the most appropriate treatment alternative for these patients.

At present, two conceptually different treatment approaches are available. The drug-free therapeutic community (TC) attempts to promote detoxification and abstinence in a structured, peer-oriented community setting, using group "encounter" therapy (see DeLeon & Rosenthal, 1979). Methadone maintenance (MM), in contrast, attempts to stabilize opiate use through substitution of the synthetic opiate methadone, simultaneously developing family, job, and social supports in that context (see Lowinson & Millman, 1979).

From a practical perspective the treatment of mixed abusers in either of these modalities has been problematic. They have the poorest posttreatment outcomes and the worst record of behavior problems during therapeutic community treatments, often disrupting treatment activities and dropping out prematurely (McLellan, Druley, & Carson, 1978; Simpson, Savage, Lloyd, & Sells, 1978). Treatment within a methadone maintenance modality is often inappropriate or illegal, since many mixed abusers have not had sufficient opiate abuse to meet FDA admission requirements. Thus, the practical treatment options available for the mixed abusers are often limited. However, from a theoretical perspective, there is reason to believe that methadone maintenance, for those mixed abusers who qualify, would be more effective than drug-free treatment. Patients who abuse nonopiate drugs often experience a greater range and severity of psychiatric problems than patients who abuse opiates alone (Grant & Judd, 1976; McLellan, MacGahan, & Druley, 1980; Zuckerman et al., 1975). The psychiatric symptoms associated with nonopiate use may respond to the potential psychotropic effects of methadone: Clinical and laboratory studies suggest that the opiates (including methadone) may have antidepressant, anti-anxiety, and antipsychotic effects (Comfort, 1977). Thus, methadone maintenance may be a more effective treatment than drug-free therapy for this population of patients. With this in mind, we compared the relative effectiveness of methadone maintenance and drug-free therapy in the treatment of mixed drug abusers *who could have legally been treated in either modality*. For purposes of comparisons we selected opiate-stimulant abusers, opiate-depressant abusers, and an opiate-only group, adopting the hypothesis that methadone maintenance would be more effective with the two mixed abuse groups than would drug-free therapy.

The subjects for this study were selected *retrospectively* from a larger pool of 382 male veterans who had been treated for drug dependence at the Philadelphia or Coatesville VA Medical Centers during 1978. All of these subjects had received a comprehensive evaluation at admission to treatment and at 6-month follow-up as part of a larger study (McLellan, Luborsky, O'Brien, Woody, & Druley, 1982). Each of the patients had received a minimum of 15 inpatient days or 15 outpatient visits in either the drug-free, therapeutic community (TC) or methadone maintenance (MM) rehabilitation programs.

3.1. Treatment Programs

The drug abuse treatment network of the Veterans Administration in the Philadelphia area consists of two inpatient drug abuse programs at the Coatesville VA Medical Center (one conventional TC, one combined alcohol and drug abuse TC). In addition, there is one outpatient methadone maintenance clinic at the Philadelphia VA Medical Center. This treatment network has enjoyed cooperative referral arrangements since 1970. Once admitted to either hospital, patients are assigned to rehabilitation treatmant based upon their personal requests, the clinical judgment of the treatment team, or administrative considerations such as bed census and patient-visit criteria.

3.1.1. Therapeutic Community

The TC program is an intensive, 90-day, 30-bed program geared primarily toward social habilitation. Individual and group (encounter) therapies, in addition to the social structure of a self-governing community, are the primary therapeutic tools. Family therapy as well as educational and vocational counseling are also offered. The program is directed by a social worker and includes a staff of two psychologists, two social workers, two nurses, and four rehabilitation technicians. Limited use of psychotropic medications is permitted when indicated by the attending physician.

3.1.2. Methadone Maintenance

The MM program is a research and treatment program that offers methadone maintenance plus individual counseling and vocational guidance to an active census of 300 patients. The use of adjunctive psychotropic medications is also permitted within this program and is supplemented by regular sessions of individual counseling by the rehabilitation technicians. The staff consists of three psychiatrists, one psychologist, two social workers, two nurse-practitioners, and ten re-habilitation technicians.

3.2. Evaluation Procedures

Since we wanted to compare treatment outcomes in subjects who *could* have been assigned to either the TC or MM programs, we selected only those subjects who met or surpassed the minimum FDA require-ments for methadone treatment. These included a minimum of three years of physical dependence on opiates, and two or more prior unsuc-cessful detoxifications. One hundred and seventeen subjects were se-lected retrospectively in this manner; thus, all subjects were legally eligi-ble for either TC or MM treatment.

These subjects were further divided into three drug-use groups as indicated, based upon the pattern of their drug use. Opiate-Stimulant (OS) subjects reported 3 or more years of regular stimulant (ampheta-mine, methylphenidate, cocaine) use, but less than 2 years of regular depressant use (the operational definition of "regular use" being a fre-quency of three or more occasions per week). Opiate-Depressant (OD) subjects reported 3 or more years of regular depressant (barbiturates, benzodiazepines, sedative-hypnotics) use, but less than 2 years of reg-ular stimulant use. Opiate-Only (OO) subjects reported greater than 3 years of opiate use and less than 2 years regular use of either stimulants or depressants. Alcohol and marijuana use were recorded but were not used in forming the three research groups. Initial analyses of these drug-use patterns indicated a tendency ($p < .12$) toward greater alcohol use by the OD and OO groups, and greater marijuana use ($p < .15$) by the OS group. However, these differences in frequency of use were not statistically significant ($p > .10$).

Once these three drug-use groups were differentiated, subjects within each were further divided into those who had been treated in TC and those who had been treated in MM. Subjects were then matched on the basis of age, race, years of drug use, number of prior treatments, and psychiatric severity. Psychiatric severity was based on the psychiatric scale of the Addiction Severity Index (McLellan, Luborsky, & Erdlen, 1980; McLellan, Luborsky, Woody, & O'Brien, 1980), an index that has been shown to be the best general index of overall impairment at treat-ment admission and a good predictor of outcome (McLellan, Luborsky, Woody, Druley, & O'Brien, 1983). Thirteen subjects were lost as a result of the matching procedure, leaving 104 subjects in the study—15 pairs of OS subjects, 15 pairs of OD subjects, and 22 pairs of OO subjects.

The study data was derived from administrations of the Addiction Severity Index (ASI; McLellan, Luborsky, Woody, & O'Brien, 1980) at admission to treatment and again at 6-month follow-up. The ASI is a structured, 30–40 minute, clinical research interview designed to assess problem severity in six areas commonly affected by addiction. It is important to note that problem severity in each area is assessed independently and six areas are included: medical, legal, substance abuse, employment, family, and psychiatric problems. In each of the areas, objective questions are asked concerning the number, extent, and duration of problem symptoms in the patient's lifetime and in the past 30 days. The patient also supplies a subjective report of the recent (past 30 days) severity and importance of the problem area. The interviewer assimilates the two types of information to produce a rating (0–9) reflecting the extent to which treatment is needed by the patient in each area. These ratings have been shown to provide reliable and valid general estimates of problem severity for both alcoholics and drug addicts, and the individual items offer a comprehensive basis for clinical and experimental assessment.

All ASI follow-up interviews were conducted by a research technician, either in person or over the phone. No information from secondary sources was used, and all data were closely monitored to preserve confidentiality. The validity of the follow-up data was maintained through built-in consistency checks within the ASI and through spot checks on subsamples of the population assessing the ASI data against urinalysis, pharmacy, and criminal justice system records.

It was important for the aims of this study to have general measures of treatment outcome, since single-item measures can be inherently unreliable (Nunnally, 1967). We therefore constructed criterion composites or factors from sets of single items within each of the ASI problem areas. Several items from each problem area were intercorrelated to exclude those that were unrelated, and the remaining items were standardized and tested for conjoint reliability using Cronbach's formula (Cronbach & Furby, 1980). A set of four to six items from each ASI problem area was selected using this procedure, and each set showed a standardized reliability coefficient of .73 or higher. Seven composites or factors (medical problems, employment, drug use, alcohol use, legal status, family problems, and psychiatric function) were constructed in this manner, and scores on each were calculated for all patients at admission and follow-up. This method has been employed effectively by Luborsky and Mintz in their studies of outcome from psychotherapy (Luborsky, 1980; Mintz & Luborsky, 1979).

3.3. Admission Comparisons

Table 7 provides information obtained at admission on the OS, OD, and OO groups assigned to either MM or TC modalities and indicate that they were well-matched in terms of demographics and general status. None of the drug use groups receiving contrasting treatment differed significantly on more than three items. Further, there were no consistent trends with regard to overall severity at the time of admission.

Comparisons among the three drug-use classes (collapsing across treatment groups) revealed a somewhat higher proportion of black subjects in the OO group than in the other two categories. This is consistent with the clinical impression shared by most workers that black drug abusers are less prone to regular, mixed drug abuse than their white counterparts. Analyses of the remaining demographic items showed no significant differences among the three drug-use classes. However, analyses of the admission status measures obtained from the ASI revealed a trend toward better overall status for the Opiate-Only group. These differences were most marked in the psychiatric status variables. No consistent differences in overall status were observed between the OS and the OD patients.

3.4. Outcome Comparisons between Treatment Programs

Comparisons of admission to follow-up *improvement*, as well as adjusted (ANCOVA) comparisons of 6-month *outcomes*, were calculated for the two treatment programs. These results are presented for each of the drug-use groups individually.

3.4.1. Opiate-Stimulant Subjects

Admission and follow-up results are presented for all OS subjects in Table 8, for both MM and TC treatment program patients. Considerable differences are seen in the performance of patients in the two treatment programs. The MM subjects showed more improvement in more criterion areas than the matched patients from the TC program. MM patients showed significant improvements in all criterion areas except medical status and family relations, while the TC patients only showed clear gains in employment and some improvement in drug use.

Analyses of covariance on the 6-month outcome status variables for the two treatment programs yielded similar results and can be found in the last column of Table 8. In each of these analyses, the admission

Table 7
Background Characteristics of Three Groups of Mixed Drug Abusers

Variable	Opiate-Stimulant drug users		Opiate-Depressant drug users		Opiate Only drug users	
	TC	MM	TC	MM	TC	MM
N	15	15	15	15	22	22
Age	32	30	31	30	33	33
Race (% black)	32	28	50	52	59	64
Yrs. education	11	12	12	12	12	12
Previous treatments	5	5	6	5	4	4
Medical severity[b]	2.8	2.6	3.2	3.0	1.9	1.8
Previous hosp.	2	3	4	3	3	2
% w/chronic prob.	33	40	34	29	40 *	23
Employment severity[b]	5.6	5.9	6.1	5.5	4.6	4.5
% w/skill or trade	80 *	67	40 *	63	68	58
% worked past 3 yrs.	65	68	42	48	55	48
Subs. abuse severity[b]	6.3	6.8	7.0	6.4	6.0	5.6
Yrs. alcohol abuse	1.5	1.8	5.0	3.4	2.6	2.2
Yrs. opiate use	9	11	8	6	12	10
Yrs. stimulant use	7	8	1	1	1	1
Yrs. depressant use	2	1	9	7	1	0
Legal severity[b]	4.0	4.2	5.0	4.2	4.0	4.4
Arrests	6	5	9	4	6	4
Months incarcerated	27	29	20 *	14	24	24
Family/social sev.[b]	5.3	4.9	4.4	3.7	3.7	3.6
% Divorced/Separated	27	20	31	33	23	27
% Never Married	40	33	27	27	36 *	18
% Living Alone	33	27	18	22	18	18
Psychiatric sev.[b]	4.5	4.4	5.1	4.7	3.3	2.3
% prev. hosp.	33	40	29	23	18	11
% sign. depressed	43	38	73	66	51	44
% suicide attempts	7	18	27	21	9	5
% hallucinated	36	22	21 *	7	0	8
Maudsley neurot.	32	34	30	27	18	20
Shipley C.Q.	88	85	76	79	88	91
Beck Dep. Inv.	11	13	21	19	17	15

[a]* = $p < .05$ by paired t-test.
[b]ASI severity scores: 0 = No problem; 9 = Maximum severity.

criterion score was used as the only covariate because the subjects were matched on virtually all other relevant variables. The follow-up outcome measures revealed better outcome in the MM subjects on all but 3 of the 21 comparisons made, and these differences were significant on 10 of those comparisons. Thus, these treatment results indicate greater and

Table 8
Opiate-Stimulant Subjects Treated in Therapeutic Community or Methadone
Maintenance[a]

Criteria[b]	Therapeutic community (N = 16)			Methadone maintenance (N = 16)			ANCOVA on 6-Mo. scores[c]
	Adm.		6-Mo.	Adm.		6-Mo.	
Medical factor	11.2		12.9	9.0		9.7	
Days med. prob.	9		10	7		7	
Employment factor	20.4	*	8.7	31.3	*	6.1	+
Days worked	3		9	2	*	12	
Money earned	54	+	277	68	*	371	+
Welfare support	62		38	71	+	47	
Alcohol use factor	15.4		16.1	20.5	+	10.6	*
Days intoxicated	4		5	7	+	2	
Money spent on alc.	7		6	8	+	3	
Drug use factor	149.1	+	61.0	197.4	*	38.2	*
Days opiate use	11	+	5	15	*	3	
Days dep. use	2		1	3		2	
Days stim. use	10	*	2	8	+	2	
Money spent/drugs	148	+	72	230	*	70	+
Legal status factor	66.2	+	21.0	86.3	*	16.4	+
Days comm. crime	6		4	15	+	6	
Illegal income	160	+	58	130	+	14	+
Family rel. factor	13.0		10.6	15.9		11.9	
Days fam. prob.	10		9	12		8	
Psychological factor	15.4		14.3	17.0	+	12.7	+
Days psych. prob.	12		10	14	+	7	+

[a] + = $p < .05$; * = $p < .01$.
[b] All data reflect the 30-day periods prior to admission (Adm.) and follow-up (6-Mo.).
[c] Admission (Adm.) score was the covariate in each analysis.

more pervasive improvements, as well as generally better 6-month outcomes, in the OS subjects who were treated in a methadone maintenance program than in matched OS subjects treated in a therapeutic community.

3.4.2. Opiate-Depressant Subjects

Table 9 presents admission and follow-up data for the OD subjects treated in each of the two rehabilitation programs. Again, there was an overall difference in the performance of the two treatment programs but

Table 9

Opiate-Depressant Subjects Treated in Therapeutic Community or Methadone Maintenance[a]

Criteria[b]	Therapeutic community (N = 18)			Methadone maintenance (N = 18)			ANCOVA on 6-Mo. scores[c]
	Adm.		6-Mo.	Adm.		6-Mo.	
Medical factor	15.1		10.3	10.2		7.1	
Days med. prob.	12		8	7		5	
Employment factor	29.2	+	4.0	21.5	+	10.6	+
Days worked	2	+	10	6		8	+
Money earned	65	+	285	117	+	211	
Welfare support	59		48	68		106	+
Alcohol use factor	25.0		15.9	16.7		9.9	
Days intoxicated	6		2	4		2	
Money spent on alc.	24	+	9	13		10	
Drug use factor	114.5	*	35.3	140.1	*	80.1	+
Days opiate use	14		8	17	*	6	+
Days dep. use	17	+	8	13		10	
Days stim. use	3		1	3		1	
Money spent/drugs	70		24	131	*	54	
Legal status factor	76.4	*	8.2	91.8	*	32.8	*
Days comm. crime	8	+	1	14	+	7	+
Illegal income	188	+	10	169		81	+
Family rel. factor	9.8		6.8	12.7		10.9	
Days fam. prob.	7		5	9		8	
Psychological factor	18.9	*	9.3	19.4		16.2	*
Days psych. prob.	20	*	6	9		8	+

[a] + = $p < .05$; * = $p < .01$.
[b] All data reflect the 30-day periods prior to admission (Adm.) and follow-up (6-Mo.).
[c] Admission (Adm.) score was the covariate in each analysis.

opposite to that seen in the OS class. The Table 9 analyses of admission to follow-up improvement indicate that 11 variables showed significant improvement in the TC subjects as compared with 7 significant changes in the MM subjects. Neither group showed evidence of significant improvement in medical condition or family relations, and only minor improvements were shown by the TC subjects in alcohol use. TC subjects improved somewhat less than the MM subjects in drug use, but showed somewhat more improvement in employment and legal status and clearly greater improvement in psychiatric function.

The between-programs analyses of covariance on the 6-month outcome data substantiated the results of the within-program analyses. As can be seen in the last column of Table 9, the follow-up status of the TC patients was better than that of the MM patients on all but 3 of the 21 measures, and significantly so on 10 variables. These results indicate generally greater improvement and better 6-month outcomes among OD subjects treated in the therapeutic community than matched OD patients treated in methadone maintenance.

3.4.3. Opiate-Only Subjects

The within- and between-program comparisons for the OO subjects are presented in Table 10. No clear differences in levels of improvement or in 6-month follow-up status were evident between the two treatment programs. Within-program comparisons indicate 11 variables that showed significant improvement for the TC subjects and 13 significantly improved criteria for the MM subjects. The MM subjects showed somewhat more improvement than the TC subjects in the area of medical status, while the TC subjects showed greater improvement in employment and family relations. The remaining areas revealed approximately equal levels of improvement between the two programs. OO subjects, in general, showed the greatest improvements in the areas of drug use, employment, and legal status, and this is consistent with previous reports from this clinic regarding the major results of drug abuse treatment (McLellan et al., 1982).

Between-program analyses of the follow-up data also indicated few significant outcome differences between the two rehabilitation programs. Thus, Opiate-Only subjects showed generally better outcomes than other drug-use subjects, but approximately equal levels of improvement and similar 6-month adjustments when treated in either the therapeutic community or in the methadone maintenance programs.

3.5. Discussion of the Findings

Prior to discussion and interpretation of these findings, it is important to clarify the limitations of the study. Perhaps the major qualification to the results is the limitation in our experimental design. The present data were collected retrospectively as part of a larger treatment evaluation study, and the patients were not randomly assigned to the two treatment programs. This raises the possibility that the treatment groups were not equal on important variables prior to treatment. We attempted to address this limitation retrospectively in two ways. First, we took care to match the patients on those variables that have histor-

Table 10
Opiate-Only Subjects Treated in Therapeutic Community or Methadone Maintenance[a]

Criteria[b]	Therapeutic community			Methadone maintenance			ANCOVA on 6-Mo. scores[c]
	Adm.		6-Mo.	Adm.		6-Mo.	
Medical factor	5.6		6.2	7.7		5.8	
Days med. prob.	4		4	6	+	3	
Employment factor	27.0	*	4.0	32.0	*	8.6	
Days worked	4	+	11	6		10	
Money earned	135	+	316	156		213	
Welfare support	34		25	96		77	+
Alcohol use factor	17.3	+	6.5	20.3	+	8.3	
Days intoxicated	4		1	5	+	1	
Money spent on alc.	10	+	2	19	+	2	
Drug use factor	164.7	*	49.9	210.7	*	41.6	+
Days opiate use	18	*	3	22	*	3	
Days dep. use	4		2	5		2	
Days stim. use	4		3	4		2	
Money spent/drugs	205	*	32	251	*	40	
Legal status factor	94.6	*	18.0	88.7	*	20.1	
Days comm. crime	15	+	4	12	+	3	
Illegal income	249	+	80	270	+	105	
Family rel. factor	7.0		5.6	12.7		14.0	+
Days fam. prob.	7		4	9		10	
Psychological factor	6.5		6.4	7.0	+	4.1	
Days psych. prob.	3		4	7		5	

[a] $+ = p < .05$; $* = p < .01$.
[b] All data reflect the 30-day periods prior to admission (Adm.) and follow-up (6-Mo.).
[c] Admission (Adm.) score was the covariate in each analysis.

ically been important in determining treatment outcome. The data indicate that, with few exceptions, the matched groups were equal in terms of demographic, background, and current status factors. Secondly, in order to adjust for incomplete matchings, we utilized analyses of covariance. While this alone is not a sufficient substitute for a randomized design, the procedure does permit a conservative analysis of outcome between groups (see Simpson *et al.*, 1978). Taken together, the matching and the analysis of covariance procedures compensate substantially for the lack of an experimental design, and permit cautious interpretation of the findings, although experimental replication is desirable in the long term.

The absence of random assignment to treatments also raises the possibility that one of the treatment groups received more favored, better prognosis patients. We do not feel this is likely in the present study for two reasons. First, despite the fact that the groups were not assigned randomly, they were assigned in the same way and in approximately the same proportion to each of the treatment programs (i.e., on the basis of their personal choice, the clinical judgment of the admission staff, administrative considerations, and simple chance). This reduces the possibility of differential assignment strategies. Secondly, if differential assignment was in effect, it does not, by itself, explain the differential performance of the three drug-pattern subjects in the two programs. The fact that neither program was generally more effective *across* all drug-pattern groups suggests that neither program received generally favored patients.

Given the present results and the design limitations, we would like to interpret the data in a manner that is consistent with certain known facts and with the clinical pictures presented by these patients.

3.5.1. Opiate-Stimulant Group

The results of the treatment comparison in the OS group are not surprising because we have studied this population of stimulant abusers extensively and, coupled with the reports of others, have concluded that the symptoms of hypersensitivity, dependence, paranoia, and mania are characteristic of these patients, often long after successful detoxification (Ellinwood, 1967; McLellan *et al.*, 1979). It is not yet clear whether the stimulants actually cause or simply exacerbate these symptoms. Nevertheless, OS patients often experience a significant range and intensity of psychiatric symptoms as they enter the TC for rehabilitation. Since the philosophy of the TC program prohibits the use of psychotropic medication, OS patients can be uncomfortable from the outset. In addition, the pressures of the new interpersonal living arrangements, the patient government, and especially the encounter therapy of the group sessions, may prove too much for many of these patients, leading to increased anxiety or early termination and an unfavorable treatment response.

The relatively better performance of the OS group in the MM program may be a function of several factors. First, the therapy provided by the counselors tended to be more individual and supportive and was thus potentially less threatening and more engaging to this group of patients than was group encounter. Secondly, the daily methadone maintenance regimen provided a regular schedule, thereby stabilizing their lives to some extent. Further, the philosophy of the program was

such that regular psychotropic medicines were provided when necessary. The daily methadone schedule provided a vehicle for their prescription (taken with or mixed in the methadone), thereby insuring compliance. Finally, there is clinical and some experimental evidence for the
direct anti-anxiety, antipsychotic actions of methadone (Berken, 1982;
Salzman, 1982). Several stimulant-abusing patients have presented with
psychotic symptoms that they (and often their families) claim are abated
during methadone stabilization. While the strength and range of methadone's neuroleptic action remains to be tested, it is well tolerated without the extrapyramidal side effects of traditional neuroleptics.

3.5.2. Opiate-Depressant Group

The differential performance shown by the OD subjects in the two
treatment programs was directly opposite to the results seen in the OS
subjects. However, the results of several studies with this population
suggest that they often present a different clinical picture than their OS
counterparts. Our experience with depressant or sedative abusers indicates that they often exhibit clear evidence of depressive spectrum disease, even following detoxification (McLellan et al., 1979). Again, it is
not possible to determine at this time whether the psychiatric symptoms
are the impetus for depressant abuse or its result, or both. Given this
clinical picture, we might have expected the methadone maintenance
regimen to show better results, under the assumption of a mood-reg-
ulative and antidepressant effect of the methadone. However, it has also
been our experience that patients who abuse depressants and sedative
hypnotics (including the benzodiazepines) continue to have serious
problems with depression even when on methadone. In these patients,
the depressant effect of the sedatives can predominate over the reported
antidepressant effects of methadone, and there may be a methadone-
sedative interaction in patients who continue to abuse sedatives. We
speculate that under these conditions, it may be especially difficult to
stabilize these patients, thus reducing the likelihood of improvement.

The inpatient and drug-free status of the TC may be particularly
important for this population of OD patients. For example, our earlier
work has indicated the significant tendency toward suicidal ideation and
suicidal attempts among the depressant users (McLellan et al., 1979).
While this may be a joint function of greater depression and the danger
of accidental overdose, we suggest that an inpatient setting may be a
more conservative and therapeutic modality in which to treat these patients. Secondly, we have evidence suggestive of cognitive impairment
in the depressant-abusing population (Grant, Adams, & Carlin, 1978;

Grant & Judd, 1976; Judd & Grant, 1975) that may require a sustained period of stable drug-free treatment in order for the recovery of cognitive functioning to take place. Again, we must emphasize that these suggestions are *post hoc* speculations, but we feel they are consistent with the data, with our clinical experience, and with relevant experimental work in the field.

3.5.3. Opiate-Only Group

As we have suggested in this paper, the clinical picture presented by the majority of our OO subjects is generally much different than that seen in either of the mixed abuse classes. The majority of our OO patients have a stable pattern of drug use, clear problems of unemployment and crime, but few psychiatric symptoms other than mild to moderate depression and sociopathy (McLellan *et al.*, 1979). As reported in previous studies (McLellan *et al.*, 1982; McLellan *et al.*, 1983), we have seen marked treatment effects in these patients in both the TC and MM programs. Also, methadone maintenance appears to provide a relatively specific and effective therapy for this type of patient. We consider these patients to have the best general prognosis, since they have often developed skills that can be applied to employment and they have few psychiatric symptoms that would interfere with interpersonal relations. Thus, we were not surprised to see substantial and relatively equal amounts of improvement in patients in both treatment programs.

4. Summary

Historically, clinicians and researchers have recognized an important association between psychiatric symptomatology and drug abuse. However, there has been considerable confusion and controversy regarding the nature of this association. Therefore, our research group has undertaken a series of studies investigating the relationships between use of particular street drugs and the development of specific psychiatric syndromes with the following results.

Chronic substance abusers often choose their drugs of abuse from families of drugs with psychologically similar effects. Chronic use of nonopiate stimulant or depressant drugs is associated with long-term and often severe psychiatric sequalae (stimulants—psychosis, mania; depressants—depression, cognitive impairment). Whether these psychiatric symptoms are *caused de novo* by the nonopiate drug use, are simply revealed or exacerbated over time as the drug abuser continues "*self-medication*" of symptoms, or are the result of some combination of these effects, is not yet known and must await a well controlled, prospective study.

Patients who primarily use opiates do *not* seem to develop significant psychiatric symptoms over time. Apparently, opiates themselves are unlikely to produce long-term psychiatric symptoms and may even be prophylactic for some symptoms of affective and thought disorder.

Patients whose drug use is limited to opiates can usually be treated successfully in either a methadone maintenance or a drug-free therapeutic community modality. Patients who chronically abuse stimulants in combination with opiates seem to fare better in a methadone maintenance program, where psychotropic medications are available for treatment of psychiatric symptoms that may have developed over time, and where methadone may itself act as a psychotropic. Opiate-sedative abusers present as a particular risk for outpatient methadone treatment due to the severity of their depression, increased suicidality, and potential for opiate-sedative overdose. These patients may be more safely and successfully treated in an inpatient, drug-free setting.

5. References

Baldassarini, R. The basis for the amine hypotheses in affective disorders. *Archives of General Psychiatry*, 1975, *32*, 14–35.

Bell, D. S. The experimental reproduction of amphetamine psychosis. *Archives of General Psychiatry*, 1973, 68–81.

Berken, G. H. *Methadone, an agent to control psychiatric rage.* Paper presented at the New York Academy of Science Conference on Opioids in Mental Illness, New York City, October 28, 1982.

Berzins, J. I., Ross, W. F., & English, G. E. Subgroups among opiate addicts: A typological investigation. *Journal of Abnormal Psychology*, 1974, *83*, 65–73.

Brown, B., & Chaitkin, L. *Use of stimulant/depressant drugs by drug abuse clients in selected metropolitan areas* (Technical Paper #27, National Institute on Drug Abuse). Washington, D.C.: United States Government Printing Office, 1980.

Carlsson, E. T., & Simpson, M. M. Medical use of morphine in the United States. *American Journal of Psychiatry*, 1963.

Comfort, A. Morphine as an antipsychotic: Relevance of a 19th-century therapeutic fashion. *Lancet*, 1977, *2*, 448–449.

Cox, C., & Smart, R. Social and psychological aspects of speed use. *International Journal of Addiction*, 1972, *7*, 16–21.

Cronbach, L. J., & Furby, L. How should we measure "change"—Or should we? *Psychological Bulletin*, 1980, *74*, 204–231.

Crowley, R. J., Chesluk, D., Dilts, S., & Hart, R. Drug and alcohol abuse among psychiatric admission. *Psychological Bulletin*, 1974, 172–177.

DeLeon, G., & Rosenthal, M. Therapeutitic communities. In R. L. DuPont, A. Goldstein,

& J. O'Donnell (Eds.), *Handbook on drug abuse*. Washington, D.C.: National Institute on Drug Abuse, 1979.

Dorus, W., & Senay, E. Depression, demographic depression, and drug abuse. *American Journal of Psychiatry*, 1980, *137*, 24–36.

Duarte-Escalante, O., & Ellinwood, E. Effects of chronic amphetamine intoxication on adrenergic and cholinergic structures in the central nervous system: Histochemical observations in cats and monkeys. In E. Ellinwood & S. Cohen (Eds.), *Current concepts on amphetamine abuse*. Washington, D.C.: National Institute on Mental Health, United States Government Printing Office, 1972.

Ellingboe, J., Mendelson, J. H., & Kuehnle, J. C. Effects of heroin and naltrexone on plasma prolactin levels. *Pharmacologic and Biochemical Behavior*, 1980, *12*, 66–72.

Ellinwood, E. Amphetamine psychosis. I. Description of the individuals and process. *Journal of Nervous and Mental Disease*, 1967, *144*, 273–283.

Ellinwood, E. Amphetamines/anorectics. In R. DuPont, A. Goldstein, & J. O'Donnell (Eds.), *Handbook on drug abuse*. Washington, D.C.: National Institute on Drug Abuse, 1979.

Ellinwood, E., & Duarte-Escalante, O. Chronic methamphetamine intoxication in three species of experimental animals. In E. Ellinwood & S. Cohen (Eds.), *Current concepts on amphetamine abuse*. Washington, D.C.: National Institute on Mental Health, 1972.

Gendreau, P., & Genderau, L. P. The "addiction-prone" personality: A study of Canadian heroin addicts. *Canadian Journal of Behavioral Science*, 1970, *2*, 206–210.

Gold, M. S., Donabedian, R. K., & Dillard, M. Antipsychotic effect of opiate agonists. *Lancet*, 1977, *2*, 398–399.

Goldberg, L. Drug abuse in Sweden. *United Nations Bulletin on Narcotics* (Vol. 2). New York: U.N., 1968.

Grant, I., & Judd, L. Neuropsychological and EEG disturbances in polydrug users. *American Journal of Psychiatry*, 1976, *133*, 44–51.

Grant, I., Adams, K., & Carlin, A. The collaborative neuropsychological study of polydrug abusers. *Archives of General Psychiatry*, 1978, *35*, 401–409.

Grant, I., Mohns, L., & Miller, M. A neuropsychological study of polydrug abusers. *Archives of General Psychiatry*, 1978, *35*, 420–427.

Greenblatt, D. J., & Shades, R. I. *Benzodiazepines in clinical practice*. New York: Raven Press, 1974.

Griffith, J. D., Cavanaugh, J., & Oates, J. A. Paranoid psychosis in man induced by the administration of d-amphetamine. *Pharmacologist*, 1968, *10*, 16–22.

Griffith, J. D., Fann, W. E., & Oates, J. A. The amphetamine psychosis: Experimental manifestations. In E. Ellinwood & S. Cohen (Eds.), *Current concepts on amphetamine abuse*. Washington, D.C.: U.S. Government Printing Office, 1972.

Hekimian, L. J., & Gershon, S. Characteristics of drug abusers admitted to a psychiatric hospital. *Journal of the American Medical Association*, 1968, *205*, 1016–1021.

Heller, M. E., & Mordkoff, A. M. Personality attributes of the young non-addicted drug abusers. *International Journal of Addictions*, 1972, *7*, 411–415.

Herl, D. Personality characteristics of a sample of heroin addicts, methadone maintenance applicants. *British Journal of Addiction*, 1976, *71*, 161–166.

Hill, H. E., Haertzen, C. E., & David, H. An MMPI factor analytic study of alcoholics, narcotic addicts, and criminals. *Quarterly Journal of Studies on Alcohol*, 1962, 230–237.

Judd, L., & Grant, I. Brain dysfunction in chronic sedative users. *Journal of Psychedelic Drugs*, 1975, *7*, 3–16.

Kaufman, E. The psychodynamics of opiate dependence: A new look. *American Journal of Drug and Alcohol Abuse*, 1974, *1*, 16–27.

Lowinson, J., & Millman, R. Clinical aspects of methadone maintenance treatment. In R. Dupont, A. Goldstein, & J. O'Donnell (Eds.), *Handbook on drug abuse.* Washington, D.C.: U.S. Government Printing Office, 1979.

Luborsky, L. Predicting the outcomes of psychotherapy. *Archives of General Psychiatry,* 1980, *37,* 471–481.

McLellan, A. T., & Druley, K. A. Non-random relation between drugs of abuse and psychiatric diagnosis. *Journal of Psychiatric Research,* 1977, *13,* 14–27.

McLellan, A. T., Druley, K. A., & Carson, J. E. Evaluation of substance abuse problems in a psychiatric hospital. *Journal of Clinical Psychiatry,* 1978, *39,* 107–116.

McLellan, A. T., Luborsky, L., & Erdlen, F. The Addiction Severity Index. In E. Gottheil, A. T. McLellan, & K. A. Druley (Eds.), *Substance abuse and psychiatric illness.* New York: Pergamon Press, 1980.

McLellan, A. T., MacGahan, J., & Druley, K. A. Psychopathology and substance abuse. In E. Gottheil, A. T. McLellan, & K. A. Druley (Eds.), *Substance Abuse and Psychiatric Illness.* New York: Pergamon Press, 1980.

McLellan, A. T., Woody, G. E., & O'Brien, C. P. Development of psychiatric disorders in drug abusers. *New England Journal of Medicine,* 1979, *301,* 1310–1314.

McLellan, A. T., Luborsky, L., Woody, G. E., & O'Brien, C. P. An improved evaluation instrument for substance abuse patients: The Addiction Severity Index. *Journal of Nervous and Mental Disease,* 1980, *168,* 26–33.

McLellan, A. T., Luborsky, L., O'Brien, C. P., Woody, G. E., & Druley, K. A. Is treatment for substance abuse effective? *Journal of the American Medical Association,* 1982, *247,* 1423–1428.

McLellan, A. T., Luborsky, L., Woody, G. E., Druley, K. A., & O'Brien, C. P. Predicting response to drug and alcohol treatments: Role of psychiatric severity. *Archives of General Psychiatry,* 1983, *40,* 620–625.

Mintz, J. A., & Luborsky, L. Measuring the outcomes of psychotherapy: Findings of the Penn Psychotherapy Project. *Journal of Consulting and Clinical Psychology,* 1979, *47,* 319–334.

Nunnally, J. *Psychometric theory.* New York: McGraw-Hill, 1967.

Nyswander, M. Drug addiction. In S. Arieti & E. B. Bowdy (Eds.), *American handbook of psychiatry.* New York: Basic Books, 1967.

Pittel, S. Psychological aspects of heroin and other drug dependence. *Journal of Psychedelic Drugs,* 1971, *4,* 16–31.

Pope, H. G. Drug abuse and psychopathology (editorial). *New England Journal of Medicine,* 1979, *24,* 391.

Salzman, B. The effect of methadone maintenance on schizophrenic opiate addicts. In K. Vereby (Ed.), *Opioids in mental illness.* New York: New York Academy of Sciences, 1982.

Schechter, M. *Anlage for schizophrenia.* Paper presented at the Annual Meeting of the American Psychiatric Association, San Francisco, May 1980.

Schildkraut, J. The catecholamine hypothesis of affective disorders: A review of supporting evidence. *American Journal of Psychiatry,* 1965, *122,* 509–522.

Schuster, C. R. Presentation at Philadelphia Veterans Administration Medical Center, August 1982.

Schuster, C. R., & Wilson, M. L. The effects of various pharmacological agents on cocaine self-administration by Rhesus monkeys. In E. Ellinwood & S. Cohen (Eds.), *Current concepts on amphetamine abuse.* Washington, D.C.: U.S. Government Printing Office, 1972.

Senay, E., Dorus, W., & Meyer, W. P. Psychopathology in drug abusers. *American Journal of Psychiatry,* 1982, *40,* 211–214.

Sharpless, S. K. Hypnotics and sedatives. I. The barbiturates. In L. S. Goodman & A. Gillman (Eds.), *The pharmacological basis of therapeutics.* Toronto: MacMillan, 1965.

Simes, L. K. Intelligence test correlates of Shipley–Hartford performance. *Journal of Clinical Psychology*, 1958, *14*, 399–404.

Simpson, D. D., Savage, L. J., Lloyd, M. R., & Sells, S. B. *Evaluation of drug abuse treatments based upon first year follow-up* (NIDA Research Monograph). Washington, D.C.: U.S. Government Printing Office, 1978.

Smith, D. E. The characteristics of dependence in high-dose methamphetamine use. *International Journal of Addictions*, 1969, *4*, 312–322.

Stein, L. Self-stimulation of the brain and the central stimulant action of amphetamine. *Federation Proceedings*, 1964, *23*, 117–119.

Tatetsu, S. Methamphetamine psychosis. In E. Ellinwood & S. Cohen (Eds.), *Current concepts on amphetamine abuse.* Washington, D.C.: U.S. Government Printing Office, 1972.

Woody, G. E., & Blaine, J. Depression in narcotics addicts: Quite possibly more than a chance association. In R. Dupont, A. Goldstein, & J. O'Donnell (Eds.), *Handbook on drug abuse.* Washington, D.C.: U.S. Government Printing Office, 1979.

Woody, G. E., O'Brien, C. P., & Rickels, K. Depression and anxiety in heroin addicts. *American Journal of Psychiatry*, 1975, *32*, 411–414.

Zuckerman, M., Sola, S., & Masterson, J. MMPI patterns in drug abusers before and after treatment in therapeutic communities. *Journal of Consulting and Clinical Psychology*, 1975, *43*, 286–296.

8

Drug-Induced Psychosis

NEUROBIOLOGICAL MECHANISMS

SAM CASTELLANI, WILLIAM M. PETRIE, AND EVERETT ELLINWOOD, JR.

1. Introduction

Bonhoeffer (1910) initially made the distinction between endogenous and exogenous psychoses, suggesting that toxic–metabolic factors can produce symptoms similar to those of functional psychoses. Drug-induced psychoses have in the last 30 years been observed with a vast array of prescribed and illicit pharmacological agents. Accordingly, extensive work has been done in the past three decades, attempting to elucidate neurobiological mechanisms that underlie both natural and drug-induced psychotic states. Indeed, the current explosive growth in biological psychiatry was heralded by early hypotheses relating molecular structures of hallucinogens to those of endogenous biogenic amines (Osmond & Smythies, 1952). We (Ellinwood & Petrie, 1976, 1979) have previously described the complex interaction between clinical and pharmacological factors that affect behavioral-symptom outcomes of abused drugs in a clinical setting. In this chapter we address only the pharmacological aspect of illicit psychotomimetic drugs currently in popular usage, with specific focus on central neuronal mechanisms that are common to both the different drug categories and the natural psychoses.

SAM CASTELLANI • Department of Psychiatry, School of Medicine, University of Kansas, Wichita, Kansas 67214.   WILLIAM M. PETRIE • Department of Psychiatry, Vanderbilt University, Nashville, Tennessee.   EVERETT ELLINWOOD, JR. • Box 3870, Duke University Medical School, Durham, N.C., 27710.

2. Symptom Patterns Induced by Psychotomimetic Drugs

The term "psychotomimetic" will be used here to refer to four illicit drug groups that are capable of inducing positive schizophreniform symptoms (hallucinations, delusions, thought disorganization) in the absence of notable impairment of sensorium: hallucinogens, stimulants, phencyclidine, and cannabis.

2.1. Hallucinogens

This class of drugs includes: indoleamines—lysergic acid diethylamide (LSD), psilocybin, psilocin, N,N-dimethyltryptamine (DMT), and 5-methoxy-DMT (5MeODMT); and amphetamine derivatives—mescaline, 2,5-dimethoxy-4-methylamphetamine (DOM), dimethoxyamphetamine (DMA), and para-methoxyamphetamine (PMA). These agents reliably produce specific perceptual, affective and cognitive changes in a single dose (Brawley & Duffield, 1972). Perceptual changes consist of heightened and altered sensitivity to sounds, shapes, and color, distortion of body image, synesthesia, and visual hallucinations. Affective experiences include elation, depression, fear, and blunting. Cognitive processes can show depersonalization, over-inclusive philosophical thought, profound meaning and significance, magical and paranoid thinking, clang associations, echolalia, and associative loosening.

Although a rapid tolerance develops to the psychic effects of hallucinogens (Balestrieri & Fontanari, 1959), prolonged psychotic reactions have been documented following both single doses and chronic usage (Cohen & Ditman, 1963; Glass & Bowers, 1970; Ungerleider, Fisher, Goldsmith, Fuller, & Forgy, 1968). These psychotic reactions are characterized by paranoid delusions, blunted affect, and depersonalization. Rorschach responses in chronic hallucinogen users show disruptive, idiosyncratic thinking and confusion of body boundary (Tucker, Quinlan, & Harrow, 1972). Flashbacks, or recurrent visual changes similar to those of the initial drug experience, can occur for weeks or months after a hallucinogen-induced episode (Horowitz, 1969).

2.2. Stimulants

Stimulants include amphetamines, cocaine, methylpenidate, and phenmetrazine. Acute usage of stimulants produces hypertalkativeness, increased arousal, euphoria, feelings of enhanced confidence and self-esteem, behavioral aggressiveness, sharpened consciousness, and feelings of significance. Chronic usage of amphetamine can result in a psychotic syndrome virtually indistinguishable from the early stages of schizophrenia: ideas of reference, pseudophilosophical ideation, paranoid delusions, visual and auditory hallucinations (Ellinwood, 1967; Ellinwood, Sudilovsky, & Nelson, 1973). An interesting aspect of chronic amphetamine-induced psychosis is the temporal evolution of symptoms observed both in street users and under experimental conditions with normal volunteers (Bell, 1973; Ellinwood, 1967). From early, predominately affective symptoms—euphoria, pleasant suspiciousness, depression—there develops increasing suspiciousness and ideas of reference to the point of a rather sudden formation of coherent, paranoid, delusional ideation. Delusional thought may persist beyond discontinuation of amphetamine drugs (Ellinwood, 1973); following long periods of abstinence after an amphetamine psychotic episode, psychotic symptoms may reappear with moderate doses of amphetamine (Bell, 1973; Ellinwood, 1973) or under stress (Utena, 1974). Psychotic states have also been described following chronic usage of cocaine and phenmetrazine (Bejerot, 1970).

2.3. Phencyclidine

Phencyclidine (PCP) is a drug with protean pharmacological and clinical effects. It possesses combined stimulant, sedative, and anesthetic properties (Domino, 1978). Low doses of PCP can produce heightened sensitivity to sounds and colors, depersonalization, sensory illusions (micropsia, macropsia), euphoria, anxiety, panic, and racing thoughts; higher doses can cause distorted body image, delusions, auditory hallucinations, mania, echolalia, disorganized thought processes, catatonia, and violence (Castellani, Giannini, & Adams, 1982a; Pearlson, 1981). Prolonged psychotic states have been described after acute PCP usage (Fauman & Fauman, 1980; Smith, 1980), and clinical reports suggest that long-term psychotic reactions may result from chronic usage (Fauman & Fauman, 1978; Smith, 1980).

A feature which distinguishes presentations of acute PCP intoxication from those of other psychotomimetic agents is the possible occurrence of sensorium impairment—drowsiness, acute and prolonged delirium, with disorientation, memory deficits, and clouded consciousness. High PCP doses can cause greater obtundation and coma (Pearlson, 1981).

2.4. Cannabis

Cannabis or tetrahydrocannabinol and derivatives (THC) in lower doses produce euphoria, relaxation, drowsiness, heightened perception of sounds, shapes, and colors, and alterations in body image and time perception (Jones, 1973). Higher doses have been reported to cause panic reactions (Weil, 1970), toxic delirioid-paranoid psychosis (Talbott & Teague, 1969; Tennant & Groesbeck, 1972), and experiences closely resembling those induced by LSD: temporal disorientation, feelings of significance, perceptual distortion, visual and auditory hallucinations, depersonalization, and fragmented thought processes (Isbell, Gorodetzsky, Jasinski, Claussen, Spulak, & Korte, 1967; Jones, 1973).

Psychotic reactions have been described following chronic high dose usage of THC (marijuana or hashish) in different cultures, with mixed symptoms of paranoid delusions, hallucinations, and delirium (Bernhardson & Gunne, 1972; McGlothin, 1972; Tennant & Groesbeck, 1972).

In summary, while all psychotomimetic agents can induce a similar vivid psychotic state with acute and chronic usage, certain qualitative symptom differences exist between the drug groups. Hallucinogens cause greater visual perceptual alterations; stimulants produce more definite systemized paranoid delusions associated with chronic usage; and PCP and, to a lesser extent, THC cause greater sensorial dysfunction.

3. Neurotransmitter Mechanisms

Considerable work has been done in the past two decades to determine CNS neurotransmitter mechanisms involved in the actions of psychotomimetic drugs. The following section presents relevant neurotransmitter changes associated with acute psychotomimetic drug administration.

3.1. Hallucinogens

3.1.1. Serotonin

Many studies have demonstrated that LSD and other hallucinogens (psilocybin, DMT, 5MeODMT, Mescaline, DOM) increase brain 5HT levels and decrease brain 5HT turnover, whereas the nonhallucinogen brom LSD (BOL) does not (Anden, Corrodi, & Fuxe, 1971; Anden, Corrodi, Fuxe, & Meek, 1974; Freedman, Gottlieb, & Lovell, 1970; Fuxe, Holmstedt, & Jonsson, 1972; Leonard, 1973). Supportive of central 5HT deficiency in hallucinogen effects are studies showing that LSD effects in man were attenuated by a monoamine oxidase inhibitor (Resnick, Krus, & Raskin, 1964) and 5-hydroxtryptophan (Pare & LaBrosse, 1963), and enhanced by reserpine (Resnick, Krus, & Raskin, 1965).

Further understanding of hallucinogen 5HT mechanisms was gained by studies of Aghajanian and colleagues, who showed that systemic and directly applied (microiontophoretic) LSD and related indoleamines (psilocin, DMT, 5MeODMT) caused profound depression of 5HT midbrain raphe neuronal firing (Aghajanian, 1972; Aghajanian, Foote, & Sheard, 1970). This effect was not produced by BOL. Further, the Aghajanian group (deMontigny & Aghajanian, 1977; Haigler & Aghajanian, 1977), using microiontophoretic techniques, demonstrated that hallucinogenic indoleamines exert much more potent inhibitory effects on pre- (raphe) versus postsynaptic sites, in contrast to 5HT, which inhibits pre- and postsynaptic sites equally. Based on these findings, the authors hypothesized that hallucinogenic effects of indoleamines are due to disinhibition of 5HT postsynaptic sites (i.e., lateral geniculate and amygdala areas).

Consistent with this hypothesis are the findings of Strahlendorf, Goldstein, Rossi, and Malseed (1982), who found that systemic and direct application of LSD to raphe neurons significantly increased visual evoked response amplitudes. The authors cite evidence that the 5HT raphe system has inhibitory actions on several forebrain areas, including the lateral geniculate nucleus, a visual relay center, and they postulate that the hallucinogenic actions of LSD may be due to depression of raphe activity.

Several groups have shown that LSD has high affinity, reversible, saturable, stereospecific (dLSD, greater) binding to brain membranes in areas with known rich serotonergic innervation (Bennett & Aghajanian, 1974; Fillion, Rousselle, Fillion, Beaudoin, Goiny, Deniau, & Jacob, 1978; Lovell & Freedman, 1976; Peroutka & Snyder, 1979). The lack of a change in LSD or 5HT brain binding following raphe lesions suggests that binding occurs at postsynaptic serotonergic receptors. The findings of different LSD affinities and pharmacological actions between 5HT and spiroperidol brain binding sites led Peroutka and Snyder (1979) to postulate the existence of two anatomically and functionally distinct central serotonergic receptors, termed $5HT_1$ and $5HT_2$. The $5HT_2$ receptor, in contrast to $5HT_1$, binds spiroperidol and shows LSD antagonist versus agonist actions. The possibility that either or both 5HT receptors mediate the psychotomimetic effects of hallucinogens is interesting, but awaits supportive evidence.

However, the CNS effects of mescaline challenge a pure 5HT hypothesis of hallucinogen action. Mescaline shows equivocal effects on central 5HT turnover (Freedman *et al.*, 1970; Tonge & Leonard, 1969); it depresses only certain midbrain raphe neurons (ventral part of the dorsal nucleus; Haigler & Aghajanian, 1973), and produces inconsistent generalization to discriminative stimulus properties of indoleamine hallucinogens (Glennon, Young, Rosecrans, & Kallman, 1980; Winter, 1978).

3.1.2. Norepinephrine

A possible common mechanism of action between mescaline and indoleamine hallucinogens is suggested by the studies of Aghajanian (1980) and McCall and Aghajanian (1980), who found that LSD and mescaline enhanced the firing of locus coeruleus norepinephrine (NE) neurons in response to external stimuli, and potentiated the excitatory effects of 5HT and NE on facial motoneurons. The latter effect was also produced by psylocin and DMT, but not by the nonhallucinogen lisuride. These data, coupled with the decrease in brain NE levels and/or increased NE turnover or release induced by LSD (Freedman, 1963; Leonard & Tonge, 1969), mescaline (Leonard & Tongue, 1969), psilocybin (Anden *et al.*, 1971), 5MeODMT (Fuxe *et al.*, 1972), and DOM (Leonard, 1973) suggest that the actions of hallucinogens may involve enhanced central NE. However, other studies found no effects of LSD (Peters, 1974), mescaline (Tilson & Sparber, 1972) or DMT (Anden *et al.*, 1971) on brain NE.

3.1.3. Dopamine

LSD (d isomer) has been shown to possess notable central dopamine (DA) agonist and antagonist effects. Agonist effects include stimulation of adenylate cyclase activity in striatal (Von Hungen, Roberts, & Hill, 1975) and limbic (Ahn & Makman, 1979; Spano, Kumakura, Tonon, Gavoni, & Trabucchi, 1975) areas, displacement of brain [^{3}H]DA (Creese, Burt, & Snyder, 1976) and [^{3}H]apomorphine (Whitaker & Seeman, 1977) binding, contralateral turning in unilateral substantia nigra-lesioned rats (Pieri, Pieri, & Haefely, 1974), and potentiation of amphetamine and apomorphine hyperactivity (Fink, Morgenstern, & Oelssner, 1979). Many of these effects are blocked by neuroleptic drugs. Dopamine antagonist effects are blockade of DA-stimulated adenylate cyclase activity (Ahn & Makman, 1979; Von Hungen *et al.*, 1975) and displacement of brain [^{3}H]haloperidol binding (Creese *et al.*, 1976).

An important consideration is whether hallucinogens other than LSD or nonhallucinogens show central dopaminergic actions. Mescaline, like LSD, stimulated adenylate cyclase in anterior limbic structures and auditory cortex (Ahn & Makman, 1979) and enhanced amphetamine- and apomorphine-induced hyperactivity (Fink *et al.*, 1979), but showed no effects on striatal adenylate cyclase (Von Hungen *et al.*, 1975) or DA receptor binding (Whitaker & Seeman, 1977). Lisuride produced contralateral turning in substantia nigra lesioned rats (Pieri, Schaffner, Pieri, DaPrada, & Haefely, 1978) and BOL showed inhibition of DA-stimulated adenylate cyclase (Von Hungen *et al.*, 1975). Thus, the data indicate that DA effects cannot sufficiently explain the central actions of hallucinogens.

Important functional interrelationships appear to exist between central 5HT and DA in hallucinogen actions. A serotonergic influence on DA actions is suggested by the findings that injection of LSD and mescaline into the median raphe nucleus potentiated apomorphine-induced hyperactivity (Fink & Oelssner, 1981). These effects can be explained by an inhibitory serotonergic mechanism from median raphe nuclei to substantia nigra neurons (Dray, Gonye, Oakley, & Tanner, 1976). By contrast, neuroleptics inhibit LSD-induced decrease in 5HT turnover (Jacoby & Poulakos, 1977), binding to the $5HT_2$ receptor (Peroutka & Snyder, 1979) and psychic effects in humans (Isbell & Logan, 1957; Moskowitz, 1971), suggesting DA antagonistic effects on hallucinogen-induced central 5HT actions.

Several investigators have proposed molecular conformational requirements for indoleamine and phenylethylamine hallucinogens, implying the existence of a functionally active central receptor common to both drug classes (Guion, 1981; Nichols, Pfister, & Yim, 1978; Smythies, Benington, & Morin, 1970; Snyder & Richelson, 1970). Similarly, Nichols (1976), based on structural similarities between LSD and apomorphine, postulated that the action of hallucinogens involves stimulation of both 5HT and DA central receptors. Such hypotheses are provocative, but need confirmatory data.

3.2. Stimulants

3.2.1. Catecholamines

A large body of evidence implicates central catecholamines, especially DA, in the actions of amphetamine and other stimulants (cocaine, methylphenidate, and amphonelic acid). These agents all cause reuptake blockade (Glowinski, Axelrod, & Iversen, 1966; Ross, 1979) and release (Heikkila, Orlansky, & Cohen, 1975; Moore, Chiueh, &

Zeldes, 1977) of central DA. Animal studies demonstrate that central DA systems mediate stereotypy and hyperactivity induced by amphetamine (Anden, 1977; Kelly, Seviour, & Iversen, 1975) and cocaine (Scheel-Kruger, Braestrup, Nielson, Golembiowska, & Mogilnicka, 1977). There is some controversy regarding the involvement of NE in locomotor stimulant actions of amphetamine, with evidence for DA only (Moore, 1977) and both DA and NE (Cole, 1978).

Important differences have been found between the mechanism by which amphetamines and nonamphetamines exert their central DA actions. Thus, nonamphetamines, in contrast to amphetamines, show less central DA releasing action (Heikkila et al., 1975; Moore et al., 1977), and their behavioral effects are inhibited by reserpine, but little or not at all by alpha-methylparatyrosine (αMPT) (Scheel-Kruger, 1971; Van Rossum & Hurksman, 1964). Therefore, amphetamines are thought to act primarily by releasing DA from newly synthesized intraneuronal pools, and nonamphetamines by reuptake blockade and making stored vesicular DA available for release (Moore et al., 1977).

3.2.2. Acetylcholine

Amphetamine and cocaine increase acetylcholine (Ach) turnover rate in the cerebral cortex (Ngai, Shirasawa, & Cheney, 1979; Robinson, Cheney, & Moroni, 1978) and the behavioral effects of amphetamine, cocaine, and methylphenidate are blocked by cholinergic agents (e.g., physostigmine) and enhanced by anticholinergics (atropine, scopolamine) (Galambos, Pfeifer, Gyorgy, & Molnar, 1967; Janowsky, El-Yousef, Davis, & Sekerke, 1972; Klawans, Rubovits, Patel, & Weiner, 1972). These findings indicate cholinergic inhibition of stimulant DA-mediated effects, and/or central anticholinergic actions of stimulant drugs.

3.2.3. Serotonin

Amphetamine has been reported to increase 5HT levels in the pons and striatum (Beauvallet, Legrand, & Solier, 1970). Cocaine was shown to decrease concentrations (Pradhan, Roy, & Pradhan, 1978) and turnover (Friedman, Gershon, & Rotrosen, 1975) of brain 5HT, and inhibit high affinity neuronal tryptophan uptake (Mandell & Knapp, 1977). Stimulant-induced behaviors appear to be inhibited by central 5HT: Reduction in central 5HT by synthesis inhibition or receptor blockade potentiates amphetamine (Breese, Hollister, & Cooper, 1977; Segal, 1977) and cocaine (Scheel-Kruger et al., 1977) behavioral effects.

3.3. Phencyclidine

3.3.1. Dopamine

Phencyclidine, like stimulants, has potent central DA actions. These include uptake inhibition (Garey & Heath, 1976; Vickroy & Johnson, 1980), release (Vickroy & Johnson, 1982), and induction of stereotyped behaviors in animals that are inhibited by neuroleptic drugs (Castellani & Adams, 1981a; Murray & Horita, 1979; Schlemmer, Jackson, Preston, Bederka, Garver, & Davis, 1978).

Many observations indicate that PCP-induced central DA actions are similar to those of the nonamphetamine stimulants. For example, compared to amphetamine, PCP shows considerably less DA releasing effects (Doherty, Simonovic, So, & Meltzer, 1980), and PCP-induced decrease in striatal DOPA accumulation (an index of tyrosine hydroxylase activity) was blocked by reserpine and gamma butyrolactone, indicating a dependence on DA storage pool and nerve impulse, respectively (Doherty et al., 1980). Also, PCP-induced hyperactivity and stereotypy were inhibited by reserpine (Meltzer, Sturgeon, Simonovic, & Fessler, 1981). These findings suggest that PCP, like nonamphetamine stimulants, mobilizes storage pool DA for impulse-dependent release.

3.3.2. Norepinephrine

Phencyclidine has been reported to block reuptake (Taube, Montel, Hau, & Starke, 1975) and decrease brain levels (Leonard & Tonge, 1969) of brain NE. Marwaha and colleagues (Marwaha, Palmer, Woodward, Hoffer, & Freedman, 1980) found that PCP, like NE, depressed the spontaneous activity of Purkinje cerebellar neurons, and this effect was blocked by destruction of NE nerve terminals. They also reported that the inhibition of Purkinje neuronal activity by two stereoisomeric forms of a PCP analog correlated with actions in the mouse rotarod test (Marwaha, Palmer, Hoffer, Freedman, Rice, Paul, & Skolnick, 1981). Taken together, these findings suggest that PCP may have important central presynaptic NE inhibitory actions.

3.3.3. Acetylcholine

Phencyclidine has shown potent anticholinergic effects in biochemical, behavioral, and electrophysiological studies, including displacement of brain muscarinic receptor binding (Vincent, Cavey, Kamenka, Geneste, & Lazdunski, 1978), inhibition of peripheral smooth

muscle contraction (Kloog, Rehavi, Maayani, & Sokolovsky, 1977), disruptive of spatial alternation behaviors (Weinstein, Maayani, Glick, &
Meibach, 1981), and depression of Ach-induced excitation of hippocampal neurons (Bickford, Palmer, Hoffer, & Freedman, 1982). Weinstein *et
al.* (1981), using quantum mechanics, determined that PCP and analogs
possess molecular properties compatible with physiological binding at
the Ach receptor. An anticholinergic contribution to PCP-induced hyperactivity and stereotyped behaviors is indicated by the effects of atropine (enhancement) and physostigmine (inhibition) (Adams, 1980;
Castellani & Adams, 1983; Murray & Horita, 1979); however, such effects were not found by all investigators (Meltzer *et al.*, 1981).

Phencyclidine also shows competitive inhibition of acetylcholinesterase and butylcholinesterase (Maayani, Weinstein, Ben-Zvi, Cohen, &
Sokolovsky, 1974), and in a recent study, Boggan, Evans, and Wallis
(1982) found that PCP increased *in vivo* brain muscarinic receptor binding. Castellani and Adams (1983) reported that physostigmine increased
PCP-induced ataxia. Thus PCP appears to possess notable central cho-
linomimetic as well as anticholinergic actions.

3.3.4. Serotonin

Serotonin effects reported for PCP include increased brain 5HT and
decreased 5HIAA levels (Tonge & Leonard, 1971) and 5HT uptake blockade in rat striatal synaptosomes (Smith, Meltzer, Arora, & Davis, 1977),
and human platelets (Arora & Meltzer, 1980). However the findings for
brain 5HT and 5HIAA levels were reversed when a different strain of
rats was used (Tonge & Leonard, 1971).

3.3.5. Opiates

A growing body of evidence indicates that PCP has important interactions with central opiate receptors. First, PCP competes with brain
binding of several opiate ligands (Itzhak, Kalir, Weissman, & Cohen,
1981; Vincent *et al.*, 1978). Second, SKF 10,047, a putative opiate sigma
receptor agonist, shows pharmacological actions in the chronic spinal
dog (Jasinski, Shannon, Cone, Vaupel, Risner, McQuinn, Su, & Pickworth, 1981) and rat discriminative stimulus test (Holtzman, 1980; Shannon, 1981) very similar to those of PCP, and selectively displaces brain
[^{3}H]PCP binding (Zukin & Zukin, 1979). Finally, Castellani, Giannini,
and Adams (1982b) found that PCP-induced ataxia was markedly enhanced by metenkephalin and morphine and inhibited by naloxone; and
PCP combined with these opiate agonists elicited the typical opioid cata-

lepsy. Together, these studies indicate important interactions of PCP with the putative central sigma, mu, and delta opiate receptors.

3.3.6. Phencyclidine Receptor

Several groups have demonstrated the presence of saturable, relatively high affinity binding of PCP to rat brain membranes (Quirion, Rice, Skolnick, Paul, & Pert, 1981; Vincent, Kartalovski, Geneste, Kamenka, & Lazdunski, 1979; Zukin & Zukin, 1979). Specificity of PCP binding was shown by high correlations between the ability of several PCP analogs to displace [^{3}H] PCP and their potency in mouse rotarod and rat discrimination tests. The validity of PCP binding has been questioned by Maayani and Weinstein (1980) due to possible artifacts of filtration methods. However, Quirion *et al.* (1981) reported stereospecific displacement of brain [^{3}H] PCP using two isomeric forms of a PCP analog that showed specificity in electrophysiological and behavioral tests (Marwaha *et al.*, 1980, 1981, *vide supra*), and methods without filtration. These data suggest the interesting hypothesis that the psychopharmacological effects of PCP may be mediated by specific brain PCP receptors.

3.4. Cannabis

3.4.1. Catecholamines

The evidence for central catecholamine effects of THC is sparse and inconclusive. Thus, increased (Singh, Mazumdar, & Prasad, 1980); decreased (Ho, Taylor, Fritchie, Englert, & McIsaac, 1972), and no change (Maitre, Stahelin, & Bein, 1970) in brain NE levels have been reported following THC. Studies showing decreased DA turnover in rat brain (Vestergaard, Rubin, Beaubrunn, Cruickshank, & Picou, 1971), and unilateral turning behavior (Hine, Friedman, Torrelio, & Gershon, 1975; Waters & Glick, 1973) suggest that THC has central DA receptor agonist effects.

3.4.2. Acetylcholine

Actions of THC on peripheral Ach are primarily anticholinergic, showing inhibition of Ach release (Layman & Milton, 1971) and smooth muscle contraction (Dewey, Harris, & Kennedy, 1972; Layman & Milton, 1971). Central effects of THC also appear to be anticholinergic:

THC produced marked decreases in brain Ach levels without changes in choline acetyltransferase (Askew, Kimball, & Ho, 1974), inhibition of habituation (Brown, 1971) and potentiation of amphetamine-induced hyperactivity (Dagirmanjian & Boyd, 1962; Garriot, King, Forney, & Hughes, 1967), the latter two effects being induced by anticholinergic agents.

3.4.3. Serotonin

Several investigators demonstrated increased brain 5HT concentrations (Holtzman, Lovell, Jaffe, & Freedman, 1969; Sofia, Ertel, Dixit, & Barry, 1971; Singh, Mazumdar, & Prasad, 1980) following THC administration. Also, Truitt and Anderson (1971) reported that THC reduces brain 5HT turnover.

3.5. Summary and Integration

The above review indicates the following general conclusions:

1. Hallucinogens have potent inhibitory effects on 5HT raphe neuronal activity and central 5HT turnover. Stimulants, PCP, and THC also show evidence for inhibition of central 5HT turnover, although the effects are less clear. Inhibition of the central 5HT system and consequent disinhibition of postsynaptic sensory relay and limbic areas may underlie the hallucinogen component of psychotomimetic effects, that is, heightened sensitivity, distortion of incoming sensory information, and visual hallucinations. Evidence showing that LSD enhances visual evoked responses, and the 5HT raphe system exerts a toxic inhibitory influence on arousal (Trulson & Jacobs, 1981) and sensory relay and limbic sites (lateral geniculate nucleus and amygdala), supports this hypothesis.
2. Enhancement of central postsynaptic DA is an effect common to all four psychotomimetic drug groups. Stimulants show more distinct DA actions and schizophreniform symptoms, suggesting that positive schizophreniform symptoms in the psychotomimetic process may be mediated primarily by increased central postsynaptic dopaminergic functioning.
3. The functional relationship between central 5HT and DA mechanisms in psychotomimetic effects suggests intriguing possibilities. Available data indicate a central 5HT inhibitory influence on hallucinogen and amphetamine-induced dopaminergic

motor effects and, conversely, substantial blocking actions by neuroleptics of hallucinogen-induced 5HT biochemical and behavioral effects in animals and psychotomimetic effects in humans. The hypothesis of a central hallucinogen receptor site with combined 5HT and DA recognition, and the role of the putative $5HT_1$ and $5HT_2$ central receptors are fertile areas for future research.

4. Central anticholinergic effects are most prominent in PCP and THC intoxication and quite likely underlie the greater sensorial impairment induced by these drugs. The hypothesis that a central DA-cholinergic balance exists such that enhancement in relative dopaminergic versus cholinergic functioning predisposes to a psychotic process (Friedhoff & Alpert, 1973) is consistent with the overall neurochemical data on psychotomimetic drugs.

5. The role of central NE in psychotomimetic actions is uncertain. The findings of enhanced locus coeruleus and facial motoneuron excitability by LSD and mescaline, and the PCP-induced noradrenergic inhibition of Purkinje cerebellar neurons suggest possible important mechanisms, but need replication and further study.

6. Two recent sets of findings for PCP are provocative, but demand further research: the central PCP receptor and PCP central opiate effects.

4. Cortical and Limbic Electrographic Effects

The different psychotomimetic drugs produce similar electrographic changes in both cerebrocortical and limbic structures. Thus, many studies found that LSD and related indoleamines (psilocybin, psilocin, DMT), mescaline, DOM, PMA, and stimulants all produce an EEG arousal pattern (i.e., low voltage, fast activity) in many species of laboratory animals (Bradley & Elkes, 1957; Brawley & Duffield, 1972; Hollister, 1968). While hallucinogens were initially not found to consistently alter the EEG in man, a desynchronizing, frequency-enhancing effect of LSD became apparent when it reversed the slowing and synchronization caused by other drugs (Fink, 1963; Hollister, 1968). Itil (1969) found that LSD administration to chronic schizophrenic patients caused EEG low voltage, fast activity accompanied by motor agitation and hallucinations; both EEG and behavioral changes were reversed with chlorpromazine. Amphetamine, methamphetamine, cocaine, and methylphenidate produce consistent low voltage, fast EEG activity in animals and humans (Wallach & Gershon, 1971; Saito, 1974). The EEG arousal pattern induced by psychotomimetic drugs, coupled with evidence for enhanced arousal in schizophrenia (Shagass, 1976; Venables & Wing, 1962; Wallach & Wallach, 1964), supports the hypothesis that abnormal hyperarousal may be an important mechanism underlying psychotomimetic psychosis and schizophrenia (see Section 5.5 in this chapter).

Phencyclidine produces EEG-alternating low voltage, fast activity and hypersynchronous slowing in the dog and monkey (Domino, 1964) and slowing in humans (Rodin, Luby, & Meyer, 1959). Similarly, THC has been reported to produce increased EEG synchrony and high voltage, slow waves and spikes in rats (Pirch, Cohn, Barnes, & Barrett, 1972) and cats (Hockman, Perrin, & Kalant, 1971). No appreciable THC effects have been found in the human EEG (Heath, 1972; Rodin, Domino, & Porzak, 1970). The slowing, synchronizing actions of PCP and THC may be due to their central anticholinergic properties, since anticholinergic agents similarly cause consistent EEG hypersynchronous slowing (Karczmar, 1976).

Psychotomimetic drugs all produce striking abnormal electrographic changes in limbic structures, often accompanied by psychotic-like behaviors in animals. Monroe and Heath (1961) reported that LSD, LSD derivatives, and mescaline caused hypersynchronous paroxysmal sharp waves in the septum and hippocampus, concomitant with agitated, catatonic behaviors in monkeys. Neither the electrographic nor behavioral effects were observed following 5HT or BOL. Adey, Bell, and Dennis (1962) found that LSD, psilocybin and psilocin caused spikes and sharp waves in the entorhinal cortex and hippocampus in cats, accompanied by bizarre "kangaroo" postures, staring and pawing at imaginary objects. Similar spike-wave activity has been observed in deep temporal and limbic (amygdala, septum, hippocampus) sites following LSD and mescaline in schizophrenic and epileptic patients, accompanied by psychotic behaviors (Monroe, Heath, Mickle, & Llewellyn, 1957; Schwarz, Sem-Jacobsen, & Peterson, 1956); both electrographic and behavioral changes were reversed with chlorpromazine.

Phencyclidine and THC cause spike and/or low frequency wave activity in several limbic structures (amygdala, septum, hippocampus, cingulate gyrus) in cats (Adey & Dunlop, 1960; Contreras, Guzman-Flores, Dorantes, Ervin, & Palmour, 1981; Hockman et al., 1971) and monkeys (Domino, 1964; Heath, Fitzjarrell, Fontana, & Garey, 1980; Martinez, Stadnick, & Schaeppi, 1972).

Stimulants induce spindling (sinusoidal waves of 20–50 Hz) in olfactory and amygdala structures in the rat (Stripling & Ellinwood,

1977), cat (Ellinwood, Kilbey, Castellani, & Khoury, 1977; Ellinwood, Sudilovsky, & Nelson, 1974) and dog (Domino & Ueki, 1960). In our laboratory, amphetamine in combination with disulfiram (Ellinwood, Sudilovsky, & Grabowy, 1973), and cocaine in high (CD50) doses (Castellani, Ellinwood, Kilbey, & Petrie, 1983) produced lower frequency amygdala spindling (10–21 Hz) and spikes seemingly paced by spindle bursts and associated with preseizure behaviors: arrest of movement, staring, bizarre stereotyped movements and limb posturing, leading in some animals to generalized motor seizures. Similarly, Jacobs and Trulson (1981) found high-frequency electrical bursts in the amygdala associated with LSD-induced arrest of movement and staring in cats. Interestingly, Eidelberg, Long, and Miller (1965) reported that several hallucinogens (LSD, mescaline, bufotenin, bulbocapnine) and PCP all markedly inhibited amygdala, olfactory bulb, and pyriform 40 Hz activity in the cat, and Domino (1964) found the same effect with PCP in the dog and monkey. However, similar suppression of olfactory-amygdala 40 Hz activity was found with 5-hydroxytryptophan, reserpine, and pheniprazine (Eidelberg *et al.*, 1965), indicating that this effect is not specific to psychotomimetic agents.

Taken together, the above data suggest that abnormal limbic electrical activity (i.e., seizure discharges and/or inhibition or abnormally low frequency of amygdala spindling) may be an important correlate of psychotomimetic drug actions. The commonality and relevance of abnormal limbic functioning in psychotomimetic and natural psychotic states are discussed in Section 5.5 of this chapter.

5. Psychotomimetic Drug Mechanisms and Natural Psychosis

5.1. Hallucinogens: Transmethylation and 5HT Hypotheses

The validity of a model psychosis begins with the assumption that phenomenological isomorphism may reflect underlying pathophysiological processes common to the model and the entity that it represents. Although differences exist between hallucinogen-induced and natural psychotic states, similarities are quite apparent. Thus, "psychedelic" symptoms (i.e., heightened sensory and affective awareness and visual hallucinations) occur in early stages of schizophrenia (Bowers & Freedman, 1966; Chapman, 1966), and schizophrenic symptoms (i.e., delusional thinking, loose associations, echolalia) emerge in LSD intoxication (Brawley & Duffield, 1972). Moreover, both hallucinogenic and natural psychotic states are effectively treated with neuroleptic drugs (Isbell & Logan, 1957; Moskowitz, 1971).

The concept that an endogenous psychotogen may have central importance in the pathogenesis of schizophrenia was first advanced in 1952 by Osmond and Smythies, who noted the similarity in chemical structure between norepinephrine and mescaline, and postulated that in schizophrenia a mescaline-like compound is synthesized by abnormal methylation of a natural biogenic amine. This hypothesis, termed "transmethylation hypothesis," was later extended to possible methylation of indoleamines. Shortly after, Gaddum and Hameed (1954) and Wooley and Shaw (1954) observed that LSD and other hallucinogens antagonized the peripheral effects of 5HT, and postulated that schizophrenia may result from a central 5HT deficiency. These hypotheses have stimulated research seeking evidence for (1) abnormal methylation and methylated compounds, and (2) abnormal 5HT metabolism in schizophrenic patients; this research has not yielded confirmatory results (Barchas, Elliott, & Berger, 1978). However, the 5HT hypothesis remains viable in light of recent findings.

Bowers (1973, 1975) found that, in a group of schizophrenic patients, cerebrospinal fluid (CSF) 5-HIAA following probenecid correlated negatively with favorable prognosis scores and motor activity, and drug-induced (mostly LSD), compared to natural psychotics, had lower CSF 5-HIAA levels and higher prognosis scores. Thus, a subgroup of schizophrenics with better prognosis and more activity may have abnormally depressed 5HT metabolism. These results are consistent with the decreased central 5HT turnover found with administration of hallucinogens and other psychotomimetic drugs (reviewed earlier).

Bennett, Enna, Bylund, Gillin, Wyatt, and Snyder (1979) demonstrated in three independent replications that [^{3}H] LSD, but not [^{3}H] 5HT binding, was reduced in the frontal cortex of schizophrenic patients, apparently unrelated to medications, postmortem changes, or demographic features. This finding poses the intriguing possibility that schizophrenia may be associated with decreased sensitivity in the putative $5HT_2$ receptor.

5.2. Stimulants: Dopamine Hypothesis

Amphetamine psychosis is regarded as a most valid model of schizophrenia for several reasons. First, its symptom picture is quite similar to that of acute paranoid schizophrenia, including both temporal development and the presence of auditory hallucinations and Schneiderian first rank symptoms (Bell, 1973; Ellinwood, 1967). Second, the treatment of choice for both amphetamine psychosis and schizophrenia is neuroleptic drugs. Third, the primary central mechanism of amphetamine appears to be enhanced postsynaptic DA, and of neuroleptic drugs, postsynaptic DA receptor blockade (Meltzer & Stahl, 1976). Such evidence provides the basic foundation for the DA hypothesis, which states that schizophrenia is associated with enhanced DA functioning at crucial central receptor sites (Meltzer & Stahl, 1976). A key drawback of the DA hypothesis is the absence of data showing increased CNS DA metabolism in schizophrenia. An alternative explanation is that central DA receptors are supersensitive in schizophrenia. This hypothesis receives some support in neuroendocrine challenge studies (Pandey, Garver, Tamminga, Ericksen, Ali, & Davis, 1977; Rotrosen, Angrist, Clark, Gershon, Halpern, & Sachar, 1978) and postmortem brain studies (Owen, Crow, Poulter, Cross, Longden, & Riley, 1978; Lee & Seeman, 1980), and becomes quite relevant in stimulant animal models of psychosis (see the following text).

5.3. Phencyclidine

The PCP model psychosis is based on the observations that normal volunteers given PCP (Bakker & Amini, 1961; Luby, Cohen, Rosenbaum, Gottlieb, & Kelley, 1959) and patients presenting with acute PCP intoxication (Castellani *et al.*, 1982a; Pearlson, 1981) exhibit schizophrenic-like symptoms: disorganized thinking, loose associations, neologisms, auditory hallucinations, delusions, and catatonia. When administered to schizophrenic patients, PCP produced a striking exacerbation of thought disorder and primitive behaviors that persisted for a month (Luby *et al.*, 1959). Further, Luby, Gottlieb, Cohen, Rosenbaum, and Domino (1962) reported that, like schizophrenic patients, normals given PCP showed a marked attenuation of psychotic symptoms under conditions of sensory isolation.

Recent studies in man shed light on neurochemical mechanisms and possible pharmacological treatment of PCP-induced psychosis. Castellani and colleagues (Castellani, Adams, & Giannini, 1982; Castellani *et al.*, 1982a; Castellani, Giannini, Boeringa, & Adams, 1982) tested the effects of intramuscular haloperidol, physostigmine, and meperidine for treating acute PCP intoxication, based on findings from animal studies. Haloperidol substantially reduced schizophreniform symptoms (delusions, disorganized thinking, auditory hallucinations); physostigmine improved symptoms of anxiety, excitement and agitation, but improved psychotic symptoms to a lesser extent; meperidine reduced schizophreniform symptoms (although haloperidol-treated patients were selected for greater psychotic symptom levels versus meperidine-treated patients). These results support CNS dopaminergic, anticholinergic, and possibly opiate antagonist mechanisms in PCP-induced psychotic symptoms. They are consistent with much animal data (reviewed above) showing potent central DA agonist and anticholinergic actions of PCP, and with DA and anticholinergic hypotheses (Friedhoff & Alpert, 1973) of psychosis.

5.4. Cannabis

Cannabis in high doses can produce schizophreniform symptoms: paranoid delusions, auditory hallucinations, and thought disorganization (see earlier discussion), which although not generally as consistently elicited nor severe as symptoms due to the other psychotomimetics, has been considered a model of schizophrenia (Jones, 1973). Although the neurochemical bases of THC-induced mental effects are not well established, the data are in accord with that of other psychotomimetic drugs, showing dopaminergic, anticholinergic, and antiserotonergic effects. Cannabis also shows profound effects on temporal processing, which, similar to PCP, is likely due to central anticholinergic effects, possibly in the hippocampus (Drew & Miller, 1974).

5.5. Arousal Hypotheses and Limbic Mechanisms

Arousal hypotheses provide a bridge between neurochemical, electrophysiological, and behavioral events in understanding both drug-induced and natural psychotic states. Thus, many studies suggest that schizophrenics are in a state of heightened arousal as measured by flicker fusion, palmar skin conductance, and saccadic eye movements (Venables & Wing, 1962; Wallach & Wallach, 1964; Zahn, Rosenthal, & Lawlor, 1968). Shagass (1976) examined consistent findings on evoked potentials in schizophrenics and concluded that the more rapid, greater amplitude and less variable early component may reflect an impairment in the normal regulating mechanism that "filters" incoming sensory information. The involvement of hyperarousal in psychotomimetic drug actions is indicated by their ability to produce EEG arousal patterns, that is, desynchronized low voltage, fast activity, and inhibitory effects on central 5HT (reviewed earlier). The 5HT raphe system exerts a general inhibitory influence on arousal and appears necessary for slow-wave sleep (Morgane & Stern, 1978; Trulson & Jacobs, 1981). Consistent with the known diminished stage 4 sleep in schizophrenics (Feinberg, Braun, Koresko, & Gottlieb, 1969), and inhibitory actions of hallucinogens on central 5HT, it has been hypothesized that inhibition of the central 5HT system may cause impairment in psychophysiological filtering and hyperaroused flooding into consciousness of perceptual-cognitive events characteristic of hallucinogen-induced and schizophrenic psychotic states (Bowers, 1975; Weil-Malherbe, 1978).

The involvement of abnormal limbic functioning in both drug-induced and natural psychotic processes is indicated by several lines of evidence. First, abnormal limbic electrical activity has been observed in both psychotomimetic drug conditions (reviewed earlier) and schizophrenics (Heath, 1962; Kendrick & Gibbs, 1957; Sem-Jacobsen, Peterson, Lazarte, Dodge, & Holman, 1955). Second, impaired habituation to repeated stimulation is seen with hallucinogen drugs (in animals; Key & Bradley, 1960) and in schizophrenics (Gruzelier & Venables, 1972). Habituation appears to be mediated in great part by the hippocampus and amygdala (Pribram & McGuinness, 1975). Third, notable similarities exist in the phenomenology of psychotomimetic states, early stages of schizophrenia and temporal lobe epilepsy, whose primary pathology is in limbic structures (Slater, 1969). Ellinwood (1974) has noted the striking similarities between symptoms in amphetamine psychosis and temporal lobe epilepsy. These include: heightened sensory awareness, feelings of novelty and portentiousness, ideas of reference and influence, temporal distortions (*deja vu, jamais vu*), distorted body image, depersonalization, and visual and auditory hallucinations.

In summary, CNS hyperarousal and limbic system dysfunction appear to be important pathological processes that underlie both psychotomimetic and natural (mainly schizophrenia) psychoses.

6. Animal Models of Psychosis

Further study of the interrelationships between neurochemical, behavioral, and electrophysiological changes—and, in addition, the temporal dimension—is provided by animal models of psychosis.

6.1. Hallucinogens

Jacobs, Trulson, and Stern (1977) discovered that LSD elicited specific behaviors in the cat (i.e., limb flick and abortive grooming), which they postulated to be a model of hallucinogenic actions in man. Trulson and Jacobs studied cats with chronically implanted electrodes and found a correlation between LSD and 5-MeODMT induction of these behaviors and onset of raphe neuronal unit activity depression (Jacobs & Trulson,

1981; Trulson & Jacobs, 1979a, 1979b). Supportive of a 5HT inhibitory mechanism, the authors found that PCPA potentiated LSD-induction of these behaviors (Trulson & Jacobs, 1976). The effects of LSD deviated from a strict temporal correlation, showing that the induced behaviors outlasted raphe depression by 2–4 hours, and a rapid tolerance developed to the behaviors, but not raphe depression. Such findings call into question these hallucinogen behaviors as a complete explanatory model in view of evidence that chronic usage may enhance the propensity to psychotic states (see the earlier discussion). Furthermore, the lack of specificity of the cat hallucinogen behavioral model is shown by the study of Marini and Sheard (1981), who found that the nonhallucinogen methysergide elicited in the cat limb flick and abortive grooming, as well as tolerance and cross tolerance to LSD for these behaviors.

Braff and Geyer (1980) examined the effects of both acute and chronic LSD on the rat startle response and found that, while tolerance developed to acute impairment of habituation, chronic LSD potentiated startle response intensity. The authors proposed that rat startle may be a valid model of LSD psychosis. Freedman and Boggan (1974) reported that chronic, compared to acute, LSD administration caused the elevation of brain 5HT to begin earlier and terminate sooner, and the decrease in 5-HIAA to be reduced, suggesting possible biochemical correlates of tolerance and negative tolerance observed to the psychic effects in man. Further, while acute LSD injections in rats caused decreased 5HT turnover in several brain areas, LSD injections for 14 days resulted in decreased 5HT turnover in midbrain only (Peters & Tang, 1977). Lee and Geyer (1980) found no differences in LSD-induced increase in raphe 5HT levels between acute and chronic treatment. Peters and Tang (1977) reported that two weeks following 14-day LSD injections, elevated midbrain 5HT levels were still present. Finally, Trulson and Jacobs (1979c) found that repeated administration of LSD for 4 days in rats produced decreased [^{3}H] 5HT and [^{3}H] LSD binding in brainstem and forebrain areas.

Together, these data suggest that tolerance to LSD-induced psychic effects may be related to diminished serotonergic actions in limbic and cortical sites, but not raphe neurons, while the sustained LSD-induced raphe neuronal inhibition, which may persist several days beyond discontinuation of LSD, could underlie the psychotic symptoms and "flashbacks" associated with chronic LSD usage.

6.2. Stimulants

The stimulant-induced behavioral model of psychosis has proved valuable for two major reasons. First, the human process which it represents, amphetamine psychosis, bears a close resemblance in symptoms and temporal development to paranoid schizophrenia. Second, it has provided a means for studying chronic changes in behavioral, electrophysiological, and neurochemical events. Chronic administration of amphetamine and cocaine leads to augmented hyperactivity and stereotyped behaviors in rats, cats, and monkeys (Castellani, Ellinwood, & Kilbey, 1978; Ellinwood, 1971; Ellinwood, Sudilovsky, & Nelson, 1972; Ho, Taylor, Estevez, Englert, & McKenna, 1977; Post, 1977; Segal & Mandell, 1974; Stripling & Ellinwood, 1977). Superimposed upon investigative stereotypies, there emerge certain behaviors, termed "end stage" behaviors: limb flicks, abortive grooming, hyperreactivity, increased startle response, and abnormal dystonic postures (Castellani *et al.*, 1978; Ellinwood & Kilbey, 1977). We have postulated that the enhanced motor activity and stereotypy and end stage behaviors may provide a model for evolving symptoms and full blown psychosis, respectively, in amphetamine psychosis and possibly schizophrenia (Castellani *et al.*, 1978; Ellinwood, Sudilovsky, & Nelson, 1973). Several neurophysiological mechanisms have been proposed to explain chronic stimulant-induced behavioral changes: (Ellinwood & Kilbey, 1980; Ellinwood & Lee, 1983). Two such hypothetical mechanisms are receptor supersensitivity and kindling.

Dopamine supersensitivity is supported by the findings that apomorphine-induced stereotyped behavior is potentiated by prior chronic amphetamine (Klawans & Margolin, 1975) and cocaine (Kilbey & Ellinwood, 1977) administration, and DA receptor binding is increased in striatal and limbic areas following chronic amphetamine (Borison, Hitri, Klawans, & Diamond, 1979; Howlett & Nahorski, 1978). Furthermore, depletion of striatal DA was observed following chronic amphetamine (Ricaurte, Schuster, & Seiden, 1980; Segal, Weinberger, Cahill, & McCunney, 1980) and cocaine (Ho *et al.*, 1977), and this could presumably result in receptor supersensitivity. However, decreases have been found in both stereotyped behaviors (Nelson & Ellison, 1978) and striatal [^{3}H] spiroperidol binding (Neilson, Neilson, Ellison, & Braestrup, 1980) following continuous amphetamine administration. Thus a supersensitivity mechanism to explain chronic stimulant behaviors is not consistently supported.

Kindling is a phenomenon whereby repeated low-level electrical stimulation of certain brain areas, most prominently the amygdala, results in increasing spread of local afterdischarges and enhanced preseizure behaviors leading to generalized seizures (Goddard, McIntyre, & Leech, 1969). We (Ellinwood *et al.*, 1977) and Post (Post, 1977) postulated that a pharmacological kindling-like mechanism may account for the augmented motor behaviors following chronic stimulant administration, and possibly the psychotic symptoms associated with temporal lobe epilepsy and schizophrenia. Consistent with a kindling-like mechanism, studies in our laboratory found enhanced amygdala spindling and spike activity accompanying the augmented stereotypy and end stage behaviors following chronic amphetamine in cats (Ellinwood *et al.*, 1974) and cocaine in rats (Stripling & Ellinwood, 1977). Chronic cocaine administration in rats produced no effect (Stripling & Ellinwood, 1977) and enhancement on the rate of subsequent amygdala kindling induced by electrical stimulation (Kilbey, Ellinwood, & Easler, 1979). However, the latter effect was directly attributable to cocaine-induced convulsions prior to kindling. Stripling and Hendricks (1981) corroborated this observation, showing that one or three convulsions induced by cocaine enhanced the rate of subsequent olfactory bulb kindling. Other studies report that prior chronic cocaine administration inhibited (increased latency) (Sato, Hikasa, & Otsuki, 1979) and did not affect (Rackham & Wise, 1979) amygdala kindling. In summary, the data suggest that while cocaine-induced convulsions may share with kindling a common neurophysiological mechanism, kindling as a mechanism contributing to chronic stimulant-induced stereotypy and hyperactivity is not established. Future studies should examine effects of kindling, given prior to stimulants on chronically induced behaviors.

Another possibility is that reduced central 5HT functioning contributes to chronic stimulant-induced behavioral changes. This is supported by several lines of evidence. First, chronic stimulant administration elicits behaviors virtually identical to hallucinogen-induced behaviors, which are thought to be due to decreased central 5HT (see earlier discussion; Ellinwood & Kilbey, 1977; Nielson, Lee, & Ellison, 1980; Trulson & Jacobs, 1979d). Second, chronic administration of amphetamine and cocaine decreases 5HT levels in brainstem, diencephelon, and striatum (Ricaurte *et al.*, 1980; Roy, Bhattacharyya, Pradhan, & Pradhan, 1978; Taylor & Ho, 1977; Trulson & Jacobs, 1979d); chronic amphetamine also decreases 5HT in neocortex, hippocampus, and amygdala (Ricaurte *et al.*, 1980; Trulson & Jacobs, 1979d). Third, LSD binding in striatum and frontal cortex was decreased following 5 days' continuous release (pellet) of amphetamine (Nielsen, Nielsen, Ellison, & Braestrup, 1980). Fourth, decreased central 5HT leads to increased locomotor activity induced by amphetamine and cocaine, possibly due to reduction in an inhibitory pathway from median raphe to substantia nigra (Dray *et al.*, 1976; see 3.2.3. above). Finally, in our laboratory, following chronic cocaine administration in cats, both cocaine plus PCPA and LSD plus PCPA elicited strong hallucinogen behaviors (limb flick, abortive grooming; Ellinwood & Kilbey, 1977).

The possible involvement of Ach in chronic stimulant behaviors is suggested by the cholinergic inhibition of stimulant behaviors (described above) coupled with the findings of marked increase in acetylcholinesterase concentrations in several brain areas following chronic amphetamine administration in cats and monkeys (Duarte-Escalante & Ellinwood, 1972). These findings suggest a cholinergic mechanism that is adaptively activated to balance stimulant-induced excitation of central catecholamines.

In summary, the chronic stimulant model psychosis has yielded data that indicate possible involvement of the following progressive CNS changes through time: increased postsynaptic DA, decreased postsynaptic 5HT, and abnormal limbic spindle-seizure activity. More hypothetically, DA receptor supersensitivity and kindling are neurophysiological processes that have been generated and tested by the model.

6.3. Phencyclidine

Like stimulants, PCP produces in animals hyperactivity and stereotyped behaviors that are blocked by neuroleptic drugs and physostigmine, and enhanced by apomorphine and atropine. Since chronic PCP usage appears to precipitate psychotic states, chronic PCP-induced behaviors in animals may provide a model for PCP psychosis in man. Chronic PCP administration in animals has been reported to increase (Castellani & Adams, 1981b; Smith, Biggs, Vroulis, & Brinkman, 1981) and decrease (Sturgeon, Fessler, London, & Meltzer, 1982) stereotyped behaviors, and show biphasic increase followed by decrease in locomotor activity (Castellani & Adams, 1981b; Sturgeon et al., 1982).

Neurochemical changes reported following chronic PCP administration are equally sparse and conflicting. Hsu, Smith, Rolsten, and Leelavathi (1980) found that acute PCP increased choline acetyltransferase and acetylcholinesterase activity in rat hippocampus, and these values returned to control levels following chronic treatment. These data suggest tolerance to the anticholinergic effects of PCP. Also, no change (Schwartz, Moerschbaecher, Thompson, & Keller, 1982) and decrease (Ward & Trevor, 1981) in muscarinic cholinergic binding has been reported following chronic PCP administration.

A DA supersensitivity mechanism in chronic PCP-induced behaviors is not supported since two studies found an absence of change in apomorphine-induced stereotypy following chronic PCP administration (Castellani & Adams, 1981b; Smith, Biggs, Vroulis, & Brinkman, 1981). Also, a decrease in DA-stimulated adenylate cyclase (Leelavathi, Misra,

Shelet, & Smith, 1980) and decreased [^{3}H] spiroperidol binding in rat striatum (Robertson, 1982) was seen following chronic PCP, suggesting a DA receptor subsensitivity mechanism underlying chronic PCP effects.

Interestingly, Smith, Leelavathi, Hsu, Ho, Tansey, Taylor, and Biggs (1981) found that following chronic PCP, 5HT uptake was increased or unaffected, in contrast to 5HT uptake blockade observed with acute PCP. Together with the properties of PCP to decrease brain 5HT turnover, these data suggest that 5HT mechanisms may contribute to the acute and chronic psychotomimetic effects of PCP.

Finally, PCP has been reported to raise the afterdischarge threshold in previously amygdala-kindled animals (Freeman, Jarvis, & Duncan, 1982) and inhibit the rate of amygdala kindling in daily injections given prior to electrical stimulation (Bowyer, 1982), indicating an inhibitory effect on amygdala kindling.

7. Summary and Conclusions

The foregoing review indicates that, while a simple unitary mechanism underlying psychotomimetic drug actions is not firmly established, the following parsimonious conclusions can be made: (1) The "hallucinogen" component (i.e., perceptual distortions and visual hallucinations) may be primarily mediated by decreased 5HT functioning at forebrain areas: limbic and sensory relay nuclei; (2) the "schizophreniform" component (i.e., systemized delusions, auditory hallucinations, disorganized thinking) may primarily involve increased central DA functioning; (3) central anticholinergic mechanisms probably reflect mainly sensorial impairment, and may also involve removal of an inhibitory control of central DA functioning; (4) disruption of limbic activity at crucial sites (i.e., amygdala, septum, hippocampus) appears to be a key pathological process common to psychotomimetic psychosis, schizophrenia, and temporal lobe epilepsy; (5) arousal concepts provide important explanatory links between neurochemical, electrophysiological, and behavioral levels for both psychotomimetic and natural psychotic states; and (6) animal models, especially the chronic stimulant model, have brought further integration to these levels and have generated research into two neurophysiological constructs, supersensitivity and kindling. Thus, for example, progressive development of decreased 5HT and increased DA functioning, which may involve gradual neuronal membrane changes at crucial limbic sites, appear necessary for the development of stimulant-induced psychosis.

Major gaps in current knowledge, which require further elucidation, are the precise relationships between psychotomimetic-induced neurochemical-membrane (receptor) changes and anatomical-electrophysiological (limbic-cortical) events.

The study of neurobiological mechanisms in psychotomimetic psychoses has deepened our understanding of the natural psychoses, especially schizophrenia. Indeed, the hallucinogen and amphetamine model psychoses have been integral to the development of the abnormal psychotogen and DA hypotheses, respectively, of schizophrenia. Factors that are intimately connected to neurobiological events and thus demand further research for a thorough understanding of psychotomimetic and natural psychoses are genetic, predisposing personality and current environmental stressor variables.

8. References

Adams, P. M. Interaction of phencyclidine with drugs affecting cholinergic neurotransmission. *Neuropharmacology*, 1980, *19*, 151–153.

Adey, W. R., & Dunlop, C. W. The action of certain cyclohexamines on hippocampal system during approach performance in the cat. *Journal of Pharmacology and Experimental Therapeutics*, 1960, *130*, 418–426.

Adey, W. R., Bell, F. R., & Dennis, B. J. Effects of LSD-25, psilocybin, and psilocin on temporal lobe EEG patterns and learned behavior in the cat. *Neurology*, 1962, *12*, 591–602.

Aghajanian, G. K. Influence of drugs on the firing of serotonin-containing neurons in brain. *Federation Proceedings*, 1972, *31*, 91–95.

Aghajanian, G. K. Mescaline and LSD facilitate the activation locus coeruleus neurons by peripheral stimuli. *Brain Research*, 1980, *186*, 492–498.

Aghajanian, G. K., Foote, W. E., & Sheard, M. H. Action of psychotogenic drugs on single midbrain raphe neurons. *Journal of Pharmacology and Experimental Therapeutics*, 1970, *171*, 178–187.

Ahn, H. S., & Makman, M. H. Interaction of LSD and other hallucinogens with dopamine-sensitive adenylate cyclase in primate brain: Regional differences. *Brain Research*, 1979, *162*, 77–88.

Anden, N. -E. Functional effects of local injections of dopamine and analogs into the neostriatum and nucleus accumbens. In E. Costa & G. L. Gessa (Eds.), *Advances in biochemical psychopharmacology*. New York: Raven Press, 1977.

Anden, N. E., Corrodi, H., & Fuxe, K. Hallucinogenic drugs of the indolealkylamine type and central monoamine neurons. *Journal of Pharmacology and Experimental Therapeutics*, 1971, *179*, 236–249.

Anden, N. E., Corrodi, H., Fuxe, H., & Meek, J. C. Hallucinogenic phenethylamines: Interactions with serotonin turnover and receptors. *European Journal of Pharmacology*, 1974, *25*, 176–184.

Arora, R. C., & Meltzer, H. Y. *In vitro* effect of phencyclidine and other psychomotor stimulants on serotonin uptake in human platelets. *Life Sciences*, 1980, *27*, 1607–1613.

Askew, W. E., Kimball, A. P., & Ho, B. T. Effect of tetrahydro-cannabinols on brain acetylcholine. *Brain Research*, 1974, *69*, 375–378.

Bakker, C. B., & Amini, F. B. Observation on the psychotomimetic effects of sernyl. *Comprehensive Psychiatry*, 1961, *2*, 269–280.

Balestrieri, A., & Fontanari, D. Acquired and cross-tolerance to mescaline, LSD-25, and BOL-148. *Archives of General Psychiatry*, 1959, *1*, 279–280.

Barchas, J. D., Elliott, G. R., & Berger, P. A. Biogenic amine hypotheses of schizophrenia. In L. C. Wynne, R. L. Cromwell, & S. Matthysse (Eds.), *The nature of schizophrenia*. New York: Wiley, 1978.

Beauvallet, M., Legrand, M., & Solier, M. Actions de l'amphetamine sur la teneur en 5-hydroxytryptamine des differentes aires du cerveau de rat. *Comptes Rendus Des Seances De La Societe De Biologie Et De Ses Filiales*, 1970, *164*, 1462–1467.

Bejerot, N. A comparison of the effects of cocaine and synthetic central stimulants. *The British Journal of Addiction*, 1970, *65*, 35–37.

Bell, D. S. The experimental reproduction of amphetamine psychosis. *Archives of General Psychiatry*, 1973, *29*, 35–40.

Bennett, J. L., & Aghajanian, G. K. D-LSD binding to brain homogenates: Possible relationship to serotonin receptors. *Life Sciences*, 1974, *15*, 1935–1944.

Bennett, J. P., Enna, S. J., Bylund, D. B., Gillin, J. C., Wyatt, R. J., & Snyder, S. H. Neurotransmitter receptors in frontal cortex of schizophrenics. *Archives of General Psychiatry*, 1979, *36*, 927–934.

Bernhardson, G., & Gunne, L. Forty-six cases of psychosis in cannabis abusers. *International Journal of the Addictions*, 1972, *7*, 9–16.

Bickford, P. C., Palmer, M. R., Hoffer, B. J., & Freedman, R. Interactions between phencyclidine and cholinergic excitation of hippocampal pyramidal neurons. *Neuropharmacology*, 1982, *21*, 729–732.

Boggan, W. O., Evans, M. G., & Wallis, C. J. Effect of phencyclidine on [³H] QNB binding. *Life Sciences*, 1982, *30*, 1193–1200.

Bonhoeffer, K. *Die Exogenen Reaktion Stypen*. Berlin: Springer, 1910.

Borison, R. L., Hitri, A., Klawans, H. L., & Diamond, B. I. A new animal model for schizophrenia: Behavioral and receptor binding studies. In E. Usdin, I. S. Kopin, & J. Barchas (Eds.), *Catecholamines: Basic and clinical frontiers* (Vol. 1). New York: Pergamon Press, 1979.

Bowers, M. B. LSD-related states as model of psychosis. In J. O. Cole, A. M. Freedman, & A. J. Friedhoff (Eds.), *Psychopathology and psychopharmacology*. Baltimore: The John Hopkins University Press, 1973.

Bowers, M. B. Serotonin (5HT) systems in psychotic states. *Psychopharmacology Communications*, 1975, *6*, 655–662.

Bowers, M. B., & Freedman, D. X. "Psychedelic" experiences in acute psychoses. *Archives of General Psychiatry*, 1966, *15*, 240–248.

Bowyer, J. F. Phencyclidine inhibition of the rate of development of amygdaloid kindled seizures. *Experimental Neurology*, 1982, *75*, 173–183.

Bradley, P. B., & Elkes, J. The effects of some drugs on the electrical activity of the brain. *Brain*, 1957, *80*, 77–117.

Braff, D. L., & Geyer, M. A. Acute and chronic LSD effects on rat startle: Data supporting an LSD-rat model of schizophrenia. *Biological Psychiatry*, 1980, *15*, 909–923.

Brawley, P., & Duffield, J. C. The pharmacology of hallucinogens. *Pharmacological Reviews*, 1972, *24*, 31–66.

Breese, G. R., Hollister, A. S., & Cooper, B. R. Role of monoamine neural pathways in d-amphetamine- and methylphenidate-induced locomotor activity. In E. H. Ellinwood & M. M. Kilbey (Eds.), *Cocaine and other stimulants.* New York: Plenum Press, 1977.

Brown, H. Some anticholinergic-like behavioral effects of trans (-)-Δ^8 tetrahydrocannabinol. *Psychopharmacologia,* 1971, *21,* 294–301.

Castellani, S., & Adams, P. M. Effects of dopaminergic drugs on phencyclidine-induced behavior in the rat. *Neuropharmacology,* 1981, *20,* 371–374. (a)

Castellani, S., & Adams, P. M. Acute and chronic phencyclidine effects on locomotor activity, stereotypy and ataxia in rats. *European Journal of Pharmacology,* 1981, *73,* 143–154. (b)

Castellani, S., & Adams, P. M. Effects of dopaminergic, cholinergic, and opiate agents on phencyclidine-induced behaviors in the rat. In J. M. Kamenka, E. F. Domino, & P. Geneste (Eds.), *Phencyclidine and related arylcyclohexylamines: Present and future applications.* Ann Arbor, Mich.: NPP Books, 1983.

Castellani, S., Adams, P. M., & Giannini, A. J. Physostigmine treatment of acute phencyclidine intoxication. *Journal of Clinical Psychiatry,* 1982, *43,* 10–11.

Castellani, S., Ellinwood, E. H., & Kilbey, M. M. Behavioral analysis of chronic cocaine intoxication in the cat. *Biological Psychiatry,* 1978, *13,* 203–215.

Castellani, S., Giannini, A. J., & Adams, P. M. Physostigmine and haloperidol treatment of acute phencyclidine intoxication. *American Journal of Psychiatry,* 1982, *139,* 508–510. (a)

Castellani, S., Giannini, A. J., & Adams, P. M. Effects of naloxone, metenkephalin, and morphine on phencyclidine-induced behavior in the rat. *Psychopharmacology,* 1982, *78,* 76–80. (b)

Castellani, S., Ellinwood, E. H., Kilbey, M. M., & Petrie, W. M. Cholinergic effects on arousal and cocaine-induced olfactory-amygdala spindling and seizures in cats. *Physiology and Behavior,* 1983, *31,* 461–466.

Castellani, S., Giannini, A. J., Boeringa, J. A., & Adams, P. M. Phencyclidine intoxication: Assessment of possible antidotes. *Journal of Toxicology: Clinical Toxicology,* 1982, *19,* 313–319.

Chapman, J. The early symptoms of schizophrenia. *British Journal of Psychiatry,* 1966, *112,* 225–251.

Cohen, S., & Ditman, K. S. Prolonged adverse reactions to lysergic acid diethylamine. *Archives of General Psychiatry,* 1963, *3,* 475–480.

Cole, S. O. Brain mechanisms of amphetamine-induced anorexia, locomotion, and stereotype: A review. *Neuroscience and Biochemical Reviews,* 1978, *2,* 89–100.

Contreras, C. M., Guzman-Flores, C., Dorantes, M. E., Ervin, F. R., & Palmour, R. Naloxone and phencyclidine: Interacting effects on the limbic system and behavior. *Physiology and Behavior,* 1981, *27,* 1019–1026.

Creese, I., Burt, D. R., & Snyder, S. H. The dopamine receptor: Differential binding of d-LSD and related agents to agonist and antagonist states. *Life Sciences,* 1976, *17,* 1715–1720.

Dagirmanjian, R., & Boyd, E. S. Some pharmacological effects of two tetrahydrocannabinols. *Journal of Pharmacology and Experimental Therapeutics,* 1962, *135,* 25–33.

deMontigny, C., & Aghajanian, G. K. Preferential action of 5-methoxytryptamine and 5-methoxydimethyltryptamine on pre-synaptic serotonin receptors: A comparative iontophoretic study with LSD and serotonin. *Neuropharmacology,* 1977, *16,* 811–818.

Dewey, W. L., Harris, L. S., & Kennedy, J. S. Some pharmacological and toxicological effects of 1-trans-Δ^8 and 1-trans-Δ^9-tetrahydrocannabinol in laboratory rodents. *Archives of International Pharmacodynamics,* 1972, *196,* 133–145.

Doherty, J. D., Simonovic, M., So, R., & Meltzer, H. Y. The effect of phencyclidine on dopamine synthesis and metabolism on rat striatum. *European Journal of Pharmacology*, 1980, 65, 139–149.

Domino, E. F. Neurobiology of phencyclidine (sernyl), a drug with an unusual spectrum of pharmacological activity. *International Review of Neurobiology*, 1964, 6, 303–347.

Domino, E. F. Neurobiology of pheychylidine: An update. In R. C. Peterson & R. C. Stillman (Eds.), *Phencyclidine (PCP) abuse: An appraisal*. Washington, D.C.: U.S. Government Printing Office, 1978.

Domino, E. F., & Ueki, S. An analysis of the electrical burst phenomenon in some rhinencephalic structures of the dog and monkey. *Electroencephalography and Clinical Neurophysiology*, 1960, 12, 635–648.

Dray, A., Gonye, T. J., Oakley, N. R., & Tanner, T. Evidence for the existence of a raphe projection to the substania nigra in rat. *Brain Research*, 1976, 113, 45–57.

Drew, W. G., & Miller, L. L. Cannabis: Neural mechanisms and behavior: A theoretical review. *Pharmacology*, 1974, 11, 12–32.

Duarte-Escalante, O., & Ellinwood, E. H. Effects of chronic amphetamine intoxication on adrenergic and cholinergic structures in the central nervous system: Histochemical observations in cats and monkeys. In E. H. Ellinwood & S. C. Cohen (Eds.), *Current concepts on amphetamine abuse*. Rockville, Maryland: National Institute of Mental Health, 1972.

Eidelberg, E., Long, M., & Miller, M. K. Spectrum analysis of EEG changes induced by psychotomimetic agents. *International Journal of Neuropharmacology*, 1965, 4, 255–264.

Ellinwood, E. H. Amphetamine psychosis: I. Description of the individuals and process. *Journal of Nervous and Mental Disease*, 1967, 144, 273–283.

Ellinwood, E. H. Effect of chronic methamphetamine intoxication in rhesus monkeys. *Biological Psychiatry*, 1971, 3, 25–32.

Ellinwood, E. H. Amphetamine and stimulant drugs. In the second report of the national commission on marihuana and drug abuse, *Drug use in america: Problems in perspective*. Washington, D.C.: U.S. Government Printing Office, 1973.

Ellinwood, E. H. Behavioral and EEG changes in the amphetamine model of psychosis. In E. Usdin (Ed.), *Neuropsychopharmacology of monoamines and their regulatory enzymes*. New York: Raven Press, 1974.

Ellinwood, E. H., & Kilbey, M. M. Chronic stimulant intoxication models of psychosis. In I. Hanin & E. Usdin (Eds.), *Animal models in psychiatry*. New York: Pergamon Press, 1977.

Ellinwood, E. H., & Kilbey, M. M. Fundamental mechanisms underlying altered behaviors following chronic administration of psychomotor stimulants. *Biological Psychiatry*, 1980, 15, 749–757.

Ellinwood, E. H., & Lee, T. H. Effect of central systemic infusion of d-amphetamine on the sensitivity of nigral dopamine cells to apomorphine inhibition of firing rates. *Brain Research*, 1983, 273, 379–383.

Ellinwood, E. H., & Petrie, W. M. Psychiatric syndromes induced by nonmedical use of drugs. *Alcohol*, 1976, 3, 177–215.

Ellinwood, E. H., & Petrie, W. M. Drug-induced psychoses. In R. W. Pickens & L. L. Heston (Eds.), *Psychiatric factors in drug abuse*. New York: Academic Press/Grune and Stratton, 1979.

Ellinwood, E. H., Sudilovsky, A., & Grabowy, R. Olfactory forebrain seizures induced by methamphetamine and disulfiram. *Biological Psychiatry*, 1973, 7, 89–99.

Ellinwood, E. H., Sudilovsky, A., & Nelson, L. Behavioral analysis of chronic amphetamine intoxication. *Biological Psychiatry*, 1972, 4, 215–229.

Ellinwood, E. H., Sudilovsky, A., & Nelson, L. M. Evolving behavior in the clinical and experimental amphetamine (model) psychosis. *American Journal of Psychiatry*, 1973, *130*, 1088–1093.

Ellinwood, E. H., Sudilovsky, A., & Nelson, L. M. Behavior and EEG analysis of chronic amphetamine effect. *Biological Psychiatry*, 1974, *8*, 169–176.

Ellinwood, E. H., Kilbey, M. M., Castellani, S., & Khoury, C. Amygdala hyperspindling and seizures induced by cocaine. In E. H. Ellinwood & M. M. Kilbey (Eds.), *Cocaine and other stimulants*. New York: Plenum Press, 1977.

Fauman, M. A., & Fauman, B. J. The psychiatric aspects of chronic phencyclidine use: A study of chronic PCP users. In R. C. Peterson & R. C. Stillman (Eds.), *Phencyclidine (PCP) abuse: An appraisal*. Washington, D.C.: U.S. Government Printing Office, 1978.

Fauman, B. J., & Fauman, M. A. Part II: Psychosis. *Psychopharmacology Bulletin*, 1980, *16*, 72–73.

Feinberg, I., Braun, M., Koresko, R. L., & Gottlieb, F. Stage 4 sleep in schizophrenia. *Archives of General Psychiatry*, 1969, *21*, 262–266.

Fillion, G. M. B., Rousselle, J., Fillion, M., Beaudoin, D. M., Goiny, M. R., Deniau, J., & Jacob, J. J. High-affinity binding of [³H] 5-hydroxytryptamine to brain synaptosomal membranes: Comparison with [³H] lysergic acid diethylamide binding. *Molecular Pharmacology*, 1978, *14*, 50–59.

Fink, M. Quantitative EEG in human psychopharmacology: Drug patterns. In G. H. Glaser (Ed.), *EEG and behavior*. New York: Basic Books, 1963.

Fink, H., & Oelssner, W. LSD, mescaline and serotonin injected into medial raphe nucleus potentiate apomorphine hypermotility. *European Journal of Pharmacology*, 1981, *75*, 289–296.

Fink, H., Morgenstern, R., & Oelssner, W. Psychotomimetics potentiate locomotor hyperactivity induced by dopaminergic drugs. *Pharmacology Biochemistry and Behavior*, 1979, *11*, 479–482.

Freedman, D. X. Psychotomimetic drugs and brain biogenic amines. *American Journal of Psychiatry*, 1963, *119*, 843–850.

Freedman, D. X., & Boggan, W. O. Brain serotonin metabolism after tolerance dosage of LSD. *Advances in Biochemical Psychopharmacology*, 1974, *10*, 151–157.

Freedman, D. X., Gottlieb, R., & Lovell, R. A. Psychotomimetic drugs and brain 5-hydroxytryptamine metabolism. *Biochemical Pharmacology*, 1970, *19*, 1181–1188.

Freeman, F. G., Jarvis, M. F., & Duncan, P. M. Phencyclidine raises kindled seizure thresholds. *Pharmacology Biochemistry and Behavior*, 1982, *16*, 1009–1011.

Friedhoff, A. J., & Alpert, M. A dopaminergic-cholinergic mechanism in production of psychotic symptoms. *Biological Psychiatry*, 1973, *6*, 165–169.

Friedman, E., Gershon, S., & Rotrosen, J. Effects of acute cocaine treatment on the turnover of 5-hydroxytryptamine in the rat brain. *British Journal of Pharmacology*, 1975, *54*, 61–64.

Fuxe, K., Holmstedt, B., & Jonsson, G. Effects of 5-methoxy-N-N-dimethyltryptamine on central monoamine neurons. *European Journal of Pharmacology*, 1972, *19*, 25–34.

Gaddum, J. H., & Hameed, K. A. Drugs which antagonize 5-hydroxy-tryptamine. *British Journal of Pharmacology*, 1954, *9*, 240–248.

Galambos, E., Pfeifer, A. K., Gyorgy, L., & Molnar, J. Study on the excitation induced by amphetamine, cocaine and α-methyl-tryptamine. *Psychopharmacologia*, 1967, *11*, 122–129.

Garey, R. E., & Heath, R. G. The effects of phencyclidine on the uptake of 3H-catecholamine by rat striatal and hypothalamic synaptosomes. *Life Sciences*, 1976, *18*, 1105–1110.

Garriot, J. C., King, L. J., Forney, R. B., & Hughes, F. W. Effects of some tetrahydrocannabinols on hexobarbital sleeping time and amphetamine-induced hyperactivity in mice. *Life Sciences*, 1967, *6*, 2119–2128.

Glass, G. S., & Bowers, M. B. Chronic psychosis associated with long-term psychomimetic drug abuse. *Archives of General Psychiatry*, 1970, *23*, 97–103.

Glennon, R. A., Young, R., Rosecrans, J. A., & Kallman, M. J. Hallucinogenic agents as discriminative stimuli: A correlation with serotonin receptor affinities. *Psychopharmacology*, 1980, *68*, 155–158.

Glowinski, J., Axelrod, J., & Iverson, L. L. Regional studies of catecholamines in the rat brain: IV. Effects of drugs on the disposition and metabolism of 3H-dopamine. *Journal of Pharmacology and Experimental Therapeutics*, 1966, *153*, 30–41.

Goddard, G. V., McIntyre, D. C., & Leech, C. K. A permanent change in brain function resulting from daily electrical stimulation. *Experimental Neurology*, 1969, *25*, 295–330.

Gruzelier, J. H., & Venables, P. H. Skin conductance orienting activity in a heterogeneous sample of schizophrenics. *Journal of Nervous and Mental Disease*, 1972, *155*, 277–287.

Guion, R. R. Possible interactions between alkoxy amphetamines and brain serotonin receptors. *Journal of Theoretical Biology*, 1981, *91*, 237–239.

Haigler, H. J., & Aghajanian, G. K. Mescaline and LSD: Direct and indirect effects on serotonin-containing neurons in brain. *European Journal of Pharmacology*, 1973, *21*, 53–60.

Haigler, H. J., & Aghajanian, G. K. Serotonin receptors in the brain. *Federation Proceedings*, 1977, *36*, 2159–2164.

Heath, R. G. Common characteristics of epilepsy and schizophrenia: Clinical observation and depth electrode studies. *American Journal of Psychiatry*, 1962, *118*, 1013–1026.

Heath, R. G. Marihuana. Effects on deep and surface electroencephalograms of man. *Archives of General Psychiatry*, 1972, *26*, 577–584.

Heath, R. G., Fitzjarrell, A. T., Fontana, C. J., & Garey, R. E. Cannabis sative: Effects on brain function and ultrastructure in rhesus monkeys. *Biological Psychiatry*, 1980, *15*, 657–690.

Heikkila, R. E., Orlansky, H., & Cohen, G. Studies on the distinction between uptake inhibition and release of 3-H-dopamine in rat brain tissue slices. *Biochemical Pharmacology*, 1975, *24*, 847–858.

Hine, B., Friedman, E., Torrelio, M., & Gershon, S. Tetrahydrocannabinol-attenuated abstinence and induced rotation in morphine-dependent rats: Possible involvement of dopamine. *Neuropharmacology*, 1975, *14*, 607–610.

Ho, B. T., Taylor, D. L., Estevez, V. S., Englert, L. F., & McKenna, M. L. Behavioral effects of cocaine: Metabolic and neurochemical approach. In E. H. Ellinwood & M. M. Kilbey (Eds.), *Cocaine and other stimulants*. New York: Plenum Press, 1977.

Ho, B. T., Taylor, D., Fritchie, E., Englert, L. F., & McIsaac, W. M. Neuropharmacological study of Δ^9-and Δ^8-l-tetrahydrocannabinols in monkeys and mice. *Brain Research*, 1972, *38*, 163–170.

Hockman, C. H., Perrin, R. G., & Kalant, H. Electroencephalographic and behavior alterations produced by Δ^1-tetrahydrocannabinol. *Science*, 1971, *172*, 968–970.

Hollister, L. E. Electroencephalographic and neurophysiological studies. *Chemical Psychoses*, Springfield, Ill.: Charles C Thomas, 1968.

Holtzman, S. G. Phencyclidine-like discrimination effects of opioids in the rat. *Journal of Pharmacology and Experimental Therapeutics*, 1980, *214*, 614–619.

Holtzman, D., Lovell, R. A., Jaffe, J. H., & Freedman, D. X. 1-Δ^9-tetrahydrocannabinol: Neurochemical and behavioral effects in the mouse. *Science*, 1969, *163*, 1464–1467.

Horowitz, M. J. Flashbacks: Recurrent intrusive images after the use of LSD. *American Journal of Psychiatry*, 1969, *126*, 565–569.

Howlett, D. R., & Nahorski, S. R. Effect of acute and chronic amphetamine administration on beta-adrenoceptors and dopamine receptors in rat corpus striatum and limbic forebrain. *British Journal of Pharmacology*, 1978, *64*, 441–442.

Hsu, L. L., Smith, R. C., Rolsten, C., & Leelavathi, D. E. Effects of acute and chronic phencyclidine on neurotransmitter enzymes in rat brain. *Biochemical Pharmacology*, 1980, *29*, 2524–2526.

Isbell, H., & Logan, C. R. Studies on the diethylamide of lysergic acid (LSD-25). II. Effects of chlorpromazine, azacyclonol, and reserpine on the intensity of the LSD-reaction. *Archives of Neurology and Psychiatry*, 1957, *77*, 350–358.

Isbell, H., Gorodetzsky, C. W., Jasinski, D., Claussen, V., v. Spulak, F., & Korte, F. Effects of $(-)\Delta^9$-transtetrahydrocannabinol in man. *Psychopharmacologia*, 1967, *11*, 184–188.

Itil, T. M. Quantitative EEG and behavior changes after LSD and ditran. In A. G. Karczmar & W. P. Koella (Eds.), *Neurophysiological and behavioral aspects of psychotropic drugs*. Springfield, Ill.: Charles C Thomas, 1969.

Itzhak, Y., Kalir, A., Weissman, B. A., & Cohen, S. Receptor binding and antinociceptive properties of phencyclidine opiate-like derivatives. *European Journal of Pharmacology*, 1981, *72*, 305–311.

Jacobs, B. L., & Trulson, M. E. An animal behavior model for studying the actions of LSD and related hallucinogens. *Science*, 1976, *194*, 741–743.

Jacobs, B. L., & Trulson, M. E. The role of serotonin in the action of hallucinogenic drugs. In B. L. Jacobs & A. Gelperin (Eds.), *Serotonin neurotransmission and behavior*. Cambridge, Mass.: MIT Press, 1981.

Jacobs, B. L., Trulson, M. E., & Stern, W. C. Behavioral effects of LSD in the cat: Proposal of an animal behavior model for studying the actions of hallucinogenic drugs. *Brain Research*, 1977, *132*, 301–314.

Jacoby, J. H., & Poulakos, J. J. The actions of neuroleptic drugs and putative serotonin receptor antagonists on LSD and quipazine-induced reductions of brain 5-HIAA concentrations. *Journal of Pharmacy and Pharmacology*, 1977, *29*, 771–773.

Janowsky, D. S., El-Yousef, M. K., Davis, J. M., & Sekerke, H. J. Cholinergic antagonism of methylphenidate-induced stereotyped behavior. *Psychopharmacologia*, 1972, *27*, 295–303.

Jasinski, D. R., Shannon, H. E., Cone, E. J., Vaupel, D. B., Risner, M. E., McQuinn, R. L., Su, T. P., & Pickworth, W. B. Interdisciplinary studies on phencyclidine. In E. F. Domino (Ed.), *PCP (phencyclidine): Historical and current perspectives*. Ann Arbor, Mich.: NPP Books, 1981.

Jones, R. T. Drug models of schizophrenia-cannabis. In J. O. Cole, A. M. Freedman, & A. J. Friedhoff (Eds.), *Psychopathology and psychopharmacology*. Baltimore: The Johns Hopkins University Press, 1973.

Karczmar, A. G. Central actions of acetylcholine, cholinomimetics, and related drugs. In A. M. Goldberg & I. Hanin (Eds.), *Biology of cholinergic function*. New York: Raven Press, 1976.

Kelly, P. H., Seviour, P. W., & Iverson, S. D. Amphetamine and apomorphine responses in the rat following 6-OHDA lesions of the nucleus accumbens septi and corpus striatum. *Brain Research*, 1975, *94*, 507–522.

Kendrick, J. F., & Gibbs, F. Origin, spread and neurosurgical treatment of the psychomotor type of seizure discharge. *American Journal of Neurosurgery*, 1957, *14*, 270–285.

Key, B. J., & Bradley, P. B. The effects of drugs on conditioning and habituation to arousal stimuli in animals. *Psychopharmacologia*, 1960, *1*, 450–462.

Kilbey, M. M., & Ellinwood, E. H. Reverse tolerance to stimulant-induced abnormal behavior. *Life Sciences*, 1977, *20*, 1063–1076.

Kilbey, M. M., Ellinwood, E. H., & Easler, M. E. The effects of chronic cocaine pretreat-
    ment on kindled seizures and behavioral stereotypies. *Experimental Neurology*, 1979,
    64, 306–314.
Klawans, H. L., & Margolin, D. I. Amphetamine-induced dopaminergic hypersensitivity
    in guinea pigs. *Archives of General Psychiatry*, 1975, *32*, 725–732.
Klawans, H. L., Rubovitz, R., Patel, B. C., & Weiner, W. J. Cholinergic and anticholinergic
    influences on amphetamine-induced stereotyped behavior. *Journal of the Neurological
    Sciences*, 1972, *17*, 303–308.
Kloog, Y., Rehavi, M., Maayani, S., & Sokolovsky, M. Anticholinesterase and anti-
    acetylcholine activity of 1-phencyclohexylamine, derivatives. *European Journal of Phar-
    macology*, 1977, *45*, 221–227.
Layman, J. M., & Milton, A. S. Some actions of Δ¹ tetrahydro-cannabinol and cannabidiol
    at cholinergic junctions. *Proceedings of the British Journal of Pharmacology*, 1971, *41*, 379–
    380.
Lee, E. H. Y., & Geyer, M. A. Persistent effects of chronic administration of LSD on
    intracellular serotonin content in rat midbrain. *Neuropharmacology*, 1980, *19*, 1005–
    1007.
Lee, T., & Seeman, P. Elevation of brain neuroleptic/dopamine receptors in schizophrenia.
    American Journal of Psychiatry, 1980, *137*, 191–197.
Leelavathi, D. E., Misra, C. H., Shelet, H., & Smith, R. C. Effects of acute and chronic
    administration of phencyclidine on dopaminergic receptors in rat striatrum. *Commu-
    nications in Psychopharmacology*, 1980, *4*, 417–424.
Leonard, B. E. Some effects of the hallucinogenic drug 2, 5-dimethoxy-4-methyl-ampheta-
    mine on the metabolism of biogenic amines in the rat brain. *Psychopharmacologia*, 1973,
    32, 33–49.
Leonard, B. E., & Tonge, S. R. The effects of some hallucinogenic drugs upon the metabo-
    lism of noradrenaline. *Life Sciences*, 1969, *8*, 815–825.
Lovell, R. A., & Freedman, D. X. Stereospecific receptor sites for d-lysergic acid di-
    ethylamide in rat brain: Effects of neurotransmitters, amine antagonists, and other
    psychotropic drugs. *Molecular Pharmacology*, 1976, *12*, 620–630.
Luby, E. D., Cohen, B. D., Rosenbaum, G., Gottlieb, J. S., & Kelley, R. Study of a new
    schizophrenomimetic drug-sernyl. *Archives of Neurology and Psychiatry*, 1959, *81*, 113–
    119.
Luby, E. D., Gottlieb, J. S., Cohen, B. D., Rosenbaum, G., & Domino, E. F. Model
    psychoses and schizophrenia. *American Journal of Psychiatry*, 1962, *119*, 61–67.
Maayani, S., & Weinstein, H. "Specific binding" of ³H-phencyclidine: Artifacts of the
    rapid filtration method. *Life Sciences*, 1980, *26*, 2011–2022.
Maayani, S., Weinstein, H., Ben-Zvi, N., Cohen, S., & Sokolovsky, M. Psychotomimetics
    as anticholinergic agents—I. *Biochemical Pharmacology*, 1974, *23*, 1263–1281.
Maitre, L., Stahelin, M., & Bein, H. J. Effect of an extract of cannabis and of some
    cannabinols on catecholamine metabolism in rat brain and heart. *Agents and Actions*,
    1970, *1*, 136–143.
Mandell, A. J., & Knapp, S. Neurobiological antagonism of cocaine by lithium. In E. H.
    Ellinwood & M. M. Kilbey (Eds.), *Cocaine and other stimulants*. New York: Plenum
    Press, 1977.
Marini, J. L., & Sheard, M. H. On the specificity of a cat behavior model for the study of
    hallucinogens. *European Journal of Pharmacology*, 1981, *70*, 479–487.
Martinez, J. L., Stadnick, S. W., & Schaeppi, U. H. Δ⁹-Tetrahydrocannabinol. Effects on
    EEG and behavior of rhesus monkeys. *Life Sciences*, 1972, *11*, 643–651.
Marwaha, J., Palmer, M. R., Woodward, D. J., Hoffer, B. J., & Freedman, R. Elec-

trophysiological evidence for presynaptic actions of phencyclidine on noradrenergic transmission in rat cerebellum. *Journal of Pharmacology and Experimental Therapeutics,* 1980, *215,* 606–613.

Marwaha, J., Palmer, M., Hoffer, B., Freedman, R., Rice, K. C., Paul, S., & Skolnick, P. Differential electrophysiological and behavioral responses to optically active derivatives of phencyclidine. *Archives of Pharmacology,* 1981, *315,* 203–209.

McCall, R. B., & Aghajanian, G. K. Hallucinogens potentiate responses to serotonin and norepinephrine in the facial motor nucleus. *Life Sciences,* 1980, *26,* 1149–1156.

McGlothin, W. H. The use of cannabis: East and west. In H. M. Pragg (Ed.), *Biochemical and pharmacological aspects of dependence and reports on marihuana research.* The Netherlands: Haarlem, 1972.

Meltzer, H. Y., & Stahl, S. M. The dopamine hypothesis of schizophrenia: A review. *Schizophrenia Bulletin,* 1976, *2,* 19–76.

Meltzer, H. Y., Sturgeon, R. D., Simonovic, M., & Fessler, R. G. Phencyclidine as an indirect dopamine agonist. In E. F. Domino (Ed.), *PCP (phencyclidine): Historical and current perspectives.* Ann Arbor, Mich.: NPP Books, 1981.

Monroe, R. R., & Heath, R. G. Effects of lysergic acid and various derivatives on depth and cortical electrograms. *Journal of Neuropsychiatry,* 1961, *3,* 75–82.

Monroe, R. R., Heath, R. G., Mickle, W. A., & Llewellyn, R. C. Correlation of rhinencephalic electrograms with behavior. *Electroencephalography and Clinical Neurophysiology,* 1957, *9,* 623–642.

Moore, K. E. The actions of amphetamine on neurotransmitters: A brief review. *Biological Psychiatry,* 1977, *12,* 451–462.

Moore, K. E., Chiueh, C. C., & Zeldes, G. Release of neurotransmitters from the brain *in vivo* by amphetamine, methylphenidate and cocaine. In E. H. Ellinwood & M. M. Kilbey (Eds.), *Cocaine and other stimulants.* New York: Plenum Press, 1977.

Morgane, P. J., & Stern, W. C. Serotonin in the regulation of sleep. In W. B. Essman (Ed.), *Serotonin in health and disease: Physiological regulation and pharmacological action* (Vol. 2). New York: Spectrum, 1978.

Moskowitz, D. Use of haloperidol to reduce LSD flashbacks. *Military Medicine,* 1971, *136,* 754–756.

Murray, T. F., & Horita, A. Phencyclidine-induced stereotyped behavior in rats: Dose response effects and antagonism by neuroleptics. *Life Sciences,* 1979, *24,* 2217–2225.

Nelson, L. R., & Ellison, G. Enhanced stereotypies after repeated injections but not continuous amphetamines. *Neuropharmacology,* in 1978, *17,* 1081–1084.

Nichols, D. E. Structural correlation between apomorphine and LSD: Involvement of dopamine as well as serotonin in the actions of hallucinogens. *Journal of Theoretical Biology,* 1976, *59,* 167–177.

Nichols, D. E., Pfister, W. R., & Yim, G. K. W. LSD and phenethylamine hallucinogens: New structural analogy and implications for receptor geometry. *Life Sciences,* 1978, *22,* 2165–2170.

Nielson, E. B., Lee, T. H., & Ellison, G. Following several days of continuous administration d-amphetamine acquires hallucinogen-like properties. *Psychopharmacology,* 1980, *68,* 197–200.

Nielsen, E. B., Nielsen, M., Ellison, G., & Braestrup, C. Decreased spiroperidol and LSD binding in rat brain after continuous amphetamine. *European Journal of Pharmacology,* 1980, *66,* 149–154.

Ngai, S. H., Shirasawa, R., & Cheney, D. L. Changes in motor activity and acetylcholine turnover induced by lidocaine and cocaine in brain regions of rats. *Anesthesiology,* 1979, *51,* 230–234.

Osmond, H., & Smythies, J. R. Schizophrenia: A new approach. *Journal of Mental Science*, 1952, *98*, 309–315.

Owen, F., Crow, T. J., Poulter, M., Cross, A. J., Longden, A., & Riley, G. J. Increased dopamine-receptor sensitivity in schizophrenia. *Lancet*, 1978, *2*, 223–225.

Pandey, G. N., Garver, D. L., Tamminga, C., Ericksen, S., Ali, S. I., & Davis, J. M. Postsynaptic supersensitivity in schizophrenia. *American Journal of Psychiatry*, 1977, *134*, 518–522.

Pare, C. M. B., & LaBrosse, E. H. A further study on alleviation of the psychological effects of LSD in man by pretreatment with 5-hydroxytryptophan. *Journal of Psychiatric Research*, 1963, *1*, 271–277.

Pearlson, G. D. Psychiatric and medical syndromes associated with phencyclidine (PCP) abuse. *Johns Hopkins Medical Journal*, 1981, *148*, 25–33.

Peroutka, S. J., & Snyder, S. H. Multiple serotonin receptors: Differential binding of [^{3}H]5-hydroxytryptamine, [^{3}H]lysergic acid diethlamine and [^{3}H]spiroperidol. *Molecular Pharmacology*, 1979, *16*, 687–699.

Peters, D. A. V. Comparison of the chronic and acute effects of d-lysergic acid diethylamide (LSD) treatment on rat brain serotonin and norepinephrine. *Biochemical Pharmacology*, 1974, *23*, 231–237.

Peters, D. A. V., & Tang, S. Persistent effects of repeated injections of d-lysergic acid diethylamide on rat brain 5-hydroxytryptamine and 5-hydroxyindoleacetic acid levels. *Biochemical Pharmacology*, 1977, *26*, 1085–1086.

Pieri, L., Pieri, M., & Haefely, W. LSD as an agonist of dopamine receptors in the striatum. *Science*, 1974, *252*, 586–588.

Pieri, M., Schaffner, R., Pieri, L., DaPrada, M., & Haefely, W. Turning in MFB-lesioned rats and antagonism of neuroleptic-induced catalepsy after lisuride and LSD. *Life Sciences*, 1978, *22*, 1615–1622.

Pirch, J. H., Cohn, R. A., Barnes, P. R., & Barrett, E. S. Effects of acute and chronic administration of marijuana extract on the rat electrocorticogram. *Neuropharmacology*, 1972, *11*, 231–240.

Post, R. M. Progressive changes in behavior and seizures following chronic cocaine administration: Relationship to kindling and psychosis. In E. H. Ellinwood & M. M. Kilbey (Eds.), *Cocaine and other stimulants.* New York: Plenum Press, 1977.

Pradhan, S., Roy, S. N., & Pradhan, S. N. Correlation of behavioral and neurochemical effects of acute administration of cocaine in rats. *Life Sciences*, 1978, *22*, 1737–1744.

Pribram, K. H., & McGuinness, D. Arousal, activation, and effort in the control of attention. *Psychological Review*, 1975, *82*, 116–149.

Quirion, R., Rice, K. C., Skolnick, P., Paul, S., & Pert, C. B. Stereospecific displacement of [^{3}H]phencyclidine (PCP) receptor binding by an enantiomeric pair of PCP analogs. *European Journal of Pharmacology*, 1981, *74*, 107–108.

Rackham, A., & Wise, R. A. Independence of cocaine sensitization and amygaloid kindling in the rat. *Physiology and Behavior*, 1979, *22*, 631–633.

Resnick, O., Krus, D. M., & Raskin, M. LSD-25 action in normal subjects treated with a monoamine oxidase inhibitor. *Life Sciences*, 1964, *3*, 1207–1214.

Resnick, O., Krus, D. M., & Raskin, M. Accentuation of the psychological effects of LSD-25 in normal subjects treated with reserpine. *Life Sciences*, 1965, *4*, 1433–1437.

Ricaurte, G. A., Schuster, C. R., & Seiden, L. S. Long-term effects of repeated methylamphetamine administration on dopamine and serotonin neurons in rat brain: A regional study. *Brain Research*, 1980, *193*, 153–163.

Robertson, H. A. Chronic phencyclidine, like amphetamine, produces a decrease in

[³H]spiroperidol binding in rat striatum. *European Journal of Pharmacology*, 1982, *78*, 363–365.

Robinson, S., Cheney, D. L., & Moroni, F. Acetylcholine turnover in specific brain areas of rats injected with various antidepressants. In S. Garattini (Ed.), *Depressing disorders*. Stuttgart, Germany: Schattaner, 1978.

Rodin, E. A., Domino, E. F., & Porzak, J. P. The marihuana-induced "social high." *Journal of the American Medical Association*, 1970, *213*, 1300–1302.

Rodin, E. A., Luby, E. D., & Meyer, J. S. Electroencephalographic findings associated with sernyl infusion. *Electroencephalography and Clinical Neurophysiology*, 1959, *11*, 796–798.

Ross, S. B. The central stimulatory action of inhibitors of dopamine uptake. *Life Sciences*, 1979, *24*, 159–169.

Rotrosen, J., Angrist, B., Clark, C., Gershon, S., Halpern, F. S., & Sachar, E. J. Suppression of prolactin by dopamine agonists in schizophrenics and controls. *American Journal of Psychiatry*, 1978, *35*, 949–951.

Roy, S. N., Bhattacharyya, A. K., Pradhan, S., & Pradhan, S. N. Behavioural and neurochemical effects of repeated administration of cocaine in rats. *Neuropharmacology*, 1978, *17*, 559–564.

Saito, M. Effects of spychotrophic drugs on the human EEG based on analog frequency analysis. *Modern Problems in Pharmacopsychiatry*, 1974, *8*, 117–130.

Sato, M., Hikasa, N., & Otsuki, S. Experimental epilepsy, psychosis, and dopamine receptor sensitivity. *Biological Psychiatry*, 1979, *14*, 537–540.

Scheel-Kruger, J. Comparative studies of various amphetamine analogues demonstrating different interaction with the metabolism of the catecholamines in the brain. *European Journal of Pharmacology*, 1971, *14*, 47–59.

Scheel-Kruger, J., Braestrup, C., Nielson, M., Golembiowska, K., & Mogilnicka, E. Cocaine: Discussion on the role of dopamine in the biochemical mechanism of action. In E. H. Ellinwood & M. M. Kilbey (Eds.), *Cocaine and other stimulants*. New York: Plenum Press, 1977.

Schlemmer, R. F., Jackson, J. A., Preston, K. C., Bederka, J. P., Garver, D. C., & Davis, J. M. Phencyclidine-induced stereotyped behavior in monkeys: Antagonism by pimozide. *European Journal of Pharmacology*, 1978, *52*, 379–384.

Schwartz, R. D., Moerschbaecher, J. M., Thompson, D. M., & Keller, K. J. Effects of chronic phencyclidine on fixed-ratio responding: No relation to neurotransmitter receptor binding in rat cerebral cortex. *Pharmacology Biochemistry and Behavior*, 1982, *16*, 647–652.

Schwarz, B. E., Sem-Jacobsen, C. W., & Peterson, M. C. Effects of mescaline, LSD-25, and adrenochrome on depth electrograms in man. *Archives of Neurology and Psychiatry*, 1956, *75*, 579–587.

Segal, D. S. Differential effects of serotonin depletion on amphetamine-induced locomotion and stereotypy. In E. H. Ellinwood & M. M. Kilbey (Eds.), *Cocaine and other stimulants*. New York: Plenum Press, 1977.

Segal, D. S., & Mandell, A. J. Long-term administration of d-amphetamine: Progressive augmentation of motor activity and stereotypy. *Pharmacology, Biochemistry and Behavior*, 1974, *2*, 249–255.

Segal, D. S., Weinberger, S. B., Cahill, J., & McCunney, S. J. Multiple daily amphetamine administration: Behavioral and neurochemical alterations. *Science*, 1980, *207*, 904–907.

Sem-Jacobsen, C. W., Peterson, M. C., Lazarte, J. A., Dodge, H. W., & Holman, C. D. Intracerebral electrographic recordings from psychotic patients during hallucinations and agitation. *American Journal of Psychiatry*, 1955, *112*, 278–293.

Shagass, C. An electrophysiological view of schizophrenia. *Biological Psychiatry*, 1976, *11*, 3–30.

Shannon, H. Evaluation of phencyclidine analogs on the basis of their discriminative stimulus properties in the rat. *Journal of Pharmacology and Experimental Therapeutics*, 1981, *216*, 543–551.

Singh, C. B., Mazumdar, S., & Prasad, G. C. Neurohumoral responses to marijuana fume inhalation. *Indian Journal of Experimental Biology*, 1980, *18*, 513–515.

Slater, E. The schizophrenia-like illnesses of epilepsy. *British Journal of Psychiatry*, 1969, Suppl. 4, 77–81.

Smith, D. E. A clinical approach to the treatment of phencyclidine (PCP) abuse. *Psychopharmacology Bulletin*, 1980, *16*, 67–70.

Smith, R. C., Biggs, C., Vroulis, G., & Brinkman, S. Effects of chronic administration of phencyclidine on stereotyped and ataxic behaviors in the rat. *Life Sciences*, 1981, *28*, 1163–1174.

Smith, R. C., Meltzer, H. Y., Arora, R. C., & Davis, J. M. Effects of phencyclidine on [^{3}H] catecholamine and [^{3}H] serotonin uptake in synaptosomal preparations from rat brain. *Biochemical Pharmacology*, 1977, *26*, 1435–1439.

Smith, R. C., Leelavathi, D. E., Hsu, L., Ho, B. T., Tansey, W., Taylor, D., & Biggs, C. Acute versus chronic administration of phencyclidine: Effects on behavior and brain biochemistry. In E. F. Domino (Ed.), *PCP (phencyclidine): Historical and current perspectives*. Ann Arbor, Mich.: NPP Books, 1981.

Smythies, J. R., Benington, F., & Morin, R. D. Specification of a possible serotonin receptor site in the brain. *Neurosciences Research Progress Bulletin*, 1970, *8*, 117–123.

Snyder, S. H., & Richelson, E. Steric models of drugs predicting psychedelic activity. In D. H. Efron (Ed.), *Psychotomimetic drugs*. New York: Raven Press, 1970.

Sofia, R. D., Ertel, R. J., Dixit, B. N., & Barry, H. The effect of Δ^1-tetrahydrocannabinol on the uptake of serotonin by rat brain homogenates. *European Journal of Pharmacology*, 1971, *16*, 257–259.

Spano, P. F., Kumakura, K., Tonon, G. C., Gavoni, S., & Trabucchi, M. LSD and dopamine-sensitive adenylate-cyclase in various rat brain areas. *Brain Research*, 1975, *93*, 164–167.

Strahlendorf, J. C. R., Goldstein, F. J., Rossi, G. V., & Malseed, R. T. Differential effects of LSD serotonin and L-tryptophan on visually evoked responses. *Pharmacology, Biochemistry and Behavior*, 1982, *16*, 51–55.

Stripling, J. S., & Ellinwood, E. H. Augmentation of the behavioral and electrophysiologic response to cocaine by chronic administration in the rat. *Experimental Neurology*, 1977, *54*, 546–564.

Stripling, J. S., & Hendricks, C. Facilitation of kindling by convulsions induced by cocaine or lidocaine but not pentylenetetrazol. *Pharmacology Biochemistry and Behavior*, 1981, *15*, 793–798.

Sturgeon, R. D., Fessler, R. G., London, S. F., & Meltzer, H. Y. Behavioral effects of chronic phencyclidine administration in rats. *Psychopharmacology*, 1982, *76*, 52–56.

Talbott, J. A., & Teague, J. W. Marihuana psychosis: Acute toxic psychosis associated with the use of cannabis derivatives. *Journal of the American Medical Association*, 1969, *210*, 299–302.

Taube, H. D., Montel, H., Hau, G., & Starke, K. Phencyclidine and ketamine: Comparison with the effect of cocaine on the noradrenergic neurones of the rat brain cortex. *Archives of Pharmacology*, 1975, *291*, 47–54.

Taylor, D., & Ho, B. T. Neurochemical effects of cocaine following acute and repeated injection. *Journal of Neuroscience Research*, 1977, *3*, 95–101.

Tennant, F. S., & Groesbeck, C. J. Psychiatric effects of hashish. *Archives of General Psychiatry*, 1972, *27*, 133–136.

Tilson, H. A., & Sparber, S. B. Studies on the concurrent behavioral and neurochemical effects of psychoactive drugs using the push–pull cannula. *Journal of Pharmacology and Experimental Therapeutics*, 1972, *181*, 387–398.

Tonge, S. R., & Leonard, B. E. The effect of some hallucinogenic drugs upon the metabolism of 5-hydroxytryptamine in the brain. *Life Sciences*, 1969, *8*, 805–814.

Tonge, S. R., & Leonard, B. E. Variation in hydroxytryptamine metabolism in the rat: Effects on the neurochemical response to phencyclidine. *Journal of Pharmacy and Pharmacology*, 1971, *23*, 711–712.

Truitt, E. B., & Anderson, S. M. Biogenic amine alterations produced in the brain by tetrahydrocannabinols and their metabolites. *Annals of the New York Academy of Sciences*, 1971, *191*, 68–73.

Trulson, M. E., & Jacobs, B. L. LSD acts synergistically with serotonin depletion: Evidence from behavioral studies in cats. *Pharmacology, Biochemistry and Behavior*, 1976, *4*, 231–234.

Trulson, M. E., & Jacobs, B. L. Effects of 5-Methoxy-N-N-dimethyltryptamine on behavior and raphe unit activity in freely moving cats. *European Journal of Pharmacology*, 1979, *54*, 43–50. (a)

Trulson, M. E., & Jacobs, B. L. Dissociation between the effects of LSD on behavior and raphe unit activity in freely moving cats. *Science*, 1979, *205*, 515–518. (b)

Trulson, M. E., & Jacobs, B. L. Alterations of serotonin and LSD receptor binding following repeated administration of LSD. *Life Sciences*, 1979, *24*, 2053–2062. (c)

Trulson, M. E., & Jacobs, B. L. Chronic amphetamine administration to cats: Behavioral and neurochemical evidence for decreased central serotonergic function. *Journal of Pharmacology and Experimental Therapeutics*, 1979, *211*, 375–384. (d)

Trulson, M. E., & Jacobs, B. L. Activity of serotonin-containing neurons in freely moving cats. In B. L. Jacobs & A. Gelperin (Eds.), *Serotonin neurotransmission and behavior*. Cambridge: MIT Press, 1981.

Tucker, H. J., Quinlan, D., & Harrow, M. Chronic hallucinogenic drug use and thought disturbance. *Archives of General Psychiatry*, 1972, *27*, 443–447.

Ungerleider, J. T., Fisher, D. D., Goldsmith, S. R., Fuller, M., & Forgy, E. A statistical survey of adverse reactions to LSD in Los Angeles county. *American Journal of Psychiatry*, 1968, *125*, 352–357.

Utena, H. On relapse-liability; schizophrenia, amphetamine psychosis and animal model. In H. Mitsuda & T. Fukuda (Eds.), *Biological mechanisms of schizophrenia and schizophrenia-like psychoses*. Tokyo, Japan: Igaku Shoin, 1974.

Van Rossum, J. M., & Hurksman, J. A. Mechanism of action of psychomotor stimulant drugs. *International Journal of Neuropharmacology*, 1964, *3*, 227–234.

Venables, P. H., & Wing, J. K. Levels of arousal and the subclassification of schizophrenia. *Archives of General Psychiatry*, 1962, *7*, 114–119.

Vestergaard, P., Rubin, V., Beaubrunn, M. H., Cruickshank, E., & Picou, D. The effect of tetrahydrocannabinols on central monoamine neurons. *Acta Pharmaceutica Suecica*, 1971, *8*, 695–696.

Vickroy, T. W., & Johnson, K. M. *In vivo* administration of phencyclidine inhibits ^{3}H-dopamine accumulation by rat brain striatal slices. *Substance and Alcohol Actions Misuse*, 1980, *1*, 351–354.

Vickroy, T. W., & Johnson, K. M. Similar dopamine-releasing effects of phencyclidine and nonamphetamine stimulants in striatal slices. *Journal of Pharmacology and Experimental Therapeutics*, 1982, *233*, 669–674.

Vincent, J. P., Cavey, D., Kamenka, J. M., Geneste, P., & Lazdunski, M. Interaction of phencyclidines with the muscarinic and opiate receptors in the central nervous system. *Brain Research,* 1978, *152,* 176–182.

Vincent, J. P., Kartalovski, B., Geneste, P., Kamenka, J. M., & Lazdunski, M. Interaction of phencyclidine ("angel dust") with a specific receptor in rat brain membranes. *Proceedings of the National Academy Science,* 1979, *76,* 4678–4682.

Von Hungen, K., Roberts, S., & Hill, D. F. Interactions between lysergic acid diethylamide and dopamine-sensitive adenylate cyclase systems in rat brain. *Brain Research,* 1975, *94,* 57–66.

Wallach, M. B., & Gershon, S. A neuropsyclopharmacological comparison of d-amphetamine, L-DOPA, and cocaine. *Neuropharmacology,* 1971, *10,* 743–752.

Wallach, M. B., & Wallach, S. S. Involuntary eye movements in certain schizophrenics. *Archives of General Psychiatry,* 1964, *11,* 71–73.

Ward, D., & Trevor, A. Phencyclidine-induced alteration in rat muscarinic cholinergic receptor regulation. *European Journal of Pharmacology,* 1981, *74,* 189–193.

Waters, D. H., & Glick, S. D. Asymetrical effect of delta-9-tetrahydrocannabinol (THC) on striatal dopamine and behaviour. *Research Communications in Chemical Pathology and Pharmacology,* 1973, *6,* 57–63.

Weil, A. T. Adverse reactions to marihuana. *New England Journal of Medicine,* 1970, *282,* 997–1000.

Weil-Malherbe, H. Serotonin and schizophrenia. In W. B. Essman (Ed.), *Serotonin in health and disease: The central nervous system* (Vol. 3). New York: Spectrum, 1978.

Weinstein, H., Maayani, S., Glick, S. D., & Meibach, R. C. Integrated studies on the biochemical, behavioral, and molecular pharmacology of phencyclidine: A progress report. In E. F. Domino (Ed.), *PCP (phencyclidine): Historical and current perspectives.* Ann Arbor, Mich.: NPP Books, 1981.

Whitaker, P. M., & Seeman, P. Hallucinogen binding to dopamine/neuroleptic receptors. *Journal of Pharmacy and Pharmacology,* 1977, *29,* 506–507.

Winter, J. C. Stimulus properties of phenethylamine hallucinogens and lysergic acid diethylamide: The role of 5-hydroxytryptamine. *Journal of Pharmacology and Experimental Therapeutics,* 1978, *204,* 416–423.

Wooley, D. W., & Shaw, E. A biochemical and pharmacological suggestion about certain mental disorders. *Proceedings of the National Academy of Science, USA,* 1954, *40,* 228–231.

Zahn, T. P., Rosenthal, S. H., & Lawlor, W. G. Electrodermal and heart rate reactions in chronic schizophrenia. *Journal of Psychiatric Research,* 1968, *6,* 117–134.

Zukin, S. R., & Zukin, R. S. Specific [^{3}H]phencyclidine binding in rat central nervous system. *Proceedings of the National Academy of Sciences,* 1979, *76,* 5372–5376.

II

PROBLEMS, EFFECTS, AND TREATMENTS

9

Interactions of Drugs of Abuse with Prescription Drugs

W. H. VOGEL

1. Overview

Over the last several decades, there has been an explosive rate of growth in the development of therapeutic drugs. These newly discovered drugs have helped significantly in reducing the threat of many diseases, in markedly improving the well-being of mankind, and in greatly prolonging life. Drugs are extremely powerful and sophisticated substances and their proper use has evolved into a specific science. One primary goal of this science is to find the smallest dose that will provide optimal benefit with the least amount of risk to a particular patient. Among the many factors that can contribute to unwanted side effects and that have to be very carefully monitored is the interaction of two or more drugs in an individual.

The administration of only one drug to a patient is already the beginning of an extremely complicated process: The drug has to be dissolved in the stomach and intestines, it has to cross cell walls and barriers to reach the blood stream where it can be adsorbed by a variety of plasma proteins, and, again, after passage through other cell walls and barriers, it must reach cell receptors with which it will react in a highly specific way. Through this interaction, it will normalize or correct a pathological state. While these processes occur, liver and kidney rec-

W. H. VOGEL • Department of Pharmacology, Thomas Jefferson University, 1020 Locust Street, Room 326, Philadelphia, Pennsylvania 19107.

ognize the drug as a foreign material and chemically change the drug in the hepatic cells and excrete the drug and its metabolites by the renal cells into the urine. All these complex processes determine the tissue levels of the drug and, clinically, the beneficial and toxic effects as well as the duration of the action of the medication.

Many of these mentioned processes are shared by different drugs, and the presence of two or more drugs can lead to either increased or decreased efficacy of the individual drugs or to the appearance of unwanted, sometimes rather serious, toxic reactions. To avoid such interactions, simple rules must be followed. First, only one drug should be used, and combinations of drugs should be avoided, if at all possible. Second, if the clinical condition of the patient warrants more than one medication, one drug should be added at a time. Third, the medical literature, the handbooks, and other health professionals should be consulted if doubt exists about possible interactions. Fourth, nobody should be afraid to ask and to consult; the multitude of drugs available and the vast number of interactions possible makes it practically impossible for any one person to completely remember all interactions.

All of the above risk factors are increased if the individual is addicted to certain drugs and chemicals. Substance abusers expose themselves to a considerable degree of risk in terms of organ and tissue damage just by using drugs or chemicals in large doses and over prolonged periods of time. In addition, they inflict additional risks if they fall ill and require medical treatment and drugs. One of these risks is a possible interaction of the drug or substance of abuse with over-the-counter or prescription drugs. This risk is increased considerably if the substance abuser keeps his/her drug habit a secret from the treating physician or starts abusing drugs after having been placed on a specific medical therapy. The abused drug or substance could then decrease the efficacy of the therapeutic agent, with minimal help to the patient, or increase its toxicity, with serious damage to the individual.

The major objective of the present chapter is to outline some of the adverse interactions that can occur when substances or drugs of abuse are taken by the individual in conjunction with medically prescribed and over-the-counter drugs. As we have implied, however, the evaluation of drug effects and interactions is often a complex matter, and it is not possible in many cases to provide a simple formula predicting the effects of any given drug combination on a particular individual. Thus, before going on to describe adverse drug interactions that may occur as a result of concomitant substance abuse, an overview will be provided on the variables and mechanisms that need to be considered within the framework of drug interactions.

2. General Considerations

The beneficial and adverse or side effects that all drugs possess or can produce depend on two major factors.

The first factor consists of the chemical structure of the drug. It is the chemical structure that will determine the pharmacological properties of drugs. It will determine, for instance, if a particular drug can correct an abnormal heart rhythm in the cardiac patient, kill certain microorganisms that cause an infection, or interact beneficially with certain brain receptors in the schizophrenic patient. It is the chemical structure that will also determine the number and severity of side effects that can be expected during therapy. And extremely small changes in the arrangement of the atoms in the chemical structure can rob the drug of its beneficial action or can drastically increase its toxicity. These chemical and pharmacological-toxicological properties can be fairly well examined, characterized, and established by analytical and scientific methods available today.

The second factor consists of the biological complexity of an organism. Although great progress has been made to understand the functioning of the human body and mind, we are still a long way from knowing the intricate details of its complexity. In addition, we are individuals differing from each other and constantly undergoing specific changes in our internal environment in response to aging and external influences. Thus, the same drug can produce slightly or markedly different effects in man as opposed to woman, in orientals as opposed to occidentals, or in young as opposed to older people. In particular, age is now recognized as an important factor, and the choice or dose of a particular drug will depend strongly on the age of the patient. In addition, currently existing pathological conditions or diseases, recognized or latent, can drastically influence drug effects. For instance, liver diseases could alter significantly the fate of a drug in the body and could lead to decreased beneficial and increased toxic effects, or a hyperactive thyroid gland can markedly enhance the toxicity of certain drugs that are relatively harmless in individuals with normal thyroid activity. Indeed, the list of factors that can affect the action of drugs in the body is rapidly growing, even changing some of our old and seemingly well-established concepts. The lethal effects of an overdose of a drug have always been associated solely with the pharmacological properties of the drug; recently, however, it was shown that the dose of heroin that is lethal to animals can be markedly influenced by the environment in which it is given (Siegel, Hinson, Krank, & McCully, 1982). The dose necessary to kill an animal is much larger if given in an environment in which the drug had been previously administered to the animal, as opposed to the smaller dose when given to the animal in a new and unfamiliar environment. This might explain some of the deaths of human heroin users by doses to which they actually should have been tolerant. Thus, it becomes readily apparent that we are far from definitively predicting the exact extent of the beneficial and/or adverse reactions that a drug may produce in an individual. At best, we can be more sure for the actions of some drugs and less sure for those of other drugs. But always, we have to be prepared for the unexpected to happen.

Given the poor understanding of the human body, it is frequently rather difficult to establish a definite beneficial or toxic action of a drug unless it is easily seen or recognized. One of the many complicating factors is the "placebo" reaction, or the reaction of an individual to an inert, pharmacologically inactive substance. The way a drug is prescribed by a physician—ranging from confidence to indifference to doubt—or the education and knowledge of a particular patient taking the drug, can produce a whole variety of beneficial or adverse signs and symptoms that are not associated with the pharmacology or toxicology of the drug at all. Such placebo reactions are not isolated episodes; signs of clinical improvement or occurence of adverse reactions may occur in 30% of all individuals taking a placebo (Vogel, Goodwin, & Goodwin, 1980). This is the reason for double-blind studies in which neither the physician nor the patient knows whether a drug or placebo is taken. Only then can one eliminate both the physician's bias and the patient's expectations, and can one truly evaluate beneficial as well as toxic effects. Although this is done with a limited number of individuals in scientific studies, it cannot be done when a drug is introduced to the general population.

3. Drug Interactions

3.1. Introduction

The complexity and uncertainty of possible adverse or toxic reactions of a drug increase considerably if the organism is exposed to two or more medications at the same time. In addition to the effects of the individual drugs on the body, drugs can interact with each other in many ways and produce a whole new set of beneficial and toxic effects. A drug interaction can then be defined as an effect that is only seen in the presence of two or more drugs.

Such drug interactions can occur at different sites outside or inside the body. A few examples, by no means exhaustive, will be cited to illustrate the many levels of possibilities.

3.2. Sites Where Drugs Can Interact

Drugs can even interact prior to the administration to patients because they can be incompatible when mixed together. A drug added to an intravenous infusion of electrolytes or nutrients can be degraded or can form potentially allergenic conjugates. This, however, is more of concern to the pharmaceutical industry and the pharmacist.

Most drug interactions occur within the body. The major sites of interaction after oral administration of a drug are in the stomach and in the intestinal tract. Here, drugs usually diffuse across the gastrointestinal walls into the blood stream. This process of passive diffusion is strongly pH dependent, and the coadministration of antacids that change the stomach pH can either increase or decrease the absorption of the drug. Certain cations such as Al^{+++}, Ca^{++}, or Mg^{++}, which are often constituents of antacids or milk, can complex with drugs such as the antibiotic tetracyclines to form insoluble complexes that cannot be absorbed into the blood stream; they will leave the body without producing a therapeutic effect (Albert, 1953). Thus, it is extremely important for the individual to know and to follow instructions on how a medication should be taken.

After a medication is successfully absorbed into the blood stream, the drug can be loosely bound to circulating plasma proteins. A fine balance develops between the bound and unbound portion of the drug. Only the free form of the drug is responsible for its further pharmacological actions. A second drug that also binds to the same sites on these circulating proteins can displace the first drug and drastically alter the bound versus unbound drug balance. Thus, a patient first stabilized on one drug can, upon receiving a second drug, develop serious toxic problems due to a disturbance of the binding equilibria of both drugs.

The duration of action of drugs is partly determined by their inactivation by special enzymes in the liver. Here, drugs are changed into other chemicals that are inactive pharmacologically and that can be excreted more easily. If used over long periods of time, some drugs can stimulate these drug metabolizing enzymes and can increase the inactivation process. Thus, subsequent doses of the same drug or other drugs will show increased metabolism or decreased levels and effects. Barbiturates, alcohol, and cigarette smoke, among many chemicals and drugs, have this enzyme induction and activation effect. This is partially the reason why tolerance develops to barbiturates; subsequent doses of these drugs are more quickly metabolized and thus produce less of an effect. Heavy smokers usually need larger doses of narcotic analgesics or minor tranquilizers because these drugs are more rapidly inactivated due to smoke-induced, accelerated drug metabolism. In certain cases, the opposite effect, increased medication effects, can be observed if the metabolism of a second, inactivating drug is inhibited by another drug, thus allowing the medication to build up to dangerously high levels.

The duration of action of drugs is also partly regulated by its elimination from the body into the feces or urine. In the kidneys, drugs and their metabolites are excreted by a variety of passive and active transport mechanisms. Since some of these transport mechanisms are shared by different drugs, the excretion rate for a particular drug can be easily influenced by another drug. Functioning of some of these transport mechanisms depends on the pH of the urine. Changes in the acidity or alkalinity can either increase or decrease the rate of excretion. Thus, a change in the urinary pH can enhance or decrease drug action.

The final action of a drug consists of its effect on specific organs or tissues. The action occurs not all over these organs or tissues but with particularly small sites on or in cells, called the receptors. The action of a drug with these receptors produces the therapeutic or toxic effects. The drug can either stimulate or inhibit receptor activity, increase or decrease cell activity. The chronic use of drugs and the continuous stimulation or inhibition of receptors can lead to certain receptor changes changing responsiveness. In the case of tolerance, overstimulation leads finally to a reduced response, or an increased amount of drug is necessary to produce the same effect. Often, several drugs can stimulate or inhibit the same receptors. Thus they can interact at the receptor and enhance or reduce their individual effects. Enhancing effects can be additive in nature, or they can potentiate each other so that the combined effect is considerably larger than the sum of the individual effects.

3.3. Classification of Drug Interactions

To begin with, most drugs do not interact, and a combination of these drugs produces neither a beneficial nor an adverse reaction. Interactions that occur can be classified in different ways.

First, drugs can be classified as they affect the clinical response of each other. Drugs can interact, and the combination is therapeutically better than the effects of the individual drugs alone. The use of drugs in combination can produce certain beneficial effects not seen with the individual medications; such combinations are frequently used in cancer chemotherapy. Drugs can also interact so that the combination decreases the beneficial effects of one or all of the drugs; the outcome ranges from a reduction to an abolishment of the therapeutic effects of one or more of the drugs used. Finally, drugs can interact to produce adverse reaction(s). These effects can be the sum (addition) or more than the sum (potentiation) of the toxic effects of the individual drugs. The interaction between two or more drugs can also produce a new, adverse effect not seen with the individual drugs alone. It should also be emphasized that our failure to recognize or detect an interaction between drugs is not conclusive evidence of the absolute absence of such interaction.

Second, drug interaction, in particular an adverse interaction, can be classified either as unavoidable or avoidable interaction. If a medical condition necessitates the use of two drugs that are known to adversely affect the patient, the physician and patient have no choice and the combination must be used. The resulting interaction is then avoidable, and one hopes that the beneficial effects produced by the combination will outweigh the toxic effects. If the adverse interaction is the result of ignorance on the part of the medical profession, and if a different, safer combination of drugs could be given, this is an example of an avoidable interaction. Such interactions can be avoided by decreasing the number of drugs prescribed, by carefully checking the existing medical literature on the drugs chosen, and by consulting with other health professionals and/or pharmacologists and pharmacists.

Third, an interaction can be classified as mild or severe as well as reversible or irreversible. Fortunately, most adverse drug interactions are mild to moderate, are transient and reversible; this does not mean, however, that drug interactions should be taken lightly.

3.4. Identification of Drug Interactions

The identification of drug interactions is often quite difficult, with the exception of the more obvious and severe interactions. The first step in the identification of a possible interaction can be investigated by using laboratory animals. Drugs in different combinations at various doses can be studied in a strict, scientific manner. However, while we would easily recognize those interactions that would lead to adverse physical effects, those having adverse psychological effects are not as readily apparent with animals. That is, cognitive or emotionally disturbing effects and "painful experiences" would not be recognized to their fullest extent because the animal either does not possess these special capabilities or cannot adequately communicate such problems to the observer. In addition, animals often handle drugs differently from man, and drug interactions may occur in animals that do not occur in man, or vice versa. Finally, the animals are usually "healthy," and drug interactions of interest might occur only or become more pronounced in a given disease state. Thus, although animal studies will detect a large number of interactions and are definitely indicated on scientific, ethical, and moral grounds, they can produce both false negative or false positive results.

For ethical reasons, studies of drug interactions in man are quite limited and are largely confined to patients who receive drug combinations for medical reasons. Recognition of an interaction, then, depends to some extent on the quality of communication between the medical practitioner and the patient. Good communication is more likely when the patient is in the hospital. With the loss of the old "family doctor," and the emergence of many specialists who see the same patient simultaneously, drug interactions are more frequent and harder to identify. Even if patient–clinician communication is good, an interaction, unless severe and occurring in all individuals, is often missed, since it might be attributed to the disease or related factors. If more than two drugs are given to the patient, it is often difficult to determine which of the administered drugs causes the adverse reaction. In addition, the interaction could be the result of many other factors, including ingestion of particular foods that interact with a specific drug.

If a new interaction between drugs is recognized, the physician can report the case in the scientific literature. In addition, he is encouraged to voluntarily report this fact to the Food and Drug Administration or FDA with copies to the pharmaceutical companies involved. In a hospital setting, the pharmacy should be notified. Although most physicians do this, some physicians have become somewhat hesitant to report an adverse reaction in their patients for fear of legal action and liability problems.

At this stage, the question arises whether the initial observation of an adverse interaction is indicative of a general phenomenon or whether it occurs in this or a very few predisposed individuals. In addition, the possibility exists that the adverse effects observed during combined drug therapies are unrelated to the medication and caused by unknown factors. The decision must now be made to continue the combination or to provide warnings about its use. This is often very difficult. If these cases are not typical, and turn out to be related to factors other than the combination, witholding the medication could harm other patients who could greatly benefit from such a "safe" combination. If these cases are typical and only the tip of an iceberg of disaster, then the combination must be withheld immediately. This is often extremely difficult to judge with foresight. It is easy to judge with hindsight.

If the FDA and pharmaceutical companies have enough information to be reasonably sure that a harmful effect is indeed the result of a drug interaction, then this information is released via federal publications and the medical literature as well as in warnings in the package inserts of the drugs involved.

3.5. Incidence, Seriousness, and Avoidance of Drug Interactions

The incidence or seriousness of adverse reactions due to drug combinations varies, of course, from combination to combination, and in many instances is still difficult to assess. Decades ago, too little attention was paid to the toxicity of medications, whereas the toxicity of drugs is overplayed today. Physicians are often overly cautious about prescribing necessary drugs, and patients are unnecessarily afraid and worried about taking their medications. No drug is absolutely harmless, but should be used to the fullest extent, if a wise benefit versus risk decision favors its use as a therapeutic agent. Many more people are killed and injured in car accidents, than are hurt by drugs; and driving a car is often for pleasure only, whereas the use of drugs may be necessary for health or life.

The incidence of adverse drug interactions is difficult to assess. One study found that the average number of drugs a patient receives in a hospital is about nine (Miller, 1973). Most pharmacologists and physicians would agree that this number is too high, particularly since many of these medications were given for minor problems such as constipation, insomnia, or mild discomfort and pain.

Puckett and Visconti (1971) reported that a group of 2,422 patients admitted to a hospital in Ohio received 160 potentially interacting drug combinations; fortunately, only 7 were clinically serious. These figures might be the same, higher, or lower for other hospitals. They are probably higher for ambulatory patients who see more than one physician. Combinations of drugs are more frequently prescribed in the older population, which suffers from more ailments but which is also more likely to show adverse reactions (Laventurier, Talley, Hefner, & Kennard, 1976). The drugs most frequently mentioned in these adverse drug interactions consisted of digitalis preparations, diuretics, tranquilizers, hypnotics, sympathomimetic stimulants, steroids, anticoagulants, MAO inhibitors, phenytoin, propranolol, and others.

Fortunately, the majority of adverse drug interactions is mild and reversible when a particular drug or drug combination is discontinued. A few interactions are indeed very serious, can be irreversible, severely disabling and/or lethal. The most commonly observed adverse interactions (35%) in one of the previously cited studies (Miller, 1973) were not life-threatening and were manifested in nausea, drowsiness, diarrhea, and vomiting. However, some life-threatening effects (less than 0.1%) occurred, including cardiac arrhythmias, bone marrow depression, liver and renal failure, and electrolyte imbalance. Although we should not underestimate the seriousness of adverse drug interactions, we should not blame a drug or a drug combination for all the ill effects seen in the patient. In a retrospective study, the results of 24 individuals thought to have died as a result of a lethal drug interaction were examined. Only nine were indeed caused by an interaction, whereas the remaining deaths were due to the disease itself (Porter & Jick, 1977).

An interesting and obviously extremely important question is as follows: If a serious drug interaction is first recognized and reported in the literature, how long will it take to label this combination as dangerous? This, of course, will vary depending on the frequency with which the combination is given and the seriousness of the interactions produced. In the case of the sympathomimetic amines and MAO inhibitors, which in combination can produce a very serious rise in blood pressure or a hypertensive crisis, it took about 4 years from the first report in the medical literature to a generalized warning by agencies of the government and pharmaceutical manufacturers not to use this combination (Wooley & Hartshorn, 1977). In other cases, it might take longer or shorter, but a considerable amount of time is always necessary to verify and to validate an adverse drug reaction. The time will increase, if the combination is given infrequently, and if the adverse effects are mild or will not appear immediately. Any delayed toxic reaction, that is, a reaction that might occur months or years after multiple drug therapy had stopped, is particularly difficult to identify.

At one time, use of drug combinations or "polypharmacy" was actually promoted. It was argued that a combination of small doses of drugs with the same therapeutic effect, but different toxicities, would result in more beneficial and less adverse effects than one large dose of a single drug. While this is still valid for isolated cases, "polypharmacy" is not recommended today. Physicians are advised to use single drugs only, and to add other medications cautiously, and only if absolutely necessary. Before each addition, possible drug interactions should be checked with the pertinent medical literature, with package inserts, or with the physician's desk reference. Consultation with colleagues or other health professionals, such as pharmacists, is strongly encouraged. If no adverse interactions are known, the second or third drug can be added to the drug regimen. If drugs that can interact have to be used, a

thorough benefit versis risk decision has to be made before the potentially harmful combination is given.

The more recent arrival of affordable computer systems in medical
and pharmaceutical fields offers a unique opportunity to recognize drug
interactions and to catalog these adverse effects for informative and
preventive purposes (Greenlaw & Zeller, 1978). In an emergency setting, a microcomputer was used to identify drug interactions in 355
patients. Adverse interactions were found in 16%. As expected, the
number and severity of the interactions increased with the number of
medications taken (Karas, 1981). To assess the cost of a computerized
drug interaction monitoring system, drug therapy of patients in a 635-
bed hospital was studied over a period of 1 year. The cost per patient for
this program was $0.42. This is indeed very small; in particular, if one
considers that the authors felt that this system prevented 341 clinically
significant drug interactions from occurring (Greenlaw, 1981). Thus,
given the tremendous memory capability of computers and the declin-
ing cost as more and more computers are used, the use of such systems
in the future should readily recognize drug interactions and prevent
them from occurring.

4. Examples of Adverse Interactions between Drugs of Abuse and Therapeutic Drugs

4.1. Introduction

The foregoing discussion of the nature of drug interactions has, if
nothing else, demonstrated the complexity of the subject. Nonetheless,
we would anticipate that the concomitant use of medical and nonmedical drugs should result in a variety of adverse effects in most individuals. However, given the wide range of patterns of substance abuse,
differences in drug dose, method of administration, duration of use, and
individual differences in drug reactions, it is not currently possible to
identify and delineate all of the multiple physical and psychological
consequences that could potentially occur. We expect that such effects
are much greater than is currently realized and that more of them will be
revealed as the involved professions (physicians, psychiatrists, psychologists, nurses, social workers, drug and alcoholism counsellors, and
pharmacologists) become aware of such possibilities, and the technology for improved identification and evaluation is developed. Thus, it

is obviously not possible, at this time, to provide exact information on the incidence or prevalence of adverse interactions between substances of abuse and medical drugs. More realistically, the objective of the upcoming section of this chapter will be to point out some of the more serious known interactions. Thus, on the following pages some specific interactions of drugs or substances of abuse with prescription or over-the-counter drugs are discussed. This discussion will not only include the actual drugs but also chemicals and substances found in "street drug" preparations as impurities or substitutes. Impurities can consist of unreacted starting materials, unremoved by-products from poorly controlled synthetic procedures, or they can be deliberately added to "cut" a preparation. Some of these impurities can be fairly constant over time in a particular preparation, like quinine in heroin on the east coast, but they can also be sporadic, like paraquat in marijuana samples after the spraying of marijuana fields with this chemical. Thus, the drug of abuse, *per se*, can lead to interactions with medically indicated drugs, or the substitute/impurity can interact with the substance of abuse as well as with the therapeutic medications.

The following sections will not be comprehensive, but are intended to demonstrate some of the adverse interactions, which are possible and typical for a particular class of drugs. Some of the interactions are obvious and based on the pharmacology of the drug of abuse; for example, if a drug has stimulatory properties it will act synergistically with all drugs that also possess stimulatory properties, or it will antagonize the sedating or hypnotic actions of CNS depressants, such as tranquilizers or sleeping aids. Some of the interactions are not obvious and might appear surprising; these will show that the potential of an interaction always exists. For complete details, the reader should consult the medical literature, the materials supplied by manufacturers of the drug, or the Food and Drug Administration. A number of books on this subject are available (*Evaluation of Drug Interactions*, 1976; Hansten, 1979; James, Braunstein, Karig, & Hartshorn, 1978; Wooley & Hartshorn, 1977).

4.2. Alcohol

Alcohol is undoubtedly one of the most frequently used and abused substances. Alcohol, or the chemical compound ethanol, is a CNS depressant that, at lower doses, depresses inhibitory pathways in the brain and thus acts as a stimulant with loss of inhibitions. At higher doses, it depresses all neural pathways and acts as a sedative and hypnotic. It impairs coordination, reflex action and higher mental functions such as proper judgment and memory. However, the boundary between stimulation and depression is different from individual to individual and depends to a great deal on the mental background of the individual and environmental circumstances.

The effects of alcohol on drug metabolizing enzymes in the liver are biphasic. Acute alcohol intake can inhibit the metabolism of various medications and can lead to excessive tissue levels and toxicities of these drugs. Chronic alcohol intake can, in contrast, activate liver enzymes and speed up drug metabolism, leading to ineffective tissue levels with reduced beneficial value. For example, acute alcohol intake can potentiate the actions of the anticoagulant coumarin, leading to hemorrhagic episodes. In contrast, chronic use of alcohol can stimulate the inactivation of the same anticoagulant, rendering it less effective (Kater, 1969).

The CNS depressant effects of alcohol can be enhanced by all compounds that also possess CNS depressant activities such as barbiturates, minor tranquilizers, antihistamines, and so forth. This enhanced depression can markedly compound the ill effects of alcohol on a person's driving ability. These effects are usually additive, but potentiation may occur in sensitive individuals.

Some antidepressant tricyclics may enhance the reduction in motor skills resulting from alcohol use, and may occasionally lead to unusual and bizarre behavior. This is particularly pronounced in patients who have been just started on an antidepressant tricyclic (Landauer, Milner, & Patman, 1969).

Use of alcohol by diabetic patients on hypoglycemic drugs can lead to hypoglycemic episodes because alcohol can lower blood sugar *per se* and can decrease the effectiveness of these drugs in chronic alcoholics by increasing the metabolism of the hypoglycemic agent (Kater, Tobon, & Iber, 1969). It has also been claimed that some of the hypoglycemic drugs (e.g., chlorpropamide) can produce an "Antabuse" reaction (Asaad & Clarke, 1976).

Alcohol can irritate the gastric mucosa, and this irritant effect is enhanced in the presence of aspirin and steroids, which also irritate the stomach. This irritant effect would counteract the beneficial effects of any anti-ulcer medication.

Alcohol renders chronic alcoholics more susceptible to the adverse effects of large doses of acetaminophen, which might be the result of alcohol-induced, accelerated acetaminophen metabolism to a hepatotoxic metabolite (Emby & Fraser, 1977).

A serious interaction can occur in patients using MAO inhibitors. In this case, the interaction is not so much due to alcohol, although alcohol's metabolism can be inhibited by MAO inhibitors, but to the presence of certain amines in the alcoholic beverage (e.g., Chianti wine).

Usually, the metabolism of these amines by the enzyme MAO is very rapid, so that they cannot accumulate in the body. This is beneficial because they have the potential to increase blood pressure and heart rate. MAO inhibitors inhibit this inactivation and, thus, lead to an accumulation of these substances. The result can be a severe hypertensive crisis with internal bleeding; any intracranial bleeding can lead to severe brain damage or death.

The interaction between alcohol and Antabuse (Disulfiram) is well known. The result is an extremely unpleasant reaction, including nausea and vomiting, headache, cardiovascular problems and, in some instances, even death. This interaction is used therapeutically in very motivated, alcohol-free alcoholics. Here, Antabuse is given to the detoxified alcoholic to prevent him or her from drinking by pointing out the unpleasantness and dangers of a single drink (Kwentus & Major, 1979).

4.3. Cigarette Smoke

As common as the drinking of alcohol is the smoking of tobacco. The inhalation of tobacco smoke delivers a multitude of chemicals to the lungs and into the body.

Nicotine is the alkaloid that seems responsible for the smoking "satisfaction" and that gives some smokers a relaxed feeling. It has slight CNS stimulatory or sedative properties, depending on dose, mood, and setting. It increases heart rate and blood pressure and it constricts peripheral blood vessels; these effects make nicotine a major risk factor in vascular diseases and myocardial infarction (U.S. Department of Health, Education and Welfare, 1964). Due to these cardiovascular effects, it can counteract the beneficial effects of many antihypertensive medications or cardiac drugs. It stimulates gastric secretion and, thus, will counteract the therapeutic effects of anti-ulcer medications. It will considerably increase the risk of thrombosis in individuals using oral contraceptives (Jick, 1974; Gilman, Goodman, & Gilman, 1980).

"Tar" consists of a wide variety of organic compounds. Some of these are irritating to the lung and probably play a role in the development of smoker's cough and emphysema. However, a specific subclass of these organic compounds is the polycyclic hydrocarbons, including benzo(a)pyrene, which have been shown to be carcinogenic and are most probably involved in the higher incidence of lung cancer found in heavy smokers. These organic compounds can stimulate the metabolism and inactivation of a number of drugs (Conney, 1967). Thus, these drugs will not achieve full therapeutic efficacy. In cases of heavy smoking, the doses of many drugs, such as the minor tranquilizers, must be increased markedly.

The next major substance of smoke is carbon monoxide or CO, which binds with hemoglobin to form carboxyhemoglobin. It lowers the oxygen carrying capability of the blood, decreases oxygen supply to the brain, and, with nicotine, has been implicated in the pathogenesis of cardiovascular diseases. Its interactions with drugs are largely unknown.

In addition to the above mentioned chemicals, smoke delivers many other compounds whose pharmacological and toxicological properties are little known and whose possible interactions with prescription or over-the-counter drugs are virtually unknown.

4.4. Cannabis Sativa

The inhalation of marijuana smoke as the most common route of use delivers, like cigarette smoke, a multitude of chemicals to the body. Marijuana contains "tar" and CO, but these compounds have already been discussed.

However, the compound most specific for marijuana is delta-9-tetrahydrocannabinol, or THC. It produces the "high" feeling, an increase in pulse rate, and reddening of the conjunctiva. At higher doses, it interferes significantly with mental functions and memory. Chronic use has been claimed to decrease immunity against infections.

Preliminary evidence suggests that marijuana may impair glucose tolerance and increase the insulin requirements in diabetics (Desser, 1970). In spite of the extremely high use of marijuana by the youth of our society, its interactions with other drugs have not been examined extensively.

The possible immune suppressant action could decrease the effectiveness of any antibiotic treatment and could add to the immunosuppressant actions of steroids.

4.5. Narcotic Analgesics

The generic names of some of the drugs of this class that are most widely abused are heroin, morphine, methadone, meperidine, nalbuphine, codeine, propoxyphene, and pentazocine. Heroin is converted in the body to morphine. Morphine and meperidine are used most frequently for their analgesic properties, and codeine for its cough-depressant effects. Methadone is used therapeutically in the withdrawal of heroin and maintenance of the abusers, but is also abused by itself.

Members of this class have some common pharmacological properties. They depress specific centers of the CNS and, thus, cause analgesia, sedation, sleep, and respiratory depression. The latter effect becomes lethal at higher doses of the drugs. They stimulate the intestinal
tract, which causes constipation, and contract the pupils, which causes
"pinpoint" pupils and visual problems. In pain-free individuals, at first
experience, these drugs may often cause nausea and vomiting. In the
body, these drugs can release histamine from body stores, which causes
flushing and itching.

The narcotic analgesics produce tolerance to some of their pharmacological actions. Analgesia shows tolerance and the lethal dose increases markedly with continuous use. Constriction of the pupils and
constipation will, however, not show tolerance and will be present even
in the chronic user. The tolerance to the analgesic effect of morphine will
result in a diminished analgesic effect if another drug of this class has to
be given for pain. Since cross-tolerance exists among most members of
this class, any narcotic analgesic will produce reduced analgesic actions.
Physical dependence occurs after use of large doses after long periods of
time. The withdrawal syndrome, although highly unpleasant, is seldom
serious or life-threatening.

Constipation, a concomitant of narcotic analgesic abuse, can interfere with the absorption of many drugs and decrease the effectiveness of
their therapeutic actions.

Interactions can occur with all drugs that similarly depress the respiratory centers in the CNS. Such CNS depressant effects can be a prominent feature of a drug such as a barbiturate or it can be a more salient,
generally less recognized, property of a substance, such as some antihistamines. Antihistamines are used for different medical reasons, but
carry the property of some degree of CNS depression. Although CNS
depressant effects of drugs are mostly additive, they can potentiate each
other in sensitive individuals.

Meperidine can interact with some patients on MAO inhibitors. The
resulting symptoms are varied and may include hypotension, hypertension, as well as CNS excitation or depression (Prosser, 1968). Some
deaths have been reported due to this interaction; however, other
patients on this combination do not seem to experience an adverse interaction. Other narcotic analgesics such as morphine are not likely to
produce these severe reactions. Therefore, they could be tried in conjunction with MAO inhibitors but with great caution.

Withdrawal symptoms in patients maintained on methadone have
been reported when given the semisynthetic antibiotic Rifampin. The
reason is thought to be stimulation of methadone metabolism by the

antibiotic with a rapid drop in its blood levels (Kreek, Garfield, Gutjahr, & Guisti, 1976).

Meperidine-induced respiratory depression can be enhanced by the presence of tricyclic antidepressants. This is particularly prominent in patients with lung diseases (Gilman *et al.*, 1980). This combination is now being abused in the street, since antidepressant tricyclics are claimed to enhance the euphoric effects of narcotic analgesics.

4.6. Sedative-hypnotics

This class (phenobarbital, secobarbital, pentobarbital, butabarbital, heptabarbital, amobarbital, and nonbarbiturates such as glutethimide or methaqualon) shows as its most prominent feature CNS depression. The depression is dose-dependent and, at small doses, manifests itself as an anti-anxiety or anxiolytic action; as the doses are increased, the individual becomes drowsy, sedated, sleepy, stuporous, comatose, and at very high doses, dies of respiratory failure.

They all produce tolerance and physical dependence. Tolerance is partially caused by the stimulation of its own metabolism in the liver. Unfortunately, the lethal dose only increases moderately as tolerance to other effects such as hypnosis develops; many chronic barbiturate users die of an overdose in the misbelief that the lethal dose has increased markedly during chronic use. Physical dependence is produced by high doses and long-term use, and abrupt cessation of the drug can lead to a serious, often life-threatening, withdrawal syndrome.

Barbiturates with their prominent CNS depressant action interact with all drugs that also depress the CNS. This is usually an additive effect but, in sensitive individuals, a potentiation of these effects can be observed. An interesting case is alcohol. Chronic use of alcohol can decrease the effectiveness of barbiturates because of stimulation of barbiturate-metabolizing enzymes. However, acute alcohol ingestion and use of barbiturates can cause excessive and prolonged CNS depression, since alcohol not only has CNS depressant activity of its own, but in an acute dose, can reduce liver enzyme activity, which slows the inactivation of barbiturates (Misra, Lefevre, Ishil, Rubin, & Lieber, 1971; Rubin, Gang, Misra, & Lieber, 1970).

The therapeutic effects of a variety of drugs (e.g., beta blockers, bishydroxycoumarin, warfarin, phenylbutazone, quinidine, phenytoin, digitoxin, antipyrine, folic acid, steroids, contraceptives, antidepressant tricyclics, etc.) can be reduced by chronic barbiturate use, because barbiturates stimulate the metabolic inactivation of these medications and, thus, decrease effective tissue levels. In the cases of contraceptives, this could actually lead to ineffective levels and, perhaps, unwanted and unexpected pregnancies (Robertson & Johnson, 1976). In contrast, if the dose of a particular drug had been adjusted to account for this increased metabolism by chronic barbiturate use, and the individual stops the use of barbiturates, the inactivation of the other drugs slows and returns to "normal," and "overdose" signs and symptoms could occur, such as severe hemorrhagic episodes in the case of warfarin.

Substances that decrease the pH of the urine, like vitamin C, retard the excretion of barbiturates, which can lead to enhanced CNS depressant effects. Substances that alkalinize the urine, such as some antacids, will enhance urinary excretion of certain barbiturates, which might prove beneficial in the treatment of a barbiturate overdose.

Phenobarbital interferes with the absorption of the antifungal agent griseofulvin, which leads to decreased effectiveness of this drug.

4.7. Stimulants

This class of drugs is very varied and contains a variety of chemically different substances such as amphetamine, methamphetamine, methylphenidate, ephedrine, cocaine, and caffeine. Caffeine is not only present in coffee, but as a mild CNS stimulant in many over-the-counter preparations and soft drinks. Also, the over-the-counter appetite suppressant phenylpropanolamine falls in this class, although its CNS stimulatory effects are relatively weak. They all have in common the stimulation of the CNS; overdose or overstimulation can lead to psychotic episodes, cardiovascular problems, and seizures. They vary markedly in their own pharmacological actions. In addition to their strong CNS stimulatory effects, the sympathomimetic amines, amphetamines, ephedrine, and methamphetamine, as well as cocaine, have considerable effects on the cardiovascular system. They increase blood pressure and heart rate and can even produce arrhythmias.

Some of the pharmacological effects of the sympathomimetic amines show tolerance such as CNS stimulation, euphoria, and the lethal dose. They do not cause physical dependence in the usual sense, but abrupt cessation of chronic usage can produce changes in mood states (depression) and long periods of sleep indicative of a definitie withdrawal symptom. Suicides are possible in this poststimulatory depression.

The stimulatory actions of these drugs are increased by all drugs that also activate the CNS. On the other hand, the stimulatory actions of these drugs antagonize the anxiolytic, sedative, and hypnotic actions of the tranquilizers, sedatives, and hypnotics. This often results in a vicious cycle—use of a hypnotic in the evening, a stimulant during the day to counteract the hypnotic "hangover," and a hypnotic again in the evening to counteract the stimulatory effect of the stimulant.

The sympathomimetic amines potentiate the cardiovascular effects of antidepressant tricyclics and the MAO inhibitors (Goldberg, 1964; Roberton & Johnson, 1976). In combination with any member of these two classes, an increase in blood pressure occurs that can range from moderate to severe, even resulting in hypertensive crisis. However, the most dangerous drugs are the MAO inhibitors, which have been associated with a number of near fatalities or deaths (Lloyd & Walker, 1965).

Since excretion of amphetamine and related drugs is pH dependent, substances that make the urine alkaline can retard excretion and enhance the effects of amphetamine. On the other hand, substances that make the urine acidic will speed up excretion of amphetamine and decrease its effects. The latter process can be used in the treatment of an overdose.

Sympathomimetic drugs can reduce the antihypertensive action of many drugs. The effects of the antihypertensive drugs methyldopa, reserpine, guanethidine, and propranolol are partially antagonized by the blood-pressure-increasing, stimulatory, sympathomimetic amines (Dollery, 1965).

Sympathomimetic drugs have potent cardiac stimulatory properties and can increase heart rate. Thus, they will partially antagonize the action of beta blockers often used to reduce and normalize irregular heart rate and rhythm.

Sympathomimetic drugs can also increase blood glucose and counteract the action of hypoglycemic agents in diabetic individuals. This can lead to increased blood sugar levels.

Furthermore, chronic use of these sympathomimetic stimulants has been reported to sensitize the heart so that later exposure to general anesthetics, in particular cyclopropane and halogenated hydrocarbons, can produce severe cardiac arrhythmias (Maickel, 1979; Smith, 1980).

Caffeine, although consumed widely, has been little studied in its interactions with other drugs. Its stimulant and anxiety-producing properties would aggravate an existing anxiety and antagonize the effects of minor tranquilizers and sleeping preparations. In effect, stopping the excessive intake of coffee or caffeine-containing beverages before bedtime often eliminates the need for sleep-inducing medications. Caffeine's stimulating actions on the stomach also decreases the efficacy of anti-ulcer medications whose main purpose it is to decrease secretion and to relax the stomach musculature.

4.8. Hallucinogens

Hallucinogens are compounds that, at relatively low doses, cause arousal, excitation, and distortion of perception, and at higher doses produce hallucinations as their most dominant feature. Many have sympathomimetic actions and can increase heart rate and blood pressure as well as decrease seizure threshold. Their use can also result in panic, anxiety, psychosis, and seizures. The hallucinogens differ widely chemically and pharmacologically. Members of this group include LSD, mescaline, dimethyltryptamine, psilocyn, phencyclidine or PCP, and others. Although widely abused, these compounds have not been studied scientifically in humans, particularly during chronic use, and many of their adverse reactions as well as drug interactions are unknown.

These compounds would most likely antagonize all drugs that are used in the treatment of anxiety, depression, and schizophrenia (i.e., minor tranquilizers, antidepressants, and antipsychotics). They are likely to reduce or abolish the effectiveness of anticonvulsant medication at higher doses.

Compounds with sympathomimetic properties will show effects similar to those seen with the sympathomimetic stimulants such as amphetamine (which can also produce hallucinations in higher doses). The same drug interactions as described earlier are therefore possible with these hallucinogens.

4.9. Solvents

A wide variety of volatile organic solvents, lighter fluids, glues, and nail polish removers are being abused, mostly by inhalation. The pure chemicals in these preparations include acetone, chloroform, carbon tetrachloride, toluene, xylene, benzene, and others. They produce exhiliration, irritation, disorientation, and distortions of perceptions, but rarely hallucinations. In large doses they markedly depress the CNS. The halogenated, organic solvents can sensitize the heart, and chronic exposure to these chemicals can produce liver and kidney failure. Some of these compounds have been shown to be carcinogenic.

Exposure to trichloroethylene followed by alcohol produces a syndrome called "degreaser's flush" and consists of a flushing, lacrimation, tachypnea, and blurred vision (Stewart, Hake, & Peterson, 1974).

Since the action of many drugs is terminated by liver inactivation and renal excretion, changes in half lives of many prescription drugs can be expected to be slower in individuals who chronically abuse these solvents and who have sustained some degree of liver or kidney damage.

Methyl ethyl ketone, if abused alone, is not a neurotoxin per se; however, this chemical can markedly potentiate the neurotoxicity of other chemicals such as hexane.

A substance that is not a solvent, but often abused, is amyl nitrite. This and similar substances produce a "flush" and feeling of euphoria in some users. In the presence of alcohol (even if the alcohol is only contained in medicinal fluids or over-the-counter preparations), an interaction could lead to hypotension with the possibility of fainting and injuries resulting from a fall.

4.10. Substitutes and Impurities

Substances bought on the street often contain other substances and/or are seldom pure unless they were samples manufactured by pharmaceutical companies (Schnoll, Cohn, & Vogel, 1972). Impurities are either added on purpose to "cut" the actual drug quantities, or are derived from unskilled or sloppy synthetic procedures of the drug samples in "basement" laboratories. Every street sample is suspect to adulteration. A few substitutes and additives and their possible interactions will be discussed below.

MAO inhibitors elevate mood even in undepressed individuals and, thus, are sometimes sold for this reason on the street under other names. As mentioned before, this substitute drug can cause the most serious interaction with prescription and over-the-counter drugs as well as food components in the unsuspected user. Even the consumption of innocent vegetables can lead to a disaster.

Phenylbutazone is an anti-inflammatory agent with slight CNS stimulatory and euphoric properties. For the latter reason, it is used on the street as a substitute for CNS stimulants. However, it can enhance gastrointestinal bleeding in certain individuals if taken with alcohol, steroids, or other anti-inflammatory agents. It can interfere with hypoglycemic drugs and produce hypoglycemia. It readily displaces anticoagulants from albumin binding sites in the blood, resulting in hemorrhagic episodes.

Substance abusers often use additional medications to counteract some of the side effects of the abused substance. A most frequently used preparation by heroin users is a laxative, such as Maalox or milk of magnesia, to bring relief from heroin-induced constipation. These over-the-counter preparations are not as harmless as often thought. Long-term and excessive use of these preparations can upset the electrolyte balance of the body, particularly in the presence of a diuretic medication. The metal ions interfere with the effectiveness of some antibiotics.

Impurities can even interact with the drug of abuse *per se*. Some unexpected deaths of heroin users have been blamed on an interaction between heroin and quinine, which is frequently added to the samples of this street narcotic on the east coast (Weisman, Lerner, Vogel, & Schnoll, 1973).

Aspirin is also frequently mixed into street preparations. Although serious drug interactions between aspirin and other drugs are rare, aspirin could interact and decrease the therapeutic effects of medications such as anti-ulcer drugs or anticoagulants. The addition of sugar or starch to street drug preparations is of no consequence in the healthy user, unless it is taken by a diabetic, where it could interfere with the insulin or hypoglycemic drug requirements of the patient.

Substitution of anticholinergic drugs such as atropine, scopolamine, or some of their derivatives, can lead to enhanced anticholinergic effects (mydriasis, cycloplegia, tachycardia, urinary retention, constipation, and dry mouth and skin, with possibility of overheating in hot weather) in conjunction with drugs that possess similar properties, such as antihistamines, antipsychotics, and antidepressants.

Presence of quinidine, procainamide, or lidocaine, as found in some heroin preparations, can lead to serious cardiac abnormalities if taken with other cardiac depressants.

5. Summary

Little is known about the delicate processes of human biochemistry and physiology. Less is known about the effects of a particular drug on these processes and its exact mechanism. Much of our knowledge, in spite of major scientific advances, still rests on empirical observations. Thus, it becomes clear that even less is known about the interactions of more than one drug with the body and among themselves.

Nevertheless, we know that the drug and its chemical and pharmacological properties determine to a large extent its therapeutic and toxic reactions. In addition, individual characteristics of the patient, such as race, age, or a particular disease, kidney and liver function, serum protein levels, urinary pH, and dietary and environmental factors can modify the action of the drug and render it more effective or toxic.

This chapter has tried to establish a basis for a better understanding of drug interactions by pointing to possible sites where drugs can interact and by citing some relevant examples of clinically significant drug interactions.

To avoid or minimize drug interactions, a series of simple steps can be followed. If at all possible, only one drug should be given to a patient. This is often possible by choosing a particular drug or changing the schedule of administration. If the clinical situation demands the use of two or more drugs, a thorough examination of their pharmacological and toxicological profiles must be performed. This might involve only the treating physician. But since the possibilities of drug interactions is so vast that no one individual can remember all of them, it is strongly recommended that a physician consult with other health professionals (physicians, pharmacologists, pharmacists) as well as with other resources (literature, computer). The effort, time, and cost are negligible in comparison to sparing a patient the effects of a detrimental drug interaction. All of these efforts may be wasted, however, when the patient is a clandestine drug abuser, or, in some cases, if he has a history of substance abuse unknown to the physician.

6. References

Albert, A. Avidity of terramycin and aureomycin for metallic cations. *Nature*, 1953, *172*, 201.

Asaad, M. M., & Clarke, D. E. Studies on the biochemical aspects of the disulfiram-like reaction induced by oral hypoglycemics. *European Journal of Pharmacology*, 1976, *35*, 301–307.

Conney, A. H. Pharmacological implications of microsomal enzyme induction. *Pharmacological Review*, 1967, *19*, 317–366.

Desser, K. B. Effects of "speed" on the juvenile diabetic. *Journal of the American Medical Association*, 1970, *214*, 2065.

Dollery, C. T. Physiological and pharmacological interactions of antihypertensive drugs. *Proceedings of the Royal Society of Medicine*, 1965, *58*, 983–987.

Emby, D. J., & Fraser, B. N. Hepatotoxicity of paracetamol enhanced by ingestion of alcohol. *South African Medical Journal*, 1977, *51*, 208–209.

Evaluation of drug interactions. Washington, D.C.: American Pharmaceutical Association, 1976.

Gilman, A. G., Goodman, L. S., & Gilman, A. *The pharmacological basis of therapeutics*. New York: Macmillan, 1980.

Goldberg, L. I. Monoamine inhibitors. *Journal of the American Medical Association*, 1964, *190*, 456–462.

Greenlaw, C. W. Cost of a computerized drug interaction screening system. *American Journal of Hospital Pharmacy*, 1981, *38*, 421–424.

Greenlaw, C. W., & Zeller, D. D. Computerized drug monitoring systems. *American Journal of Hospital Pharmacy*, 1978, *35*, 1031–1032.

Hansten, P. D. *Drug interactions*. Philadelphia: Lea and Febiger, 1979.

James, J. D., Braunstein, M. L., Karig, A. W., & Hartshorn, E. A. (Eds.). *A guide to drug interactions.* New York: McGraw-Hill, 1978.

Jick, H. Smoking and clinical drug effects. *Medical Clinicians of North America,* 1974, *58,* 1141–1148.

Karas, J. Drug interactions in an emergency room. *Annals of Emergency Medicine,* 1981, *10,* 627–630.

Kater, R. M. H. Increased rate of clearance of drugs from the circulation of alcoholics. *American Journal of the Medical Sciences,* 1969, *258,* 35–39.

Kater, R. M. H., Tobon, F. T., & Iber, F. L. Increased rate of tolbutamide metabolism in alcoholic patients. *Journal of the American Medical Association,* 1969, *207,* 363–365.

Kreek, M. J., Garfield, J. W., Gutjahr, C. L., & Guisti, L. M. Rifampin-induced methadone withdrawal. *New England Journal of Medicine,* 1976, *294,* 1104–1106.

Kwentus, J., & Major, L. F. Disulfiram in the treatment of alcoholism. *Journal of Studies on Alcohol,* 1979, *40,* 428–446.

Landauer, A. A., Milner, G., & Patman, J. Alcohol and amitriptyline effects on skills related to driving behavior. *Science,* 1969, *163,* 1467–1468.

Laventurier, M. F., Talley, R. B., Hefner, D. L., & Kennard, L. H. Drug utilization and potential drug–drug interactions. *Journal of the American Pharmaceutical Association,* 1976, *16,* 77–81.

Lloyd, J. T. A., & Walker, D. R. H. Death after combined dexamphetamine and phenelzine. *British Medical Journal,* 1965, *2,* 168–169.

Maickel, R. Acute amphetamine abuse, problems during general anesthesia for neurosurgery. *Anesthesiology,* 1979, *34,* 1016–1019.

Miller, R. R. Drug surveillance utilizing epidemiologic methods. *American Journal of Hospital Pharmacy,* 1973, *30,* 584–592.

Misra, P. S., LeFevre, A., Ishil, H., Rubin, E., & Lieber, C. Increase of ethanol, meprobamate and phenobarbital metabolism after chronic ethanol administration in man and in rats. *American Journal of Medicine,* 1971, *51,* 346–351.

Mond, E., & Mack, I. Cardiac toxicity of iproniazid (marsilid): Report of myocardial injury in a patient receiving levarterenol. *American Heart Journal,* 1960, *59,* 134–139.

Porter, R. S., & Jick, H. Drug related deaths among medical inpatients. *Journal of the American Medical Association,* 1977, *237,* 879–881.

Prosser, C. D. G. The use of pethidine and morphine in the presence of monoamine oxidase inhibitors. *British Journal of Anesthesia,* 1968, *40,* 279–282.

Puckett, W. H., & Visconti, J. A. An epidemiological study of clinically significant drug interactions in a private community setting. *American Journal of Hospital Pharmacy,* 1971, *28,* 147–153.

Robertson, Y. R., & Johnson, E. S. Interactions between oral contraceptives and other drugs: A review. *Current Medical Research Opinions,* 1976, *3,* 647–679.

Rubin, E., Gang, H., Misra, P. S., & Lieber, C. S. Drug metabolism by acute ethanol intoxication. *American Journal of Medicine,* 1970, *49,* 801–806.

Schnoll, S. H., Cohn, R. D., & Vogel, W. H. A rapid TLC screening procedure for various abused psychotropic drugs. *Journal of Psychedelic Drugs,* 1972, *5,* 75–78.

Siegel, S., Hinson, R. E., Krank, M. D., & McCully, J. Heroin "overdose" death: Contribution of drug-associated environmental cues. *Science,* 1982, *216,* 436–437.

Smith, D. S. Amphetamine abuse and obstetrical anesthesia. *Anesthesia and Analgesia,* 1980, *59,* 710–711.

Stewart, R. D., Hake, C. C., & Peterson, J. E. Decreaser's Flush. *Archives of Environmental Health,* 1974, *29,* 1–6.

U.S. Department of Health, Education and Welfare. *Smoking and health: Report of the Adviso-*

ry Committee to the Surgeon General of PHS. Washington, D.C.: U.S. Government Printing Office, 1964.

Vogel, A. V., Goodwin, J. S., & Goodwin, J. M. The therapeutics of placebo. *American Family Physician,* 1980, *22,* 105–109.

Weisman, M., Lerner, N., Vogel, W. H., & Schnoll, S. H. Quality of street heroin. *New England Journal of Medicine,* 1973, *289,* 698.

Wooley, B. H., & Hartshorn, E. A. (Eds.). *Drug interaction-induced adverse reactions.* Miami: Symposia Specialists, 1977.

10

Identification and Treatment of Substance Abuse Problems in the Emergency Room

EDWIN ROBBINS, STEVEN B. KATZ, AND MARVIN STERN

1. Overview

While it may not be necessary for a culture to be involved with substance abuse, few groups have remained absolutely free (Cohen, 1982a; Nahas, 1982). In reviewing substance abuse, it becomes evident that there is considerable waxing and waning in the popularity of substances chosen and the groups that abuse. In Ireland, there was a fad with ether in the late 1800s, which resulted in passage of laws curtailing its use. In America, we have witnessed a change in opiate addiction, from middle-class women abusing tonics, to lower-class youths, primarily male, becoming involved with heroin. The recent resurgence of cocaine, especially among the more affluent, has become a matter of national concern. An excellent review of drug abuse can be found in Nicholi (1983).

Endeavoring to understand the attractiveness of intoxicants, researchers have offered them to animals on an *ad lib* basis (Jaffe, 1980). Cocaine is frequently taken to the point of inanition or death. Some animals have pushed a lever 4,000 times to obtain a cocaine reward.

EDWIN ROBBINS, STEVEN B. KATZ, AND MARVIN STERN • New York University Medical Center, 560 First Avenue, New York, New York 10016.

Opiates are less attractive. Animals gradually increase their intake over several months before learning how to avoid gross toxicity or withdraw.

The richness of drug experience has been described in the popular (Castanada, 1968; Huxley, 1963; Leary, Metzner, & Alpert, 1964) and scientific (Blum & associates, 1964; Cohen, 1965; Masters & Houston, 1966) literature.

In this chapter, the major groups of substances abused by humans will be considered. Attention will be paid to physiologic characteristics, psychological changes, and the management of acute intoxication, chronic conditions, withdrawal, and overdose.

1.1. Identification of the Abuser

Since so many people seeking help in a psychiatric setting may also be involved with substance abuse, questions regarding use of alcohol and drugs should be asked routinely in the emergency room (ER) and again after admission. The most common physical indicators of substance abuse are drowsiness and nodding, pinpointed or dilated pupils, flushed, moist, or dry skin, ecchymoses, burn marks or needle tracks, fractures, and coordination difficulties. The most pronounced behavioral changes are silliness, pressured speech, emotional lability, boisterousness, and aggressiveness. Many of the *acutely disturbed people* in ER's today are intoxicated, not schizophrenic. Psychological changes may be egosyntonic and include euphoria, tranquility, peacefulness, heightened awareness, enhanced sensory facilitation, and increased social relatedness. Egodystonic responses can be irritability, anger, depression, paranoid ideation, panic, anxiety, jitteriness, poor judgement, impulsiveness, decreased intellectual performance, and confusion.

Because of the emphasis placed upon the enhancement of self-understanding when psychedelics were first becoming popular, we asked abusers, primarily hospitalized as a result of adverse reactions, if they noted psychological benefits (Frosch, Robbins, Robbins, & Stern, 1967). The few who said they did were neurotics, concerned with "finding themselves." We felt they were rationalizing their drug abuse. Today, many of these people are in the group seeking to find their innermost selves through the use of marijuana. The majority of the abusers freely acknowledged that their motivation was to become high. It was of interest to note that most of them were quite limited in their description of their feelings or impressions while intoxicated, and often seemed unable to distinguish between the effects of the various compounds they had taken. This is in accord with the difficulty that abusers report in identify-

ing what substances they have taken when other compounds are substituted by dealers.

1.2. Laboratory Testing

Laboratory testing is of great value when a person is suspected of
being an abuser. In Table 1, substances that most laboratories are readily

Table 1
Substances Tested in Toxicology Laboratories[a]

Common name	Blood or urine[b]	Urine only[c]	Length of time detected in urine[d]
Antidepressants			
Elavil	—	x	1 day
Tofranil	—	x	1 day
Hypnotics and sedatives			
Barbiturates	x	—	5 days
Doriden	x	—	2 days
Librium	—	x	1 day
Miltown, equanil	—	x	2 days
Quaalude	—	x	1 day
Valium	—	x	1 day
Opiates and synthetics			
Codeine	—	x	3 days
Demerol	—	x	1 day
Heroin	—	x	5 days
Methadone	—	x	2 days
Morphine	—	x	5 days
Phenothiazines			
All	x	—	2 days
PCP	x	—	1 day
Stimulants			
Amphetamine	x	—	1 day
Cocaine	—	x	1.5–2 hours
Miscellaneous			
Alcohol	x	—	1 day
Quinine	x	—	5 days
Salicylates	x	—	3 days

[a]These substances are tested in most laboratories. Other commonly abused substances, such as marijuana, hallucinogens, ritalin, and medications such as other tricyclics or haloperidol cannot be tested
in many places.
[b]Quantitative analysis available for blood. Except for sedatives, most substances may not be isolated
after several hours have lapsed.
[c]Qualitative analysis only.
[d]Length of time (Done, 1968). Information supplied by Mr. Slepstein of Bellevue Hospital Toxicology
Laboratory.
Table adapted from Robbins, Agus, & Stern (1976).

able to identify from samples of blood, urine, or gastric juices are listed. Few laboratories are equipped to screen all admissions routinely, and many cannot perform highly sensitive tests for marijuana. In recent years, more sensitive tests for marijuana have been developed (Dakis, Pottash, Annitto, & Gold, 1982) so that it is possible to measure THC in the urine, but there is no correlation between a positive finding and intoxication. Efforts to correlate saliva levels or breath levels with the "stoned" state are in process (Cohen, 1983). On occasions it is advantageous to identify specific substances that are in the possession of the user. The local department of health laboratory is usually helpful. If it cannot perform the analysis, some commercial laboratories may be of assistance.

1.3. Working Definitions

Substance abuse is characterized by *a period of abuse of more than one month; complications* such as *blackouts or overdoses; inability to control, reduce, or stop* intake in spite of multiple efforts to gain control; *need for daily intake* in order to function adequately; and *impairment in social or occupational functioning* as a result of drug abuse. *Intoxication* is *maladaptive behavior secondary to substance abuse.* There are changes in perception, wakefulness, attention, thinking, judgement, emotional control, and psychomotor behavior. *Recreational use* that does not result in maladaptive behavior is not considered intoxication even though high blood levels of the substance are present. *Patterns of use* may be continuous, episodic, periodic, or in remission. *Frequency* of use, *dosage,* and *time elapsed* since the last dose all play a role in determining the course of a drug experience.

Addiction, dependence, and tolerance describe different aspects of the drug-using experience. *Addiction* is the practice of using a habit-forming substance to such an extent that cessation of use results in severe trauma, such as withdrawal symptoms. *Dependence* may be psychic or physical. In the case of psychic dependence, there is a feeling or belief that it is necessary to partake of a substance in order to achieve a "normal" feeling state. In the case of physical dependence, there are physical changes in the body that lead to withdrawal symptoms if the substance is withheld. *Tolerance* is the body's adaptation to the effects of a substance.

For substances characterized by significant tolerance, such as barbiturates and heroin, an increasing amount must be taken to achieve the early psychological effects. As tolerance increases, the habit may become too expensive or the margin of safety too limited, leading some abusers to detoxify in order to reduce their tolerance. When abuse is resumed, an overdose may occur if an error is made in judging the extent of tolerance. For some substances, such as the hallucinogens, there is an extensive refractory period, which makes it impossible for the second dose of the drug to have a comparable effect to the first. Because stimulants do not have a refractory period, abusers have learned that it is possible to enjoy a succession of highs (runs) and postpone the moment of letdown and depression. For some, this leads to compulsively taking all available drugs.

1.4. Clinical Syndromes

Delirium is an acute, disordered state with an altered, fluctuating level of consciousness, often associated with fever. The syndrome is usually associated with systemic illness, hormonal or metabolic dysfunction, or withdrawal from drugs or alcohol. It has been clinically observed that former alcoholics are more likely to become delirious in conjunction with infectious illness than are nonalcoholics.

Altered level of consciousness is manifested by defects in recent memory, difficulty in learning new material, increased concern to the point of preoccupation with the environment, and misinterpretation of events. Attention span is short and patients may "wander off" while being interviewed in response to internal or external distractions, thus making it necessary to interrupt and redirect them. Poor judgement and sensory misinterpretations are present. There may be expression of lurid tales, sometimes with paranoid content. Through careful questioning and the liberal use of imagination, it is possible to reconstruct the events that led to the patient's distortions. Visual hallucinations are most common, followed by auditory and then other somatic phenomena. Among motor manifestations are restlessness, frequently manifested by picking at skin or bedclothing, or tremulousness. Some people are quiet and withdrawn so that it is not evident that they are delirious. Myoclonus is uncommon. *Grand mal* seizures may occur in the early stages of alcohol or sedative withdrawal. If they do, electroencephalograms may be characterized by slow wave patterns, but these are not always present.

Amnestic syndromes may be acute (self-limited) or chronic (irreversible). The essential feature is impairment of short- and long-term memory in a state of normal consciousness. If there is clouding of consciousness, delirium is present. If there is a general loss of intellectual capacity, dementia is present. In addition to the problems of learning new material and recalling old, there may be disorientation, confabulation, lack of insight, denial of deficit, apathy, lack of initiative, blandness, and flattening of affect. Accompanying physical symptoms in alcoholics may include peripheral neuropathy, cerebellar ataxia, opthalmoplegia, and nystagamus as well as the common gastrointestinal and hepatic complications.

Dementia involves memory loss, impairment of abstract thinking, poor impulse control, deficits in judgement, and disturbances in higher cortical function. When language is affected, speech may be vague, stereotyped, imprecise, or aphasic. Personality deterioration is present, though there may be efforts to conceal or compensate for losses early in the course of the illness.

Other syndromes associated with substance abuse are *prolonged psychoses, flashbacks, blackouts,* and *development of amotivational syndromes.*

A *prolonged psychosis,* similar to schizophrenia, may emerge imperceptibly after a period of acute intoxication or chronic abuse of many substances, including alcohol, bromides, LSD, and PCP (Cohen & Ditman, 1963; Frosch, Robbins, & Stern, 1965; Rainey & Crowder, 1975). It may persist long after the symptoms of withdrawal should have remitted. Among explanations offered for prolonged psychosis are personality instability, a drug-induced metabolic alteration, a learned phenomenon, and a flight from reality followed by a desire to remain psychotic.

Flashbacks are the spontaneous reappearance of thoughts, mood, or perceptual changes that occurred while under the influence of drugs. Some abusers welcome them because of their pleasant connotations. Others find the experience disconcerting or frightening, and a few respond with terror or massive anxiety. Individual episodes are of short duration and, if there is no pattern of continued abuse, cycles generally do not persist for more than several months. The etiology of the flashback is uncertain. It is not likely that it is a persistent biochemical phenomenon. Sometimes, it has been associated with stress or anxiety, and may be a learned reaction, precipitated by subliminal stimuli. Flashbacks also follow use of drugs, including marijuana.

Blackouts are periods in which there is no awareness of functioning. The lapses may be of short (seconds) or relatively long (hours) duration. They occur in response to heavy use of many substances, including alcohol, amphetamines, and phencyclidine (PCP).

The *amotivational syndrome,* which is especially deleterious in adolescence, occurs when abusers become diverted from reality-based goals in favor of drug-induced fantasies (Robbins & Robbins, 1977; Robbins, Robbins, Frosch, & Stern, 1967). Considerable time may be spent in being high. Other effects are negative changes in personality, a reduction in the number and quality of friends, and alterations in the sleep-wake cycle, especially during periods of amphetamine abuse.

Idiosyncratic effects or paradoxical reactions, which are the opposite of the usual clinical effect of a medicine, may also occur. In clinical practice, physicians use these reactions to therapeutic advantage. One example is the use of amphetamines to calm hyperactive children. Barbiturates, which paradoxically may excite children or the elderly, are exploited by the abuser, who learns how to minimize their sedative effects while enhancing their excitatory ones. Paradoxical reactions also occur during withdrawal. Behaviorally, opiate and barbiturate addicts may become excited, rather than sedated. Amphetamine abusers "crash" with exhaustion, rather than remain stimulated. Paradoxical physical changes in opiate abusers include dilated, not pinpoint, pupils in coma and diarrhea instead of constipation during withdrawal.

In the remainder of this chapter we will discuss the symptomatology associated with each of the popularly abused substances, and the recommended treatment course for acute reactions and overdose.

2. Alcohol

2.1. Introduction to Symptoms of Alcohol Abuse/Alcoholism

Alcoholics can present in a myriad of ways. The chronic alcoholic is classically described as having an acneform face, bulbous red nose, hepatic palms, staggering gait, "down-and-out" appearance, neurologic deficits, or the smell of alcohol on his/her breath. Few people in the ER will appear this way. Many alcoholics seen in public hospitals will be "down and out" in their luck and will be poorly dressed, sometimes dirty, though some will be neat and clean. Behavior can range from being highly agitated to withdrawn. Many times depression is obvious and suicidal ideation may be present. Some will come to the hospital with an awareness that they have been out of control, while others will be seen for an unrelated illness and never mention that they drink excessively. Sometimes people in the latter group are not aware of their dependence, and they may enter into withdrawal if hospitalized for treatment of a medical condition. In a hospital, it is advisable to consider alcoholism or drug abuse in the differential diagnosis when septicemia, meningitis, pneumonia, gastritis, hepatitis, pancreatitis, cirrhosis, peripheral neuritis, or spinal cord disease are diagnosed. Alcoholism may be overlooked unless specific questions regarding its use are asked. When an alcoholic who has a fever of unknown origin is examined, it is necessary that he/she be observed closely because serious systemic illness such as meningitis or pneumonia, which has not yet become clinically apparent, may be present.

2.2. Symptomatology and Treatment of Major Psychiatric Syndromes

2.2.1. Acute Intoxication (Simple Drunkenness)

Acute intoxication is the most common reason for which alcoholic people are taken to an ER. They may have become boisterous, assaultive, destructive, or have had a driving accident and the police wish them to be observed. The correlation between behavior and ethanol blood level can be described with some consistency. By the time a level of 50 mg% is achieved, cortical controls have lessened and a sense of tranquility or drowsiness is present. Muscular coordination begins to be lost above 50 mg%, and by 100 mg% reaction time is prolonged, driving is dangerous, and people are considered to be legally drunk. The clinical diagnosis of acute intoxication rests upon at least one of these physiological signs: slurred speech, incoordination, unsteady gait, nystagamus or flushed face; and one of these psychological signs: mood change, irritability, loquacity, or impaired attention (Spitzer, 1980). When the blood level rises above 200 mgm%, temperature regulation is lost and amnesia, hypothermia, and early anesthesia occur. Though coma, respiratory failure, and death are common above 400 mgm%, a few people have survived above levels of 700 mgm% (Goldfrank, Flomenbaum, Lewin, & Weisman, 1982).

Treatment of acute intoxication is symptomatic. In many instances, providing a supportive environment will suffice. If patients are assaultive and disturbed, benzodiazepines and restraints may be indicated. Lethargic people should be permitted to sleep off their drunk. Sometimes, it may be necessary to rouse them to prevent stupor from developing. The use of caffeine sodium benzoate (0.5 mgm intramuscularly, or dextroamphetamine, 5 mgm orally) may be helpful (Salamon, 1980). In semistuporous people, vital signs should be observed at half-hourly intervals. The patient should lie on his/her side to minimize the likelihood of aspirating vomitus. The respiratory passage should be monitored to make certain that no obstruction to breathing develops.

2.2.2. *Toxic Psychotic States and Alcohol Idiosyncratic Intoxication*

Toxic psychotic states and alcohol idiosyncratic intoxication are to be considered when there is an acute brain syndrome manifested by confusion, bizarre thought content, severe agitation, hallucinations and/or delusions. Pathologic excitement is most frequent in young males between the ages of 18 and 25. During these periods, there can be uncontrollable assaultiveness. Generally, there is amnesia for the episode. While these symptoms may be associated with alcohol withdrawal states alone, they also may be the result of the interaction of alcohol and other substances. The duration of drug-induced psychosis is time related. Diagnosis is made by history and toxicologic studies.

Treatment of toxic psychotic states is primarily symptomatic and supportive and similar to that of all other alcohol-induced syndromes. Restraints may be necessary if there is danger to the patient or to others. When symptoms cannot be controlled by sedation with a benzodiazepine, the use of a major tranquilizer is indicated. Oral administration is preferable, but if the patient is uncooperative, intramuscular (IM) medication is necessary. We prefer to use low doses, of no more than 5 mgm per injection for haloperidol or 25 mgm for chlorpromazine. The drug should not be administered more frequently than every 30 minutes. By 2 hours, nearly all patients should have responded and be willing to accept medication orally.

2.2.3. *Withdrawal Syndrome*

Alcohol withdrawal usually does not occur until a person has been drinking heavily for 5–10 years. It is unusual below the age of 30. Most often, withdrawal follows abrupt cessation of drinking, either because of a conscious attempt to stop or because of an inability to retain alcohol secondary to gastritis or other medical conditions. Symptoms generally remit between 5 and 7 days. In its simplest form, withdrawal is characterized by tremulousness of hands, tongue, and eyelids, which begin within 8 hours after the last drink. By 24 hours (though there may be an occasional delay up to 8 days), marked tremors, which may be accompanied by insomnia, hyperactivity, nightmares, or illusions develop. Hallucinations are uncommon in simple withdrawal. If they occur, they are vague in contrast to those of alcoholic hallucinosis and alcohol withdrawal delirium. At least one of the following psychological signs— mood change, irritability, loquacity, or impaired attention—and one of these physiological signs—slurred speech, incoordination, unsteady gait, nystagamus, or flushed face—must be present for the diagnosis to be made (Spitzer, 1980). Other physical concomitants are gastritis or pancreatitis and autonomic nervous system involvement—tachycardia, sweating, or elevated blood pressure. Tremor is a common symptom and must be distinguished from idiopathic tremor, tremor induced by medication (amphetamine, epinephrine, lithium, phenothiazine, phenytoin, reserpine, thyroid hormone), substances of abuse (amphetamine, cocaine, PCP), or industrial substances (arsenic, lead, mercury, phosphorous) (Goldfrank, Flomenbaum, Lewin, & Weisman, 1982). Carbon monoxide, which may cause tremor, nystagmus, and occasionally blindness, also must be considered in the differential diagnosis, especially if there has been a history of unconsciousness.

While we prefer to *treat* people who are withdrawing within the hospital and initiate a "detoxification" therapeutic process for selected patients who are undergoing mild withdrawal, it is acceptable to initiate an identical treatment regimen on an outpatient basis. Patients should be instructed to return to the hospital if symptoms progress or if "horrors" or signs of physiological crisis develop. A benzodiazepine is the drug of choice at present. Phenothiazines do not prevent the development of delirium tremens and should be used only when benzodiazepines cannot control behavior.

As the patient responds to treatment, the benzodiazepines should be tapered so that they can be discontinued when the patient is fully recovered. This will reduce the likelihood of dependence.

Additional measures consist of assessing the need for replacement of fluids and electrolytes. People in mild withdrawal may be overhydrated, while those in more severe withdrawal may be dehydrated (Schenker, 1982). Electrolyte studies should be done at the time of initial assessment, and replacement therapy should be provided as indicated. If electrolytes are administered, the severity of deprivation and the patient's clinical status determine the route of administration. Monitoring of blood levels and observation for side effects should be done assiduously.

In cases of severe withdrawal, it may be necessary to provide glucose intravenously for one or two days to spare endogenous protein breakdown. Proponal (Inderal) reportedly has been helpful in controlling alcoholic tremor, but this treatment has not received FDA approval.

Withdrawal seizures ("rum fits") generally do not appear before 12 to 48 hours and are more likely to occur in epileptics than in people with normal EEG's. At times, it is necessary to distinguish between people who are having true *grand mal* seizures and those who are simulating in the hopes of obtaining diazepam or another sedative. Epilepsy may complicate withdrawal. If there is a past history of seizures during withdrawal, phenytoin (Dilantin) should be given prophylactically. This treatment should be sustained if the patient is also an epileptic. Otherwise, it may be discontinued after several days because the likelihood of seizure is low.

Status epilepticus responds to intravenous diazepam, 5–10 mg, which may be repeated at 10 to 15 minute intervals until 30 mg has been given (Rall and Schleifer, 1980). If status epilepticus does not respond, intravenous phenytoin, 18 mg/kg, mixed in a volutrol with 100 mg of normal saline and administered at a rate of no more than 50 mg per minute, should be given (Mack, 1983).

2.2.4. Alcohol Hallucinosis

Alcohol hallucinosis occurs in people who have drunk heavily for more than a decade. Symptoms generally appear between 3 and 5 days, but may not appear before 2 weeks have passed since the reduction or cessation of drinking. Hallucinations usually remit within one week, but occasionally a prolonged psychosis occurs. Vivid, mostly auditory, hallucinations are the most prominent symptom. When the voices are distinct, they most often are accusatory or threatening and speak of the patient in a derogatory manner. Frequently, people remain awake all night in a state of terror. Symptoms are usually short-lived. People who continue to hallucinate after a week may develop a prolonged psychosis essentially indistinguishable from paranoid schizophrenia.

Although the sensorium is described as being clear, there may be some minor memory deficits, which are secondary to other symptoms of alcoholism, frequently head trauma.

Treatment depends upon the patient's anxiety and level of disturbance. Rest and medication, most often a benzodiazpine, are indicated. Markedly anxious patients should be hospitalized. Others may be treated overnight in a holding unit or discharged for outpatient treatment.

2.2.5. Alcohol Withdrawal Delirium (Delirium Tremens)

Alcohol withdrawal delirium is a serious complication of alcoholism. Mortality figures, which in the past approached 15%, are considerably lower now and generally reflect the prognosis of the underlying medical or surgical disease.

Psychological symptoms of withdrawal delirium include clouded consciousness with fluctuating levels of attention, disorientation and memory impairment, and changes in perceptual functioning (illusions, misinterpretations, and visual or other hallucinations). Restlessness and insomnia are most pronounced during periods of anxiety, secondary to hallucinations and misperceptions.

Physically, changes include tremor and autonomic hyperactivity, such as tachycardia, dilated pupils, sweating, and hypertension. Temperature must be monitored carefully. Medical care is necessary if it rises above 101°F. and an emergency exists above 103°. Temperature may be a function of either the delirium, the use of restraints, or an underlying medical disease, which must be diagnosed and treated.

Treatment of withdrawal states has several aspects. The environment should be warm and supportive. This will allay anxiety and enable recovery to be smoother. A companion and soft lighting at all times enables better environmental contact, reducing the likelihood for the appearance of perceptual distortions, and decreasing the need for medication and restraints. When agitation occurs, benzodiazepines, either orally or intramuscularly, can be of benefit. Chloral hydrate, in addition, may be helpful in promoting sleep. Restraints should be avoided since they will be misunderstood, and, as the patient struggles against them, exhaustion and increased temperature may develop. However, they must be used if there is a danger to the patient or to others. Fever requires aggressive treatment. Alcohol rubs, ice water, and other preventative measures must be vigorously initiated.

2.2.6. Alcoholism and Depression

The risk of suicide is great in depressed alcoholics, especially if they also utilize barbiturates. Significantly depressed alcoholics should be admitted to an inpatient service. Although there are instances in which individuals feign suicidal thoughts in an effort to seek admission, if there is uncertainty about their being manipulative, it is advisable to err on the side of caution.

Treatment of suicidal patients consists of observation—either 1 to 1 for acute risks or half-hourly for people judged to be less at risk. Medical treatment is similar to that of other alcoholic patients—supportive, with vitamin supplements and benzodiazepines and restraints, if indicated. Antidepressants are not useful since their onset of action is slow, and the patients may show significant resolution of their depression within several days. Efforts should be made to engage the patient in a detoxification program, psychotherapy, and/or an Alcoholics Anonymous group by the time of discharge.

2.2.7. Alcohol Amnestic Disorder (Korsakoff's Syndrome, Werniecke's Encephalopathy)

Alcohol amnestic disorder is the name for a group of symptoms occurring in chronic alcoholics, symptoms that formerly were differentiated into separate illnesses. Impairment of recent and remote memory in the presence of a clear sensorium is the diagnostic key to this disorder. Symptoms include disorientation, confabulation, impaired judgement, and poor reality testing. Concomitant opthalmoplegia, ataxia, nystagamus, and othe' neurologic signs are an indication of mid-brain damage.

Treatment should include vitamin therapy, especially with thiamine and other members of the B complex group. If treatment is vigorously initiated, the ophthalmoplegia is rapidly reversed, and other neurological symptoms may also respond, although more slowly. There will be limited recovery from the cortical deficits.

2.2.8. Fasting Alcohol Hypoglycemia

Fasting alcohol hypoglycemia was first observed by Brown and Harvey in 1941 (Lieber, Gordon, & Southren, 1982). It develops between 6 and 15 hours following alcohol ingestion in a malnourished or fasting individual who has drunk the equivalent of between 50 to 150 ml of absolute alcohol. Children are particularly susceptible and may become ill after an alcohol rub (Lieber *et al.*, 1982). The illness is not uncommon in adults and has been reported to occur in 0.1% of the patients admitted to Bellevue's adult emergency service (Goldfrank, Flomenbaum, Lewin, & Weisman, 1982).

Patients enter the ER in a stuporous or comatose state. They also may have extensor rigidity of the extremities, unilateral or bilateral Babinski reflexes, convulsions, and conjugate deviation of the eyes. There may be hemiparesis, which often becomes permanent. Lieber and colleagues (1982) suggest that trismus, coma, and hypothermia should suggest this condition. It appears to be caused by inhibition of gluconeogenesis. Blood sugar levels are below 40 mg% and may drop below 30 mg%. Measuring plasma insulin levels permits differentiation between alcohol hypoglycemia and hypoglycemia secondary to insulin overdosage.

Treatment for alcohol- or insulin-induced hypoglycemia consists of immediate intravenous administration of 100 cc of 50% glucose in water. Additional supportive measures, as indicated, should be done.

2.3. *Summary of Treatment for Alcoholism*

Several therapeutic pathways have to be pursued simultaneously in the ER: need for admission, for psychiatric treatment and, at times, for medical treatment. It is advisable to withhold psychotropic medication until a complete physical examination has been performed unless the patient is so unruly that he becomes dangerous to himself or to others. Under these circumstances, intramuscular or intravenous medicine must be given, perhaps prior to or concomitant with the application of restraints.

In the treatment of an alcoholic, outpatient care is feasible if the patient is suffering from mild withdrawal or is able to cooperate with a detoxification program. People who are acutely or pathologically intoxicated may respond to a period of sedation and treatment in a holding unit for several hours until their acute condition passes. Then they can be evaluated for inpatient treatment or discharge. Alcoholics who are suffering from severe withdrawal reactions, serious deficits associated with chronic brain syndromes, and psychoses should receive inpatient care.

Several groups of medication are available. Most useful are the benzodiazepines, major tranquilizers, and vitamin supplements. Intravenous glucose is reversed for cases of alcoholic hypoglycemia.

Choice of psychotropic medication and frequency and route of administration depend upon several factors, including the acuteness of illness, responsiveness to medication, and whether the treatment is to be initiated in the emergency room and carried out at home or to be done in an inpatient setting. If benzodiazepines are to be used parenterally (for acutely disturbed patients), the patient should be carefully observed after the initial dose is given and, when indicated, repeated doses should be administered at half-hourly intervals for a total of no more than three additional doses before changing to oral medication. Both diazepam and chlordiazepoxide may be given intramuscularly or intravenously. Dosage for the former is 10 mg initially and 5 to 10 mg in subsequent doses. For the latter, 50 to 100 mg may be given and repeated. It is advisable that patients lie down after receiving parenteral medication. If oral medicine is given, the dose range may be the same initially and gradually changed to divided doses of 25 mg *qid* for chlordiazepoxide and 10 mg *qid* for diazepam. When indicated, the patient can be "titrated" with larger doses initially, though every effort should be made to reduce medication gradually so that the patient is medication-free at the time of discharge. Other benzodiazepines may be substituted. There is a wide margin of safety with the benzodiazepines.

Major tranquilizers should be reserved for the treatment of patients who do not respond to benzodiazepines and continue to be a danger to themselves or others because of their highly agitated state. Intramuscular medication should be given in relatively low doses (2 to 5 mg of haloperidol or 25 to 50 mg of chlorpromazine every half hour for no more than 3 doses). Afterwards, oral medication should be used.

During periods of heavy drinking, most alcoholics neglect their diets. As a result, vitamin deficiency is very likely. Thus, all alcoholics should receive vitamins as part of their treatment regimen. If there is severe gastritis, parenteral administration for the first two days of treatment is advisable. A daily dosage schedule might be: thiamine, 50–200 mgm; folic acid, 1–5 mgm; and vitamin C and niacin in the form of Berroca C, 2–4 ml. If prolongation of prothrombin time exists, vitamin K should be given even though there are anecdotal reports of paradoxical lengthening of prothrombin time in alcoholics who have received vitamin K (Greenblatt & Shader, 1975).

Antabuse has no role in the acute care of the alcoholic and should be reserved for the patient who has demonstrated an ability at total abstinence under supervision. Patients treated with Antabuse, who recently have ingested alcohol, should be observed in a medical emergency room, regardless of their psychiatric status, because an Antabuse reaction can be fatal.

We do not recommend the use of barbiturates in the treatment of alcoholics, largely because their interaction with alcohol can lead to respiratory depression and possibly inadvertent death (see Chapter 9 in this volume). In addition, their addicting potential is quite high. Another old standby, paraldehyde, continues to have some advocates, but it is not as effective as benzodiazepines and can cause sterile abscesses if injected.

Once the acute problem has been resolved, the possibility of *long-term* treatment should be considered. Perhaps the most critical step is to differentiate those who are likely to remain within a therapeutic program from those who will resume drinking within a short period after discharge. It is not within the province of this chapter to discuss the many therapies available for the outpatient treatment of alcoholism. We would like to emphasize that the majority of patients treated in public hospitals also experience social and environmental difficulties, including lack of housing, shortage of funds, impoverishment of social networks, despair, loneliness, and difficulty in obtaining work, often because of restricted vocational training. People will do best if referred to alcohol detoxification units as a first step to reentry. Patients may be introduced to ways of spending time in the company of others without drinking, and progress from there to vocational training and ultimate employment. Within the context of these transitional programs, it is possible to utilize the case manager system to help the person surmount the social and environmental problems that frequently make rehabilitation impossible. The role of Alcoholics Anonymous in the rehabilitation program cannot be overemphasized.

3. Anticholinergics

3.1. Introduction

Four groups of medicines have anticholinergic properties: (1) antidepressants; (2) antihistamines; (3) antispasmodics; and (4) antipsychotics of the phenothiazine group. At least 28 plants with anticholinergic properties have been identified (Goldfrank, Lewin, Flomenbaum, & Howland, 1982). The more common are Jimson weed, henbane, matrimony vine, nightblooming jasmine, angel's trumpet flower, and potato and tomato vines. Some are occasionally taken by abusers.

Three groups of people with anticholinergic side-effects may be seen in the ER. First are the younger abusers. Many of these unknowingly take antispasmodics (atropine or scopolamine), believing they purchased LSD, PCP, or TCP. Next is the group, usually middle-aged or elderly, that develops restlessness, confusion, and agitation in the course of treatment with a tricyclic antidepressant, phenothiazine, or antispasmotic. Finally, there is the group that has taken an anticholinergic, usually an antidepressant or phenothiazine, in a suicide attempt (Crome, 1982).

3.2. Clinical Observations and Treatment Regimens

3.2.1. Psychological Responses

Though most anticholinergic effects are dose related, there can be a range of individual susceptibility. Weiner (1980) observed toxicity to follow administration of eye drops. The earliest signs of overdose are restlessness, drowsiness, irritability, and confusion. These are followed, if the dose is sufficiently large, or if the person is unduly sensitive, by delirium. Clouded sensorium and fearful visual hallucinations may be present. Auditory hallucinations are less frequent. People who are being treated with phenothiazines, tricyclics, or antispasmodics are less likely than others receiving anticholinergics to have visual hallucinations.

3.2.2. Physical Changes

Anticholinergic side effects are mediated by the parasympathetic nervous system. Weiner (1980) has summarized the effects of atropine's dosage upon clinical response. At 0.5 mg, there is slight cardiac slowing, dryness of the mouth, and inhibition of sweating. At 1.0 mg, dryness of the mouth becomes annoying, thirst is marked, pupillary dilitation occurs, and tachycardia (occasionally preceded by bradycardia) is present. At 2.0 mg, cardiovascular sensitivity is present and palpitations occur. Near vision is blurred. All of these are accentuated at a dosage level of 3.0 mg. In addition, there is difficulty in urinating and swallowing. The skin is hot, dry, and flushed, especially about the face. Bowel sounds may be absent. Early signs of delirium are present. Above 10 mg ataxia exists, the skin is scarlet, the iris almost obliterated, delirium and restlessness so marked that restraints may be necessary. Cardiopulmonary failure may occur.

3.2.3. Treatment of Acute Intoxication

Symptoms are dose related and time limited. As a result, for all but emergencies caused by cardiovascular or respiratory collapse, treatment can proceed calmly and logically with the realization that the patient is going to recover. Restless patients are calmed as they receive attention. Alcohol/ice-water rubs are effective for temperature control and should be repeated as indicated. Restlessness and possibly seizures can be controlled by moderate doses of benzodiazepines. Diazepam (Valium) can be given in doses up to 20 mg every four hours, either orally or parenterally.

If it is indicated, 0.5 to 4.0 mg of physostigmine can be administered intravenously at a rate of 1 mg per minute. Response occurs within 15 minutes. Because the drug is rapidly metabolized, its effects are short-lived and the patient should be observed for recurrence of symptoms. If there is no response to the first dose, consider that the offending agent was not an anticholinergic. The differential diagnosis includes hypoxia, acidosis, or carbon monoxide poisoning. After alternative diagnoses have been investigated, the physostigmine test may be repeated at least twice to rule out an anticholinergic etiology. Side effects from physostigmine can be serious. Those following too rapid administration include bradycardia, hypersalivation, and emesis, which may lead to respiratory complications and possibly a seizure. Other side effects are diarrhea, nausea, and excessive urination or sweating. Most effects can be controlled by careful regulation of dosage. Those that necessitate stopping administration include marked bradycardia, convulsions, emesis, and hypersalivation, especially if there are respiratory difficulties. People with illnesses susceptible to autonomic nervous system stimulation—asthma, cardiovascular disease, thyroid disease, glaucoma, gastrointestinal or urogenital problems—should not receive physostigmine. If a toxic reaction to the drug develops, administering 0.5 mg of atropine sulfate for each mg of physostigmine will reverse its action (Hall, Feinsilver, & Holt, 1981).

4. Arylcyclohexlamines (PCP)

4.1. Introduction

In recent years, phencyclidine (PCP) has become one of the most popular substances of abuse, both because of the wide range of side effects that it can produce as well as the fact that it is readily available.

PCP was introduced as a general anesthetic for humans (Sernyl) and animals (Sernylan). Because of the high incidence of postoperative delirium and agitation, it is no longer used for humans.

4.2. Clinical Observations and Treatment Regimens

4.2.1. Psychological Effects

Cohen (1981) relates symptoms to dosage as follows:

a. *Disinhibition* (1–5 mg). Disinhibition is similar to the relief of anxiety produced by relatively low amounts of alcohol. In addition, there can be feelings of euphoria, or emotional lability. At times, numbness, comparable to the paresthesias of alcoholism, is felt. Illusions, pseudohallucinations, and synesthesias may occur.

b. *Toxic psychosis* (5–15 mg). Depersonalization, inner turmoil, confusion, disorientation, and disorganization of thinking are present and may remain for 4 to 8 hours. Behavior varies between agitation and withdrawal, occasionally to the point of muteness. Anxiety, panic, and feelings of isolation are reported. Because amnesia for the toxic psychosis can occur, the likelihood of learning from an adverse experience is lessened.

c. *Schizophreniform psychosis* (20 mg). The resemblance to schizophrenia is sufficiently similar to lead to the consideration that

PCP is one of the drugs capable of producing a "model psychosis" (see Chapter 8 in this volume). Catatonic stupor or excitement, often with paranoid ideation, are the predominant features. During this phase, the person is out of contact and cannot be "reached" verbally, so that "talking down" is not an effective therapeutic maneuver. Prolonged psychosis may follow the schizophreniform psychosis.

Other chronic changes are similar to those of drug abuse in general. Flashbacks may be more frequent and of longer duration than those seen with hallucinogens or cannabis. Among those who have written about clinical effects of PCP are: Liden, Lovejoy, and Costello, 1975; Fauman, Aldinger, Fauman, and Rosen, 1976; Luisada and Brown, 1976; Khantzian and McKenna, 1979; Smith, 1980.

4.2.2. Physical Changes

Physical changes include ataxia, dizziness, drooling, dysarthria, flushing, hypertension, increased deep tendon reflexes, mydriasis, nystagamus, sweating, and tachycardia. Nausea, bloody vomiting, diarrhea, and coma may be the result of a contaminant, l-piperidinocyclohexane-carbonitrile.

4.2.3. Treatment of Acute Intoxication

Panic and catatonic excitement are the responses most frequently encountered in the emergency room. Sedative medication, most often diazepam (10–20 mg) administered intravenously, permits regaining control. Phenothiazines may bring about a serious hypotensive reaction because of their anticholinergic effects and are contraindicated. Haloperidol (2.5–5 mg IM repeated at half-hourly intervals for 3 doses) is preferred. Reducing environmental stimuli may be helpful in calming the patient. Sometimes a dimly lit room acts as a sedative.

When force is required to subdue the patient, it is most likely that he/she is out of contact, not amenable to reason, and apt to try to inflict serious damage upon others. Staff should be well trained in techniques of subduing violent patients, and should work as a team. At least four people should approach the patient to minimize the likelihood of injury to patient or staff member. Restraints are not recommended in the treatment of acutely intoxicated people, because the patient may become exhausted in fighting against them. They do have a role if the patient is dangerous to himself or to others, and should be used until medication is effective.

Treatment of PCP ingestion (overdose) is usually not undertaken in a psychiatric emergency room, but is done in a general hospital ER. Immediate measures include gastric lavage because PCP is rapidly absorbed from the intestine, but slowly from the stomach. To minimize the effects of enterohepatic recirculation, continuous gastric suction is recommended. Clearing PCP from the body by acidifying the urine with ammonium chloride is no longer recommended for several reasons. PCP is lipid soluble, and only a small amount will be removed from the urine at any time. If death occurs from PCP overdose, the most common cause is renal failure associated with rhabdomyoglobinurea, which is best treated with alkalinization of the urine, though this treatment is contraindicated with PCP overdose. It is conceivable that acidification could precipitate rhabdomyoglobinurea, which certainly would complicate an existing condition. Finally, for acidification to be successful, the blood pH must be reduced to 7.1–7.2, which causes still other problems (Weisman, 1983).

Treatment of patients with a prolonged psychosis is similar to that of any other chronic condition. Phenothiazines are not contraindicated once the acute intoxication has passed.

5. Cannabinoids

5.1. Introduction

Marijuana has become a widely abused substance whose use cuts across all socioeconomic classes. Classifying it as a hallucinogen is not accurate because there is no cross tolerance with hallucinogens. It is more correct to consider cannabinoids in a separate class, even though many psychological effects are shared with the hallucinogens (Jaffe, 1980).

5.2. Clinical Observations and Treatment Regimens

5.2.1. Psychological Effects

Psychological effects are dose- and setting-related. Sensory changes, alteration of visual and color perception, heightened acuity, time distortions, subjective feelings of aesthetic enhancement, and increased self-awareness may be present. Many users report a feeling of increased empathy. With lower doses, marijuana eases tension and produces a relaxed state in which it is possible to reflect. Recognition of marijuana's effectiveness in producing sleep has been known for more than a century.

With larger doses, psychological disturbances are likely. Perceptual distortions, especially in unfavorable surroundings, may lead to paranoid judgements about the behavior and intent of others. Anxiety may reach panic proportions, resulting in conflict with people in the area and, rarely, hospitalization. With high dosage or heightened individual susceptibility, a toxic psychosis similar to that seen with alcohol, bromides, hallucinogens, or PCP may occur. Prolonged psychoses could follow the toxic psychosis.

5.5.2. Physical Effects

Physical changes are minimal. The reddened conjunctivae of the smoker are familiar to all. Mild stimulation of the cardiovascular system is manifested by increased pulse rate and systolic blood pressure. Bronchitis and asthma are the only demonstrable pulmonary side effects at this time. The middle finger of the hand holding the "joint" may be stained. Reports of chronic brain disease have not been substantiated. It is difficult to assess the long-term effects of cannabis because few people use this drug exclusively.

Treatment of acute reactions is largely symptomatic and resembles that for the hallucinogens (see Section 6). Marijuana does not appear to be ingested in such quantity as to produce symptoms of stupor or unconsciousness. There do not seem to be withdrawal effects following marijuana use.

6. Hallucinogens

6.1. Introduction

The serendipitous discovery of the hallucinogenic properties of lysergic acid diethylamide (LSD) by Hoffman (Stoll & Hoffman, 1943) was followed by its rapidly becoming a popular drug of abuse (Alpert, Cohen, & Schiller, 1966). Once discovered, LSD was found to be a naturally occurring substance and was isolated in a morning glory seed, romantically named "Heavenly Blue," among other compounds. Hallucinogens may be used directly as they occur in plants (peyote) or in purified form (mescaline). The widespread use of this group of substances began to wane in the late 1960s, partially because of the long refractory period following exposure, partially because of the discovery of PCP, and partially because of the increased availability of cocaine.

The mind-expanding qualities of the hallucinogens are more intense than those of cannabinoids and have been compared to mystical supernatural experiences.

6.2. Clinical Observations and Treatment Regimens

6.2.1. Psychological Effects

Barr, Langs, Holt, Goldberger, and Klein (1972) studied the psychological and physiological responses to LSD in the laboratory setting. Five response patterns were described. Those for LSD are somewhat slower in onset than for other hallucinogens.

Types 1 and 2 began within half an hour of ingestion. Individuals initially reported feeling like a different person, with impending loss of control and body image changes (parts small and disconnected). Type 1 was of short duration and passed into Type 2, which lasted for approximately 8 hours. Depersonalization, alienation and multiple somatic effects were characteristic of this stage. Type 3 reactions were intense by 2 hours after ingestion, and peaked by 5. These were highly varied and included most of the commonly experienced feelings, such as being carefree, elated, or silly. Other changes attributed to Type 3 were distortions of time, body image, and perception, especially visual feelings of unreality and of seing oneself apart from others. Type 4 reactions, also lasting about 8 hours after ingestion, included impairment of thinking, characterized by difficulty in concentration and slowing of thoughts, loss of meanings, impaired time sense, reluctance to talk, and externally directed anger. Type 5 reactions, which peaked at 5 hours, were both pleasant and distressing. Among the distressing reactions were anxiety, depression, time retardation, blankness of mind, disconnectedness of events, fear of going crazy, and uncertainty about reactions of others, while loss of thought inhibition and increased pace of thinking, opening of new horizons, gaining of insights, and pressure of speech were generally considered pleasant. LSD users have also emphasized increased somatic-aesthetic awareness, intensification of colors, preoccupation with the color blue, feeling music more intensely, and feeling one sensation through a different sensory modality (synesthesias).

Psychological effects leading to ER intervention may be characterized as symptoms of acute intoxication and panic reactions, prolonged psychosis, and flashbacks. Occasionally, schizophrenics attribute feelings of depersonalization or unreality to having been given LSD surreptitiously.

Panic reactions, as the name indicates, are awesome in their ability to frighten the abuser. They appear to be triggered by setting, interaction with others, and mood. The thought content is usually paranoid. Other symptoms characteristic of intoxication are impairment of thinking, loss of meanings, looseness of associations, muteness, withdrawal, and depression.

Prolonged psychosis is a blending of the symptoms of the LSD experience into those of a psychotic episode. It appears as if the mentally ill person never quite comes down from the LSD trip, but continues to have affective problems or disorganized thinking. It is often difficult to determine the extent to which symptoms of mental illness were present prior to the use of the hallucinogen and to what extent the present episode is a release phenomenon (Vardy & Kay, 1983). Flashbacks have been described previously (Section 1.4).

6.2.2. Physical Effects

Physical symptoms are primarily related to stimulation of the sympathetic nervous system. Activity is greatest in the first hours after ingestion. The chemically refined hallucinogens have milder action than the naturally occurring, where other compounds act as impurities. Among changes noted are elevation in blood pressure, tachycardia, gastrointestinal distress with occasional vomiting, dilated pupils, and hyperactive deep tenden reflexes.

6.2.3. Treatment

Once it was recognized that the majority of symptoms associated with a "bad trip" were time limited and responded to tranquilizers, abusers did not come to the ER for treatment unless they were seriously ill.

6.2.3.1. Treatment of Intoxication or Panic. Treatment of intoxication or panic is largely supportive. A calm atmosphere and friendly, empathic people capable of "talking the patient down" may suffice. The ambience should be similar to that utilized for most delirious reactions. Medication may either be a standard dose of a benzodiazepine, phenothiazine, or haloperidol, given orally or parenterally. We have preferred one of the major tranquilizers, largely because of our experience with them during the 1960s, when hallucinogenic reactions were at their peak in the emergency room. Generally, oral doses suffice, but in an acute episode, one to three intramuscular injections of a major tranquilizer at half-hourly intervals may be necessary.

6.2.3.2. Treatment of Chronic Psychosis. Treatment of chronic psychosis, complicated by drug abuse, is always a difficult issue because two serious problems must be independently assessed and addressed therapeutically. Unless there is an effort to help the abuser change his or her way of life, drug abuse problems will persist and undermine psychotherapeutic or psychopharmacologic efforts (see Chapter 13 in this volume).

6.2.3.3. Treatment of Flashbacks. Treatment of flashbacks is through reassurance, medication, and possibly interpretation of the symbolic meaning of the flashback. Emphasis must be placed upon the fact that the frightened person is suffering from inability to control thoughts, rather than from a chemical alteration of the brain. The trigger might be a "learned" experience, similar to that experienced when high. Anxiety may be a stimulus to the appearance of flashbacks. While exploratory therapy can be helpful, it is not likely to be as effective in the short term as antianxiety medication.

7. Inhaled Vapors: Miscellaneous Substances or Household Hallucinogens

7.1. Introduction

Organic solvents or inhaled vapors are one of the first substances seriously abused by many younger people, most often males between 7 and 17. The most popular vapors are acetone found in model airplane glue, nail polish, paint thinners, and aerosols and anesthetic agents such as ether or alcohol. Sniffers rarely enter the emergency room, but can readily be recognized as they weave ataxically down the street. While most suffer from an acute brain syndrome, an occasional abuser develops a chronic brain syndrome (see Chapter 11 in this volume).

Abusers have learned that many commonly used substances have the capacity to "turn them on." Cohen (1981) has categorized the substances according to the place in which they are found: *kitchen cabinet* (mace, nutmeg); *laundry room* (aerosols, anesthetics, commercial solvents); *medicine chest* (antihistamines—often with ephedrine—amyl nitrite); *garden* (morning glory seeds, Hawaiian wood rose seed; and *alcohol containing substances* (aftershafe lotions, cooking extracts, Listerine).

Identification of toxic substances can be difficult when commercial products are used. The local poison control center or the manufacturer can be of help. When proprietary medicines are used, it may be possible to initiate specific treatment. Most often, treatment must be nonspecific.

7.2. Clinical Observations and Treatment Regimens for Aerosols

7.2.1. Psychological Effects

Psychological effects are immediate because absorption from the lungs is rapid. The typical effects of organic brain syndromes—irritability, confusion, impulsiveness, excitement, panic, and possibly delirium may be seen.

7.2.2. Physical Effects

Physical symptoms are primarily those of an acute intoxication, including ataxia, slurred speech, nystagamus, and other symptoms of cerebellar dysfunction.

Respiratory distress is a dangerous, life-threatening complication. It may be aerosol-induced through laryngeal edema, freezing of the uvula or larynx, or coating of the bronchioles. Suffocation may also result from placing a plastic bag over the head or from stuffing a vapor-saturated rag in the mouth and falling unconscious without releasing the rag.

7.2.3. Treatment of Acute Intoxication

Treatment of acute intoxication has two aspects, psychological and physical. Drug-induced psychological symptoms are time limited and require little treatment other than a supportive environment and control of agitation with a benzodiazepine. Barbiturates, which may depress respiration further, should not be used.

In emergencies, intubation or a tracheotomy may be indicated. When aerosols coat the respiratory system, oxygen may be of help until respiratory specialists can be contacted. Cardiac complications are primarily arrythmias. If specific toxins are present, there may be renal, hepatic, or neurologic damage. During the period in which treatment is provided, base line bloods should be drawn to assess electrolyte, liver, and renal functioning, and urinary output should be measured. Patients should be observed for several hours before discharge to make certain that renal functioning is not impaired.

If there are indications of decreasing urinary output, creatinine and blood urea nitrogen (BUN) levels should be repeated. If they have risen, additional hospitalization is indicated. Tubular necrosis is the major complication. Uremia, manifested by hyperkalemic acidosis and encephalopathy, indicates progressive disease. Hemodialysis is the treatment of choice followed by renal transplant.

Hepatic failure is indicated by progressively increasing jaundice, increasing confusion, tremor, stupor, and coma. Treatment involves support of vital signs, elimination of protein from the diet, and prevention of infection.

8. Narcotics

8.1. Introduction

Narcotics are physiologically safe substances since they do not affect the major organ systems, nor do they cause central nervous system damage as do many sedatives and alcohol. They are damaging psychologically because of the ennervating consequences of dependence. Addicts' anxiety about experiencing withdrawal contributes to their focusing much of their lives toward obtaining the next "fix." Abusers who desire to work regularly have learned to time their "fixes" to correspond with the working day. Drug substitution programs (heroin in England and Methadone in the United States) have been effective in making it possible for former addicts to sustain themselves within the community, and for some, to work or attend rehabilitation programs. Drug abuse, however, continues for a small number, in spite of their enrollment in a Methadone program. Methadone dosage is an indicator of ability to tolerate a drug-free existence. Many people remain in programs for years, often at a high level of dosage. Some report that they are unable to tolerate a drug-free state, others that they enjoy the feeling of a relatively high level of Methadone. Of course, if people have been in a program for more than 5 years and are still taking more than 80 mg of Methadone, one can suspect that they have learned to control their withdrawal symptoms and are selling their weekend rations in order to earn extra money.

Multiple drug abuse is the rule, rather than the exception, at the present time. All abusers who seek help in an ER should be evaluated and have a laboratory drug screen in order to see if they are abusing other substances. Two patterns of abuse exist: One is to modify heroin's effect with alcohol or a sedative; the other is to extend the high by augmenting it with amphetamine or cocaine.

A small number of people abuse narcotics other than heroin. Though available narcotics fall into five major groups (Adriani, 1979), we prefer to think of them as being in two classes. They are morphine-like drugs, which include heroin and medicines, dolophine (Methadone), meperidine (Demerol), and morphine itself. The second group consists of codeine and its two congeners, pentazocine (Talwin) and propoxyphene (Darvon), which can be taken parentally or orally. Most often, people who abuse medically available narcotics have access, as doctors and nurses, or have become addicted in the course of medical treatment.

8.2. Clinical Observations

8.2.1. Psychological Effects of Morphine-like Drugs

The psychological effects of morphine-like drugs are described in limited terms by addicts. They especially enjoy the "rush" following intravenous administration of heroin. This sensation lasts for a few moments and is followed by relaxation, often accompanied by drowsiness ("nodding"). Other feelings, secondary to mental clouding, include lessening of concern with regard to suffering, tension, depression, and physical pain. Delirium is not a usual side effect of drugs in this group, though an occasional patient, heavily medicated with meperidine postoperatively, may develop an acute brain syndrome. Addicts frequently report a loss of sex drive after being addicted for several years, but this may partially be a function of the individual and his focus upon drugs, since many addicts are married and have intercourse and children.

8.2.2. Physical Effects of Morphine-like Drugs

Physical changes associated with *daily use* of heroin are minimal, especially in the lower dose ranges. Nausea and vomiting are common with first exposure. These symptoms rapidly disappear until the only gastrointestinal complaint is chronic constipation. Other physical changes—drowsiness, contracted pupils, marred veins, and discolored arms and possibly backs of hands secondary to injection of impurities, are familiar symptoms.

Tolerance develops slowly to the analgesic effect, which makes it possible to utilize opiates medically. It does not develop to the constipating effect, which poses a severe problem to addicts, nor to the miosis, which leads many addicts to wear sun glasses. Controlling of tolerance is a problem for addicts who are faced with maintaining dosage within physiologically safe bounds and containing the cost of each high. Some seek detoxification so that they may reduce their dosage, while others attempt to withdraw themselves.

8.2.3. Physical Effects of Codeine and its Congeners

Symptoms of codeine abuse include initial nausea, vomiting, dizziness, hypotension, and tachycardia. More serious reactions following overdose are similar to those for morphinelike drugs, though there may be a higher incidence of seizures in the codeine family.

8.2.4. Clinical Effects of Narcotics

The clinical effects of narcotics may be considered as *acute reactions, overdose, secondary illness,* and *withdrawal.* Any may bring the addict to the ER seeking help.

8.2.4.1. *Acute Reactions.* By and large, addicts will not seek treatment for acute reactions until street remedies have failed, so they may be critically ill when they come to the hospital. Acute reactions may be coma, pneumothorax or acute pulmonary edema, and respiratory failure. *Pneumothorax* occurs when an addict who is inexperienced at locating the subclavian vein perforates the apical pleura. When *pulmonary edema* is present, there may be frothing at the mouth and nasal orifices. This symptom frequently is a concomitant of opiate overdose, but may occur when overdose is not present, as an acute reaction to the opiate (Garay, 1983).

8.2.4.2. *Overdose.* Overdose is an acute reaction that may lead to death if it is not treated promptly. In cases of overdose a distinction should be made between overdose related to heroin and to complications of multiple drug abuse. If characteristic symptoms of other drugs are identified, specific therapy should be initiated.

Symptoms of *heroin overdose* include drowsiness, stupor, and coma, with respiratory and cardiovascular depression. If pupils are dilated, rather than contracted, mid-brain involvement has taken place. See Section 11 for more detailed consideration of coma.

In *meperidine overdosage,* symptoms appear to be a combination of atropine (dilated pupils) and opiate overdosage. Confusion, muscular tremors, hallucinations, and occasionally *grand mal* seizures can occur

(Jaffe & Martin, 1980). Severe headaches, nausea, vomiting, respiratory depression, and cardiovascular collapse are other possible symptoms.

8.2.4.3. Secondary Illness. Secondary illness may be transmitted by sharing work or injecting under nonsterile conditions. Illnesses that have been reported include hepatitis, malaria, septicemia, vasculitis, endocarditis, and more recently AIDS, though the method of transmission for the last is uncertain.

8.2.4.4. Withdrawal. Withdrawal symptoms from *heroin* begin approximately 8 hours after the last dose is taken, and gradually increase in severity over the next 2 days, until withdrawal is complete, about 48 hours after the first symptoms appear. *Psychological changes* most commonly seen are anger, irritability, depression, and pleading for an opiate or medication to terminate the withdrawal. Restlessness, the most common psychomotor change, is manifested by fidgeting, inability to sleep, and pacing. *Physical symptoms* include chills, piloerection ("cold turkey"), yawning, abdominal discomfort (cramps, nausea, vomiting) and secretory excess (rhinorrhea, lacrimination, diaphoresis). Withdrawal from other morphine-like drugs has a similar pattern to heroin.

Withdrawal from *codeine-like drugs* is more akin to that of barbiturates. Restlessness, agitation, confusion, euphoria, and perceptual distortions (usually visual hallucinations) may occur and be the first indication of abuse. Some abusers are aware of their addiction, especially when using Elixir of Terpin Hydrate with Codeine, while others who ingest an excessive amount of medication for pain or relief of anxiety may not be aware of the developing addiction, nor of its implications.

8.3. Treatment Regimens

8.3.1. Acute Reactions

Treatment of acute reactions usually requires a highly trained medical team that can work smoothly together (Goldfrank, Flomenbaum, Lewin, & Weisman, 1982). Treatment of acute medical emergencies is beyond the scope of this chapter, though some of the highlights of diagnosis and treatment are presented in the following text and in Section 11.

Acute brain syndromes are rarely caused by narcotics. When an addict presents with one, consider the likelihood of multidrug abuse. If it is possible to identify the etiologic agent, specific treatment, as discussed with each of the substances in this chapter, may be undertaken. Otherwise, management of an anxious or delirious person with a time-limited disorder should be undertaken. A supportive environment, with appropriate use of a benzodiazepine, is generally adequate.

Occasionally, an addict may seek help because of marked anxiety. It may be possible to provide immediate care and discharge him/her from the ER before withdrawal sets in. If this is not possible, it will be necessary to initiate treatment of withdrawal, possibly for several substances simultaneously.

8.3.2. Opiate Overdose

Treatment of opiate overdose has two components. The first is a general treatment that is given to all patients who enter the ER in a semistuporous, stuporous, or comatose condition. This consists of a rapid evaluation, initiation of life-supporting treatment, performing a thorough physical examination, and directing specific treatment for the etiologic agent and affected systems. This is discussed more fully in Section 11. Naloxone is the antidote for opiate overdosage and can reverse respiratory depression. In addition, it serves as a useful test when opiate addiction is suspected because it can produce withdrawal symptoms. It is administered intravenously, 0.4 mg every 15 minutes for up to three doses. Response is rapid if a narcotic has been taken, making nalaxone an excellent therapy for complications of overdose. If a person experiences withdrawal following a low dose, it is unlikely that there has been serious addiction. When nalaxone has been used to reverse respiratory depression, the patient should be observed for at least 24 hours to make certain there is no relapse of coma or respiratory difficulty. This is especially necessary when long-lasting drugs such as meperidine have been taken. Nalaxone is less useful as a test for pentazocine addicts, since this drug has some antagonistic qualities.

8.3.3. Secondary Illnesses

Treatment of secondary illnesses requires diagnosis and treatment of the disease as well as complications secondary to dependence.

8.3.4. Withdrawal

The treatment of withdrawal may be directed toward detoxification or substitution of Methadone for heroin. Detoxification, a drug-free state, may be achieved by substituting Methadone and gradually reducing the dosage, or by controlling most side effects with clonidine, acupuncture, or phenothiazines. The choice of inpatient or outpatient setting depends upon the method chosen. Most clinicians try to achieve a drug-free state in a closed setting, but some administer Clonidine or acupuncture in their private offices or a clinic. Transfer to Methadone maintenance usually is performed in a clinic.

When Methadone is used, the first therapeutic steps are identical for those who are ultimately to become drug free or remain on Methadone maintenance. In order to be safe, give 10 to 20 mg of the drug and observe the patient for signs of respiratory changes, drowsiness, or the appearance of withdrawal. With this approach, the initial dose cannot be lethal. When people seeking Methadone state they are in a program, verification should be obtained, as well as the date the last dose was taken and the amount given. If there is any reason to suspect that the patient had not been participating, a test dose of 0.2 mg of Nalaxone can be given. If there are signs of withdrawal, the person has been opiate free and has no further need for Methadone.

In a Methadone detoxification program, the dosage should be adjusted to a level that is able to control withdrawal symptoms. This level should be maintained for one or two days in divided doses (*qid*) and then given once daily. Reduction of dosage should be planned so that detoxification is completed by a week. In a Methadone-maintenance clinic, dosage should be adjusted in the first two days so that the patient is symptom free and feels at ease. Generally, a level of 80 to 100 mg accomplishes this goal. The dose is reduced gradually over a period of several years. Some people are able to become drug free, while others find reassurance in continuing on a lower dose.

Clonidine can relieve symptoms in patients addicted to heroin, Methadone, or other opiates (Abromowicz, 1979). About 80% of the patient's distress is relieved by this drug, which resembles opiates to the extent that it can cause mild dependence and withdrawal symptoms in hypertensive patients. When clonidine is substituted for Methadone by administering 6 mcg/kg initially, followed by 17 mcg/kg/day in divided doses, most withdrawal symptoms are suppressed. Another method of withdrawal is to utilize 0.5 mg/day for 10 days, which reduces but does not abolish withdrawal symptoms. Clonidine is not without danger, since it is an antihypertensive agent and is capable of producing hypotensive reactions in some normals.

Acupuncture can also be used to control withdrawal symptoms (Smith, Squires, Aponte, Rabinowitz, & Bonilla-Rodriquez, 1982). We have spoken with patients in Smith's clinic, who were relaxed during the withdrawal process and who said they were symptom free.

In facilities where it is not possible to obtain Methadone, it may be necessary to utilize phenothiazines in order to assist a patient who is undergoing "cold turkey." The drug should be given in sufficiently high dosage to control gastrointestinal symptoms and alleviate anxiety. The process is less painful at present than in the past, when heroin was less adulterated than it currently is.

People who are to be withdrawn from codeine, propoxyphene, or pentazocine should not receive Methadone, but the substance they have been abusing. The method of administration is similar to that outlined for Methadone or barbiturates, in which a dose that holds symptoms in check is given initially and gradually reduced over the course of a week.

9. Sedative-Hypnotics

9.1. Introduction

Most of the substances in the sedative-hypnotics group have found favor with abusers throughout the years. Although their primary pharmacologic purpose is sedation, they can secondarily provide feelings of relaxation and cortical disinhibition and, for abusers, a sense of release that can be exhilarating.

The most commonly abused substances in this group are shorter acting barbiturates (Amytal, Nembutal, Seconal, Tuinal), benzodiazepines (Librium, Valium), chloral hydrate (Noctec), ethchlorvynol (Placidyl), glutethimide (Doriden), meprobamate (Equanil, Miltown), methaqualone (Quaalude), methyprylon (Noludar), and paraldehyde. While many abusers will take any pills that are available, some try to restrict themselves to one or several groups of drugs.

The nonbarbiturate sleeping medications have become popular on the street because of their availability and similarity in action to the barbiturates. Benzodiazepines, which have led to the mild addiction of many patients, are also popular street drugs and have been discovered by many Methadone users to give a slight push to the high of the Methadone. Diazepam (Valium), which has a relatively long half-life, is the preferred drug in this group. Meprobamate, more popular when it was first introduced, received bad press on the street because of the severity of withdrawal reactions, and has been supplanted by other drugs, notably glutethimide and methaqualone. The latter, which can produce even more severe withdrawal reactions, has been withdrawn from the ethical marketplace. Chloral hydrate and paraldehyde are prescribed less frequently now than in the past and are not widely abused. Paraldehyde was primarily preferred by alcoholics who became addicted when the drug was used as a sedative during the "drying-out" process.

In addition to being used in their own right, many of these substances are taken in conjunction with other drugs of abuse, especially alcohol, amphetamines, cocaine, and heroin. They serve to extend the high of heroin and to control the peak of the stimulants. Alcoholics use barbiturates to help control "the jitters." Sometimes, unable to relax, they may use barbiturates in order to sleep. Because of the alcohol-induced confusion, as well as their inability to sleep quickly, they may lose count and take too many pills. Respiratory depression ensues and may be followed by death. Unfortunately, the combination of barbiturates and alcohol may lead to respiratory depression and death. Barbiturates, through their hepatosomal enzyme interaction, may increase sensitivity to estrogens, oral contraceptives, and warfarin. They may also lead to exacerbation of intermittent porphyria. Withdrawal from unsuspected addiction to a sedative or alcohol may precipitate the first episode of tonic clonic seizure in a person who has no history of prior epilepsy. This reaction usually follows hospitalization for another illness, often a surgical procedure.

Tolerance develops quickly. The margin between a lethal dose and that necessary to produce a psychological response gradually narrows for all drugs in the group, with the exception of the benzodiazepines, where the margin remains great. It is not widely recognized that dependence develops with low doses also, so that people who take pills for a short time may develop mild physical and psychological discomfort (tremulousness, jitteriness, anxiety) upon cessation of use, in addition to the well recognized sleep rebound effect.

9.2. Overview of the Effects of Individual Drugs

While the drugs in the sedative-hypnotics group have many characteristics in common, they also differ in several respects. In this overview, the highlights of the drugs, especially in the production of acute intoxication, will be discussed. (For a fuller description, see Harvey, 1980.)

Barbiturates are the standards by which other sedative-hypnotics are measured. Though they have lost some of their popularity to benzodiazepines in the past years and are no longer as widely used in suicide attempts as they were, they continue to be popular sleeping pills and drugs of abuse. Whether they produce excitement or depression is a function of dose level and individual susceptibility. For some, this is illustrated by a lightening of mood and possibly a sense of jocularity in the hours following awakening.

Major dangers are acute intoxication, withdrawal, and overdose.

Some long-time abusers develop a chronic brain syndrome (see Chapter 7 in this volume). In the period of acute intoxication, the major dangers are from falling while ataxic, from falling asleep while stuporous and becoming injured because of inability to feel pain, or from having an accident when driving while intoxicated and sleepy. Ecchymoses about the face and eyes and broken noses and limbs are diagnostic clues. Withdrawal, with its acute brain syndrome and *grand mal* seizures, is another source of danger and discomfort. The most serious complication is overdose, with associated respiratory and cardiovascular collapse. Tolerance develops more quickly to somnifacient effects than to respiratory, reducing the safety margin.

Symptoms of other drugs in this group parallel those of barbiturates. In some cases, they are not as severe, in others the dangers are equally as great.

Benzodiazepines have the greatest safety margin between overdose and lethality. Withdrawal reactions from benzodiazepines are less severe than those of barbiturates and are relatively uncommon, especially when one considers how prevalent their use is. There have been no reported successful suicides following the ingestion of diazepam alone. Cardiovascular effects include hypotension, increased peripheral resistance, and increased heart rate.

Chloral hydrate is a safe drug that may be taken for considerable periods of time without serious side effects. Few instances of the "chloral habit" are seen today. However, there may be lapses or breaks in tolerance, which stem from a failure of the detoxification mechanism due to hepatic damage or which may be secondary to overdose. In either case, coma and respiratory paralysis can occur, with death as the outcome.

Ethchlorvynol produces a characteristic mintlike aftertaste that may be of help diagnostically. Symptoms of overdose are nausea, vomiting, dizziness, hypotension, and facial numbness. When Placidyl is used in conjunction with tricyclic antidepressants, delirium may occur.

Gluthemide has marked antimuscarinic effects, characterized by ileus, atony of the urinary bladder, mydriasis, dryness of the mouth and skin, and hyperpyrexia. Respiratory depression is usually less severe than that of the barbiturates, but cardiovascular symptoms are equally severe. Chronic use may be associated with osteomalacia. Muscular spasms may occur during withdrawal.

Meprobamate enhances the CNS effects of most drugs. Approximately 40% of serious abusers have a withdrawal reaction. Milder reactions may occur after withdrawal from lower doses.

Methaqualone is popular among abusers who believe that it has aphrodisiac powers and that it produces a "high" which is akin to that of heroin. The drug possesses antispasmodic, local anesthetic, and weak antihistaminic properties. It has comparable antitussive activity to codeine and enhances its analgesic qualities. Methaqualone is more likely to produce side effects than others in the group. Most common are somnambulism and excessive dreaming. Paresthesias, which may persist for years, can develop. With overdosage, restlessness, excitement, and myoclonus are likely to occur. Cardiac and respiratory complications of overdose are not as severe as those seen with barbiturates. When death takes place, it is most commonly a result of an ethanol–methaqualone combination. This drug is no longer manufactured in the United States.

Methyprylon has similar effects to those of barbiturates.

Paraldehyde has mildly sedating qualities. It was the drug of choice for treating alcoholics, but has been supplanted by benzodiazepines, which are more effective and do not have paraldehyde's disagreeable odor or propensity to cause abscesses when injected intramuscularly. A small number of alcoholics became addicted to paraldehyde.

9.3. Clinical Observations and Treatment Regimens

Psychological and physical effects of sedatives can be classified as *acute intoxication, chronic brain syndromes,* and *withdrawal.*

9.3.1. Acute Intoxication

The clinical picture of an acute brain syndrome is relatively the same, no matter which toxic agent is responsible. Although alcohol is the most frequent cause, other chemicals such as barbiturates, ethchlorvynol, gluthemide, methaqualone, and methyprylon should be considered whenever an acute brain syndrome is diagnosed. Personality changes, increased irritability, memory gaps, difficulty in concentrating, confusion, and ataxia are some of the diagnostic clues. In the ER, the patient's behavior may vary in behavior from cooperative to assaultive, from withdrawn to overly friendly and intrusive.

Physical effects primarily involve the central nervous and gastrointestinal systems. Of the central nervous effects, cerebellar changes are most noticeable, especially discoordination, ataxia, and nystagamus. Other effects are amblyopia, double vision, and scotomata. Hyperreflexia may be present. Gastrointestinal complaints include gastric irritation, nausea, and vomiting. At times, the level of confusion accompanied by lethargy and temperature may lead to a mistaken diagnosis of meningitis, especially if a rash is present. With coma, there can be respiratory and circulatory collapse. Thrombocytopenia, aplastic anemia, or leukopenia are rare complications.

9.3.2. Chronic Brain Syndromes

Chronic brain syndromes are most likely to occur in people who have been addicts for several years. Symptoms are similar to other organic conditions and include confusion, memory loss, irritability, and personality deterioration. Some of the physical findings listed under acute intoxication may be present. Barbiturates are most likely to be associated with an organic syndrome, though any in the group, with the exception of the benzodiazepines, may cause a chronic brain syndrome (see Chapter 7 in this volume).

9.3.3. The Withdrawal Syndrome

The withdrawal syndrome of the chronic, heavy user is a medical emergency. Symptoms of barbiturate withdrawal (which is the model for this group) begin shortly after the last pill is ingested, but become pronounced by 8–10 hours. They include anxiety, jitteriness, weakness, gastrointestinal distress, lethargy, tremor, and ataxia. Some confusion may be present. These symptoms are quite similar to those of the acutely intoxicated state and often can best be distinguished by history and course over time. Symptoms gradually increase over the next 24–48 hours, and headache, nausea, vomiting, and a gross tremor develop. Seizures may occur within 16–48 hours after withdrawal from the shorter acting barbiturates and up to a week if the person is addicted to a longer acting drug such as phenobarbital. *Status epilepticus* and head injury are the two serious physical complications of seizures. Delirium may begin in 16–48 hours. Visual hallucinations and illusions, which usually persist for several days, are the predominant sensory disturbances. During this period, hypothermia and cardiac collapse can occur.

9.3.4. Treatment of Intoxicated States

Treatment of intoxicated states caused by sedative-hypnotics is similar to that of other intoxications. The illness is time limited. in a protected environment. Restraints should be used only if there is assaultiveness or danger to self. Because sedatives are generally used to quiet acutely disturbed people, the therapeutic armamentarium is restricted when an acutely disturbed person comes to the ER as a result of abusing a sedative. A phenothiazine or haloperidol, administered cautiously by mouth or intramuscularly, are the drugs of choice. Dosages should be relatively low in order to avoid the possibility of respiratory or cardiovascular complication.

If pills have been ingested in a *suicide attempt,* or if *stupor* or *coma* are present, the methods for treatment of *overdose* discussed in Section 11 should be followed.

9.3.5. Treatment of Withdrawal

Treatment of withdrawal is frequently initiated by evaluating sensitivity to a test dose of 200 mgm of phenobarbital in a drug-free, fasting patient. People who show no sign of sensitivity to the test dose within an hour of administration probably have been taking 900 or more mgm of barbiturate daily. Those who become drowsy have taken less than 200 mgm. Coarse nystagamus, positive Romberg's sign, and drooping eyelids suggest a daily level of 500 mgm. A mild dysarthria may also be discernible. With 600 mgm, only nystagamus and dysarthria remain. At usage of 800 mgm, only nystagamus is present.

An alternative method of estimating the initial therapeutic dose is to halve the quantity of medication the patient reports to have taken.

Treatment consists of administering the estimated daily amount of barbiturates in four divided doses, using a short-acting drug. The key to successful treatment lies in careful observation of the patient. People who become drowsy and ataxic are receiving too much medication, while those who become increasingly anxious are receiving too little. When the correct dosage level is found, withdrawal should be planned so that treatment terminates between 7 and 10 days. If the patient is addicted to a sedative other than barbiturates, a similar program of withdrawal should be initiated. While some people prefer to administer barbiturates no matter which sedative has been taken, we feel that it is preferable to administer the drug that the patient has been using. Many barbiturate addicts are chronically depressed and should be evaluated for possible suicidal ideation when withdrawal is complete.

10. Sympathomimetics

10.1. Introduction

Amphetamines and cocaine are the two sympathomimetics most commonly abused. A third, methylphenidate (Ritalin), may occasionally be taken, especially by individuals who have been treated for hyperactivity as a child or adolescent. Its physiological and psychological effects are similar to cocaine and amphetamines, but are less marked.

Medical uses of cocaine and amphetamine have been curtailed in recent years. Amphetamines presently are approved for treatment of hyperactive children, for short-term treatment of obesity, for the differential diagnosis of depression, and for mood elevation in narcolepsy. Cocaine has limited application as a nasal anesthetic. Its attractiveness to abusers lies in the rapid onset of euphoria, coupled with a feeling of mastery and power. The duration of the effect is short lived. Many abusers frequently take a succession of "fixes" ("runs") in order to sustain the effect and postpone the letdown of mood and return to a normal state. Amphetamine has a longer duration of activity than cocaine, is more plentiful and cheaper, but is less exhilarating. Routes of administration for both are inhaling, "skin-popping," and intravenous "shooting."

10.2. Clinical Observations of Amphetamines and Cocaine

10.2.1. Psychological Effects of Amphetamine

Psychological effects of amphetamine may be classified as acute or subacute if the drug has been taken for less than one month. Chronic abusers develop considerable tolerance and may accomodate up to 1,250 mgm per day without serious side effects (Angrist, Lee, & Gershon, 1974; Connell, 1958). Following a psychosis, there may be increased sensitivity to the psychological effects of amphetamines, but tolerance to the physiological effects may not be altered.

Acute or subacute reactions are dose related. The most pronounced effect is a feeling of wakefulness and alertness, which has been exploited by student users and by long-distance truck drivers. Higher doses are associated with feelings of euphoria, omnipotence, and enhanced self-confidence. Pressure of speech and overtalkativeness, which are characteristic of many stimulants, including caffeine, are also present. Enhancement of motor activity, coupled with feelings of power, have led some athletes to use amphetamines prior to games. Occasionally, people in psychotherapy give themselves a fix before a session. The therapist, who is not alert to the characteristics of stimulants, may not recognize that the patient is high. Hallucinations, illusions, and paranoid ideation may occur in the presence of a clear sensorium, which has led some to consider amphetamine psychosis as a "model psychosis." In "speed freaks," paranoid ideation and enhanced energy may lead to violent behavior. Occasionally, blackouts may occur.

Chronic reactions resemble paranoid schizophrenia, hypomania, or a manic psychosis. Hallucinations involving any of the senses may be present. Paranoid symptoms are dose related, and nearly all serious abusers exhibit them. Delusions generally are not well systematized. When paranoid psychotic features predominate, paranoid schizophrenia will be mimicked. Marked pressure of speech, overactivity, and flight of ideas suggest hypomania or a full-blown manic psychosis. Because the sensorium is often clear, diagnosis can be difficult. Amphetamine psychosis should be suspected when there is pressure of speech, evidence of weight loss, or cachexia, a high level of excitement or exhaustion, accompanied by paranoid ideation. When an organic brain syndrome is present, the diagnosis is simpler because drug abuse is more likely to be included in the differential diagnosis. Quite often, abusers will deny using amphetamines during the paranoid/acutely psychotic period and may be mistakenly diagnosed as schizophrenic or manic. Laboratory studies may be necessary to confirm the diagnosis.

10.2.2. Physical Effects of Amphetamines

Physical effects of amphetamines are the result of sympathetic nervous system stimulation. Gastrointestinal complaints include dryness of the mouth, a metallic taste, nausea, vomiting, cramps, and diarrhea. Anorexia, to the point of cachexia, is prominent in heavy users. Adrenergic crisis, primarily involving the cardiovascular system, is the most serious side effect. Arrhythmias may lead to cardiac arrest. Hypertension may be the precursor of a cerebrovascular accident manifest by coma, convulsions, and death. The pupils of amphetamine abusers are dilated and nystagamus may be present. Libidinal changes are reported by men. Initially, there may be an increased desire, often with sustained erections, but long-term use may lead to impotence. Motor system changes include tremulousness, bruxism, and stereotypical movements, often with picking of the skin. Deep tendon reflexes are increased. Diaphoresis may be marked when moderately high doses are taken. If amphetamine is snorted, there may be slight constriction of blood vessels about the nasal orifice. Nasal septal perforation occurs, but is unusual.

Exhaustion ("crashing") following sprees is associated with depression, apprehension, delirium, confusion, headache, dizziness, palpitations, and jitteriness. Dysregulation of the autonomic nervous system can develop.

10.2.3. Psychological Effects of Cocaine

The effects of cocaine are, with a few exceptions, similar to those described for amphetamine. The psychological effects of cocaine are short lived because the drug is rapidly detoxified by the body. As a result, tolerance does not develop; nor are there withdrawal symptoms. Cocaine is more likely than amphetamines to produce an acute brain syndrome. Visual hallucinations, often Lilliputian, and frequently of animals, are not uncommon. Paranoid, agitated states were seen in the first years of cocaine's popularity, but are rare today. The crazed behavior of some acutely intoxicated cocaine addicts gave rise to the expression "dope fiend."

10.2.4. Physical Effects of Cocaine

Physical changes associated with cocaine use are minimal. Some abusers who "mainline" may have distorted and scarred veins because of adulterants in the cocaine. Nasal perforation, once felt to be a *sine qua non* of cocaine abuse, is rarely seen today largely because "mainlining" and "free-basing" have become so popular.

10.3. Treatment

In the ER it is unusual to see the short-lived, acute reaction to cocaine. An occasional amphetamine abuser may come for help when acutely intoxicated. Usually sedatives suffice to calm the patient. When an acute or chronic psychosis is present, a major tranquilizer such as haloperidol (Haldol) is indicated. If depression occurs, tricyclics are useful. Inpatient treatment for compulsive cocaine abuse has been implemented in some centers. Individual and group psychotherapy, milieu and occupational therapy, and tricyclic antidepressants, as indicated, form the basis of the program. Monitoring for continued drug abuse is also done.

The major physical dangers of stimulant overdose are adrenergic crisis and cardiac or pulmonary distress. The treatment is outlined in Section 11. Exhaustion and physical inanition, seen in amphetamine abuse, can be treated by providing ample diet, vitamin supplements, and rest.

11. Evaluation and Treatment of Overdose

11.1. Introduction

Treatment of a semistuporous, stuperous, or comatose patient is a medical emergency, best performed by a trained staff in a well-equipped area. In this section, we will present an approach to the evaluation and treatment of people who have ingested a psychoactive drug.

All people who are in a semistuporous, stuperous, or comatose state should be considered as suffering from an overdose or insulin shock until a more definitive diagnosis can be made. Clues may be found by examining the patient's property to see if there are medical alert cards, doctors' names, or informational tags. The presence of an empty container near the person or of a suicide note are generally accurate indicators of intent. However, a history obtained from a friend or relative may be misleading and a thorough medical examination must be performed.

A properly organized emergency room is essential if comprehensive, rational treatment is to be offered. Staff should be well trained and retrained annually in cardiopulmonary resuscitation (CPR). Literature should be readily available (Effron, 1982; Standards of JAMA, 1980; White, 1982). The center's life-sustaining equipment should be regularly maintained and readily accessible. A fully stocked medicine cabinet with antidotes to poisons should be available, as well as a table reminding staff of how to use the antidotes.

The remainder of this chapter is organized to follow the progress of a semistuporous, stuperous, or comatose person through the course of an emergency. The following topics will be discussed: the differential diagnosis of coma; initial evaluation and initiation of treatment; and detailed evaluation, with specific emphasis upon findings in the physical examination relevant to substance abusers.

11.2. Differential Diagnosis of Coma

The differential diagnosis of coma requires consideration of many conditions. Among the more common are: acidosis (alcoholic, diabetic ketoacidosis, uremic); accident (cerebrovascular, including subarachnoid hemorrhage, rupture of middle meningeal artery, subdural hematoma, thrombosis; trauma to the skull and brain laceration or contusion); adrenal insufficiency (Addison's disease); decompression sickness; epilepsy, lowered threshold for seizure (following use of lithium, monoamine oxide inhibitors, or MAOI's, withdrawal from sedatives or alcohol, phenothiazines, tricyclics); embolus; eclampsia; expanding brain lesion; intoxication (acute overdose from any substance); infection (meningitis, encephalitis, pneumonia); infarction (cerebral, myocardial), hepatic disease; hyper- or hypo-calcemia, -glycemia, -natremia, -thermia, -thyroidism; hysteria; hypertension; syncope and terminal event of illness.

Distinguishing between coma induced by toxic and/or by structural damage is of critical importance. The presence or absence of the pupillary light reflex is of help since the reflex is usually not lost until near the point of death in toxic states, unless opiate, anticholinergic, or glutethimide overdose is present (Mack, 1983). In toxic states, there may be slight divergence of the eyes, but this is minimal compared to the extreme divergence or failure of the eyes to return to the midline following rotation of the head in cases where there is structural damage.

Semistuporous people who are agitated and negativistic may be confused with acutely psychotic people who refuse to answer questions. It is critical that physical examinations be performed and that efforts be made to see if any confusion exists.

11.3. Physical Examination

An initial brief, physical examination should be performed immediately. The focus is to highlight potential or actual problems. In summary:

1. Skin—color, dryness, turgor, temperature, burns, needle-track or hypodermic marks.
2. Skull—injury, bleeding, fluids leaking from orifice.
3. Pupils—size, regularity, diversion, and responsiveness to light.
4. Cardiovascular (see Section 11.5).
5. Respiratory (see Section 11.4).
6. Abdomen—injury, tenderness.
7. Extremities—injury, correct anatomic position.
8. Neurological examination—signs of cerebral accident, pain sensitivity, and localization.
9. Mental status—orientation to person, place, time, history of recent events, recent or chronic illness, medication taken, substance abuse, suicidal intent.

11.4. Initial Evaluation of the Respiratory System

In the initial evaluation of the respiratory system, the first assessment is to identify life-threatening problems and rectify them immediately. Although the evaluation appears to be done in consecutive steps, with experience most of the observation is done simultaneously. A quick look at the patient provides information regarding *skin* and *lip* color, cyanotic or cherry red lips being suggestive of carbon monoxide poisoning. Observation of *respiratory patterns* may show slow, deep respirations suggestive of narcotic overdose or rapid, shallow breathing more characteristic of barbiturates. Irregular rhythm, Cheyne-Stokes or Kussmaul breathing, is characteristic of central-nervous-system problems. Obstruction in the respiratory tract may be identified by hearing stertorous breathing. Examination of the airway should include listening to breath counts, noting if the tongue has been swallowed, and searching for aspirated vomitus or mucous plugs. Confirming patency of the airway includes removal of false teeth, holding the tongue in place, if necessary, with a plastic pharyngeal airway, and, if the patient is comatose, inserting a cuffed endotracheal tube.

Other therapeutic measures include administration of oxygen when breathing is sufficiently labored or when the patient does not have normal color. While a catheter is acceptable, an Ambu bag is more useful during the first hours of treatment. It facilitates administering artificial respiration at half-hourly intervals in order to clear the lungs and prevent atelectasis from developing. Another measure is to turn the patient from side to side each half hour. Bronchial suction is recommended after respirations have been established, and the gastric tube is in place. The pH of the aspirate should be done with nitrazine paper. If it is below 3, gastric juices have been aspirated. Bronchial suction should be done, and steroids given.

If the patient is comatose and respiration is labored, the possibility of opiate overdose should be considered. A drug screen should be done, and a test of naloxone (0.4 mg, IV or IM) may be given and repeated at 2- or 3-minute intervals for no more than three doses. If respirations do not improve by then, opiate coma can be ruled out. If opiate coma is present, the patient should be treated until response occurs and then observed for the next 24 hours to make certain that relapse does not occur. If it does, additional naloxone is indicated (Levin, 1983). Naloxone can cause severe withdrawal symptoms. It should not be given too vigorously as a test to nonaddicts and should be given with caution to addicts, since they, too, may suffer from withdrawal.

When an overdose of propoxyphene is being treated, larger doses of naloxone may be necessary (Moore, Rumack, Conner, & Peterson, 1980).

11.5. Cardiovascular Evaluation

Cardiovascular evaluation initially consists of preparing the veins with an indwelling catheter to assure patency, starting an IV of 5% Dextrose/water, and drawing blood for toxicological examination and baseline studies of CBC, electrolytes, SMA 12, and pH. The circulatory system is examined by monitoring the pulse for rate and rhythm and by listening to the heart and recording an EKG. If there are conduction problems or arrhythmias, the EKG recording should be continuous. Immediate intervention is indicated if ventricular fibrillation or excessive tachycardia is present.

When the initial assessment is completed, emergency treatment of blood pressure changes and heart failure can be initiated. Hypotension should be treated with the usual physical techniques for controlling shock. Blood volume is controlled with 5% Dextrose/water, and, if there is no response, a pressor such as dopamine (Intropin), norepinephrine (Levophed), or phenylephrine (Neo-synepherine) should be administered intravenously. Blood pressure must be monitored carefully every 2 minutes until a normotensive range is stabilized, and every 5 minutes thereafter. To avoid the development of ischemia, make certain that the IV flows freely and that there is no leakage to the surrounding tissue. These drugs are potentiated in the presence of MAO inhibitors and tricyclic antidepressants.

Drugs that may cause a hypertensive reaction are stimulants, such as amphetamine, or MAOI inhibitors, when dietary regulations have not been followed. The elevated pressure should be controlled by intravenous infusion of phentolamine (Regitine), nitroprusside (Nipride), trimethaphan (Arfonad), or propranolol (Inderal). All must be administered slowly, with monitoring of the blood every minute. A continuous EKG is recommended. Cardiac failure is treated by usual methods of rapid digitalization and careful monitoring of cardiac status.

11.6. Other Immediate Therapeutic Measures

Other immediate therapeutic measures are determined by the clinical condition of the patient and are generally supportive. Some believe that all comatose people should be treated immediately with a bolus of 50 cc of glucose in 100 cc of water, naloxone, 2–5 mg, and

Thiamine (10 mg), while others, more conservatively, would await the results of laboratory data and make a definitive diagnosis. Once a diagnosis of overdose is made, it is necessary to clear the stomach, if the patient has not already vomited because some residual drug may be present (Flomenbaum, 1983).

The decision to perform renal dialysis is a difficult one (Stern, 1983). The subject is discussed in some detail in Seyffart (1977) and Braxter (1982).

11.6.1. Clearing the Stomach

Induction of vomiting is indicated when patients are alert, have active gag reflexes, and have not vomited subsequent to taking their overdose. Syrup of Ipeccac and Apopmorphine are the agents of choice to induce vomiting. The former is gentler; the latter, quicker (onset of action 15 minutes in contrast to 30). Neither may be effective when dopamine blocking agents (phenothiazines or tricyclics) have been ingested. Gastric lavage is then the treatment of choice.

Dosage for Syrup of Ipeccac is 30 cc in two to three glasses of water. It may be repeated once. For Apomorphine, 5 mg, subcutaneously, is the standard dose, but the range is between 2 and 10 mg. It should not be readministered since it can produce serious side effects, including respiratory depression and occasionally death. Goldfrank et al. (1982a) recommend administration of nalaxone when vomiting is completed.

Antimetics are contraindicated when there is impending shock, or ingestion of sharp objects, corrosive poisons, strong alkalis or acids, petroleum distillates or strychine, or when there is CNS depression (following opiates, alcohol or sedatives) Goldfrank, Flomenbaum, Lewin, and Weisman (1982) additionally suggest that cirrhosis, thrombocytopenia, and active seizures are also contraindications.

When lavage is utilized, a large bore tube, such as an Ewald or Orogastric, should be passed through the mouth in order to protect the nose from injury. Lavage is performed by introducing 150–200 cc of water or saline through the tube and withdrawing the fluid. The process should be repeated until the gastric contents are clear, which generally occurs within 20 minutes. After the stomach is cleared, *activated charcoal* should be given.

A thin watery mixture (slurry) of activated charcoal, 1 gm/kg of body weight, is administered orally or passed through the tube. Activated charcoal is prepared by mixing 30 gm of activated charcoal with a glass of water. It inactivates a wide variety of substances, including most psychoactive drugs, with the exception of glutethimide (Flomenbaum,

1983). Next, one of the *ionic cathartics* is administered. This does not interact with charcoal, and it helps evacuate the GI tract.

The cathartics are: magnesium sulfate (Epsom salts), sodium sulfate, or phosphosoda. Contraindications to their use are: adynamic ileus, severe diarrhea, abdominal trauma, intestinal obstruction, renal failure (for which magnesium sulfate is contraindicated), cardiac failure (for which sodium sulfate is contraindicated), and ethylene glycol (for which phosphosoda is contraindicated).

11.6.2. Specific Antidotes

Specific antidotes should be available in every emergency room. In addition to a posted list of drugs and antidotes, a procedural manual defining the medical indications, contraindications, and complications for each antidote is helpful. In preparing the manual and lists, the experience of the medical staff and, possibly, the local poison control center can be drawn upon.

11.7. Detailed Evaluation

Detailed evaluation begins as soon as the immediate life-saving steps are completed. In performing this physical examination, the direct and indirect effects of toxic substances upon the body must be considered. They control the function of vital organs through effects upon the medulla and midbrain as well as by local effects upon the contractibility and conduction of the heart and permeability of blood vessels. The permeability of the lungs is also affected, leading to noncardiac pulmonary edema.

It is helpful to understand the effects of the autonomic nervous system (ANS) upon the body (Mayer, 1980) because it controls most bodily functions. The parasympathetic system is primarily responsible for mucous excretion, pupillary contraction, salivation, and tearing and vasodilation of the blood vessels of the head. It also slows the heart and reduces gastrointestinal motility. The sympathetic nervous system has the opposite effects, constricting blood vessels, dilating pupils, erecting hair, increasing heart rate, reducing secretions, and stimulating respirations. Table 2 summarizes these effects by system.

These effects are enhanced or blocked by different substances of abuse. Both anticholinergic (AChe) and adrenergic (Ad) drugs cause pupillary dilitation. Neither effects accomodation. Both will increase heart rate, but adrenergic effects are more likely to increase blood pres-

Table 2
Comparison of Sympathetic and Parasympathetic Responses

System involved	Sympathetic response	Parasympathetic response
Skin	moist	dry
	pale	flushed
	cool	hot
Vascular	contracted	dilated vessels
	tachycardia	bradycardia (but release phenomena can occur)
	hypertension	hypotension
Temperature	hyperpyrexia	hyperpyrexia, 2° to inability to lose heat
Visual		
Pupils	dilated	dilated
Reflexes	active	inactive
Diploplia	—	present
Mocous membranes	moist	dry
GU	—	retention
GI	nausea, vomiting, diarrhea, active bowel sounds	constipation, absent bowel sounds, thirst

sure. AChe's are more likely to cause decreased secretions, dryness of skin, flushing of face, hyperpyrexia, increased gastrointestinal motility, and stopping of urination. Scopolamine will have more of an effect upon the eyes and secretory and sweat glands than atropine, but less on the GI tract. Scopolamine, more than atropine, is apt to have CNS excitatory reactions, including a feeling of euphoria and hallucinations.

12. Summary

This chapter discusses the clinical manifestation of substance abuse, emphasizing the complications most likely to be seen in a psychiatric emergency room. The different groups of substances are discussed in some detail, with special attention given to the psychiatric and physical responses to average and large doses. Treatment of short-term side effects, psychotic responses, and the initial assessment and resuscitation of people who have taken an overdose are highlighted.

ACKNOWLEDGMENTS

Dr. Charles Rohrs assisted in the preparation of Section 2 on alcoholism. Dr. Bernard Salzman assisted in the preparation of Section 8 on narcotics.

13. References

Abromowicz, M. Clonidine (Catapras) an antihypertensive agent has been used to provide nonnarcotic detoxification. *The Medical Letter on Drugs and Therapeutics*, 1979, *21*, 100.
Adriani, J. Narcotic Poisoning. In H. F. Conn (Ed.), *Current drug therapy*. Philadelphia: W. B. Saunders, 1979.
Alpert, R., Cohen, S., & Schiller, L. *LSD*. New York: New American Library, 1966.
Angrist, B., Lee, H. K., & Gershon, S. The antagonism of amphetamine-induced symptomatology by a neuroleptic. *American Journal of Psychiatry*, 1974, *131*, 817–819.
Barr, H. L., Langs, R. J., Holt, R. R., Goldberger, L., & Klein, G. S. *LSD: Personality and experience*. New York: Wiley-Interscience, 1972.
Blum, R., & Associates. *Utopiates*. New York: Atherton Press, 1964.
Braxter, D. C. *Handbook of drug use in patients with renal disease*. Lancaster, Texas: Improved Therapeutics, 1982.
Castanada, C. *The teachings of Don Juan: The Yagui way of knowledge*. Berkeley: University of California Press, 1968.
Cohen, S. *The beyond within*. New York: Atheneum, 1965.
Cohen, S. *The substance abuse problem*. New York: Hawsorth Press, 1981.
Cohen, S. Coming of age in America—with drugs: Contemporary adolescence. *Drug Abuse & Alcoholism Newsletter*, 1982, *9*, 9.
Cohen, S. Marijuana use detection: The state of the art. *Drug Abuse & Alcoholism Newsletter*, 1983, *12*, 3.
Cohen, S., & Ditman, K. S. Prolonged adverse reactions to lysergic acid diethylamide. *Archives of General Psychiatry*, 1963, *8*, 475–480.
Connell, P. H. *Amphetamine psychosis* (Maudsley Monograph No. 5). London: Oxford University Press, 1958.
Crome, P. Antidepressant overdosage. *Drugs*, 1982, *23*, 431–461.
Dakis, C. A., Pottash, A. L. C., Annitto, W., & Gold, M. S. Persistence of urinary marijuana levels after supervised abstinence. *American Journal of Psychiatry*, 1982, *139*, 1196–1198.
Done, A. K. Inhalation of glue fumes and other substances. In *Abuse practices among adolescents*. Washington, D.C.: U.S. Government Printing Office, 1968.
Effron, D. M. *Cardiopulmonary resuscitation (CPR)*. Tulsa, Okla.: CPR Publishers, 1982.
Fauman, B., Aldinger, G., Fauman, M., & Rosen, P. Psychiatric sequelae of phencyclidine abuse. *Clinical Toxiology*, 1976, *9*, 529–538.
Flomenbaum, N. The GI front. *Medical Emergencies*, October 15, 1983, pp. 152–164.
Frosch, W. A., Robbins, E. S., & Stern, M. Untoward reactions to lysergic acid diethylamide (LSD) resulting in hospitalization. *New England Journal of Medicine*, 1965, *273*, 1235–1239.
Frosch, W. A., Robbins, E., Robbins, L., & Stern, M. Motivation for self-administration of LSD. *Psychiatric Quarterly*, 1967, *41*, 56–61.

Garay, S. Pulmonary aspects. *Medical Emergencies*, October 15, 1983, pp. 187–195.

Goldfrank, L. R., Flomenbaum, N. E., Lewin, N., & Weisman, R. S. *Toxicologic emergencies: A comprehensive handbook in problem solving*. New York: Appelton-Century, Crofts, 1982.

Goldfrank, L., Lewin, N., Flomenbaum, N., & Howland, M. A. The pernicious panacea: Herbal medicine. *Hospital Physician*, 1982, *18*, 64–68.

Greenblatt, D. J., & Shader, R. I. Treatment of the alcohol withdrawal syndrome. In R. I. Shader (Ed.), *Manual of psychiatric therapeutics*. Boston: Little, Brown, 1975.

Hall, R. C. W., Feinsilver, D. K., & Holt, R. E. Anticholinergic psychosis: Differential diagnosis and management. *Psychosomatics*, 1981, *22*, 581–587.

Harvey, S. C. Hypnotics and sedatives. In A. G. Gilman, L. S. Goodman, & A. Gilman (Eds.), *The pharmacologic basis of therapeutics* (6th ed.). New York: Macmillan, 1980.

Huxley, A. L. *The doors of perception: Heaven and hell*. New York: Harper & Row, 1963.

Jaffe, J. H. Drug addiction and drug abuse. In A. G. Gilman, L. S. Goodman, & A. Gilman (Eds.), *The pharmacologic basis of therapeutics* (6th ed.). New York: Macmillan, 1980.

Jaffe, J. H., & Martin, W. R. Opioid analgesics and antagonists. In A. G. Gilman, L. S. Goodman, & A. Gilman (Eds.), *The pharmacologic basis of therapeutics* (6th ed.). New York: MacMillan, 1980.

Khantzian, E. J., & McKenna, G. J. Acute toxic and withdrawal reactions associated with drug use and abuse. *Annals of Internal Medicine*, 1979, *90*, 361–372.

Leary, T. F., Metzner, R., & Alpert, R. *The psychedelic experience: A manual based upon the Tibetan Book of the Dead*. New Hyde Park, N.Y.: University Books, 1964.

Levin, R. I. The cardiac aspects. *Medical Emergencies*, October 15, 1983, pp. 172–186.

Liden, C. B., Lovejoy, F. H., & Costello, C. E. Phencyclidine: Nine cases of poisoning. *Journal of the American Medical Association*, 1975, *234*, 513–516.

Lieber, C. S., Gordon, G. G., & Southren, A. I. The effects of alcohol and alcoholic liver disease on the endocrine system and intermediary metabolism. In C. S. Lieber (Ed.), *Medical disorders of alcoholism*. Philadelphia: W. B. Saunders, 1982.

Luisada, P. V., & Brown, B. L. Clinical management of phencyclidine psychosis. *Clinical Toxicology*, 1976, *9*, 539–545.

Mack, B. J. The CNS slant. *Medical Emergencies*, October 15, 1983, pp. 165–171.

Masters, R. E. L., & Houston, J. *The varieties of psychedelic experience*. New York: Holt, Rinehart and Winston, 1966.

Mayer, S. E. Transmission and the Autonomic Nervous System. In A. G. Gilman, L. S. Goodman, & A. Gilman (Eds.), *The pharmacologic basis of therapeutics* (6th ed.). New York: Macmillan, 1980.

Moore, R. A., Rumack, B. H., Conner, C. S., & Peterson, R. G. Naloxone: Underdosage after narcotic poisoning. *American Journal of Diseases of Children*, 1980, *134*, 156–160.

Nahas, C. G. Hashish in Islam in the 9th and 10th century. *Bulletin of the New York Academy of Medicine*, 1982, *58*, 814–831.

Nicholi, A. M., Jr. The nontherapeutic use of psychoactive drugs: A modern epidemic. *New England Journal of Medicine*, 1983, *308*, 925–933.

Rainey, J. M., & Crowder, M. K. Prolonged psychosis attributed to phencyclidine: Report of three cases. *American Journal of Psychiatry*, 1975, *132*, 1076–1078.

Rall, T. W., & Schleifer, L. S. Drugs effective in the treatment of the epilepsies. In A. G. Gilman, L. S. Goodman, & A. Gilman (Eds.), *The pharmacologic basis of therapeutics* (6th ed.). New York: Macmillan, 1980.

Rappolt, R. T., Gay, G. R., & Farris, R. D. Emergency management of acute phencyclidine intoxication. *Annals of Emergency Medicine*, 1979, *8*, 68–76.

Robbins, E., Agus, B., & Stern, M. Drug abuse. In R. A. Glick, A. T. Meyerson, R. Robbins, & J. A. Talbott (Eds.), *Psychiatric emergencies*. New York: Grune & Stratton, 1976.

Robbins, E., Robbins, L., Frosch, W. A., & Stern, M. Implications of untoward reactions to hallucinogens. *Bulletin of the New York Academy of Medicine*, 1967, 43, 985–999.

Robbins, L., & Robbins, E. Ganja in Jamaica: A medical anthropological study of chronic marijuana use (book review). *American Journal of Psychiatry*, 1977, 121, 238–244.

Salamon, I. Alcoholism. In H. F. Conn (Ed.), *Current therapy*. Philadelphia: W. B. Saunders, 1980.

Schenker, S. Effects of Alcohol on the brain: Clinical features, pathogenesis and treatment. In C. S. Leiber (Ed.), *Medical disorders of alcoholism*. Philadelphia: W. B. Saunders, 1982.

Seyffart, G. Dialysis and haemoperfusion in poisonings. *Poison Index*. Bad Homberg: Fresnius Foundation, 1977.

Smith, D. E. A clinical approach to the treatment of phencyclidine (PCP) abuse. *Psychopharmacology Bulletin*, 1980, 16, 67–70.

Smith, M., Squires, R., Aponte, J., Rabinowitz, N., & Bonilla-Rodriquez, R. Acupuncture treatment of drug & alcohol abuse: 8 years experience emphasing tonification rather than sedation. *British Journal of Acupuncture*, 1982, 5, 9–10.

Spitzer, R. L. *Diagnostic and statistical manual of mental disorders* (3rd ed.). Washington, D.C.: American Psychiatric Association, 1980.

Standards and guidelines for Cardiopulmonary Resuscitation (CPR) and Emergency Cardiac Care (ECC). *Journal of the American Medical Association*, 1980, 244, 453–509.

Stern, L. The renal angle. *Medical Emergencies*, October 15, 1983, pp. 196–215.

Stoll, A., & Hoffman, A. Partialsynthese von alkaloiden vom typus des ergobasins. *Helvetica Chimica, Acta* 1943, 26, 944–965.

Vardy, M. M., & Kay, S. R. LSD-psychosis or LSD-induced schizophrenia? *Archives of General Psychiatry*, 1983, 40, 877–883.

Weiner, N. Atropine, scopolamine and related drugs. In A. G. Gilman, L. S. Goodman, & A. Gilman (Eds.), *The pharmacologic basis of therapeutics* (6th ed.). New York: Macmillan, 1980.

White, R. B. *Basic life support: A Ciba symposium*. West Caldwell, N.J.: Ciba, 1982.

11

Neuropsychology of Alcohol and Drug Abuse

Igor Grant and Robert Reed

1. Introduction

Human interest in substances that have the capacity to alter his intrapsychic state must be as ancient as humanity itself. Certainly, the desirable effects and many of the toxic properties of ethanol, cannabis, opium, and some of the naturally occurring central stimulants have been known from antiquity. In the late 19th century, progress in organic chemistry led to the ready availability of purified, naturally occurring psychoactive substances and their synthetic analogues. Although the chemistry and pharmacology of at least some representatives of most of the drug classes with which we shall be concerned in this chapter were already appreciated in the first half of the 20th century, it was not until the coming of the psychosocial revolution of the 1960s that any but a few scientists and small groups of addicts were aware of most of them (with the exception of alcohol). The sixties brought with them an interest in altering the psychological internal milieu. This philosophical disposition toward inner exploration was powerfully reinforced by an increasing popular understanding of some of the neurochemical bases of brain function, and by clinical psychiatry's demonstration that powerful drugs could be used to alter mood, thinking, and behavior. Thus, a sociopolitical climate that stressed freedom, the uniqueness of the individual, flexibility,

Igor Grant and Robert Reed • San Diego VA Medical Center and University of California, San Diego, School of Medicine, Department of Psychiatry (M-003), La Jolla, California 92093. This work was supported by the Medical Research Service of the Veterans Administration.

and the importance of self-knowledge coalesced with advances in neuropsychopharmacology to create a moral climate that was tolerant of self-medication, and that considered such practices in some instances to be legitimate, even fashionable.

As experimentation with various classes of psychoactive drugs began touching larger and larger segments of populations of Western countries, so it became obvious that an increasing minority of people were unable to confine their nonmedical drug use to the experimental, occasional, or even recreational.

The gradual emergence of a substantial number of nonmedical users of psychoactive drugs naturally raised concerns that the target organ of such drug use—the brain—might suffer long-term, perhaps irreversible harm. The abuse of alcohol was already seen as a model: Even before nonmedical drug use became a socially important phenomenon in the mid-1960s, Fitzhugh and associates (Fitzhugh, Fitzhugh, & Reitan, 1960, 1965) had already demonstrated that many alcoholics, selected from a population free of obvious neurological disease or intellectual impairment, nevertheless had subtle impairments in cognitive function suggestive of generalized cerebral impairment.

In 1975 our group reviewed some of the neuropsychological research then extant concerning the long-term cerebral consequences of the abuse of alcohol, marijuana, opiates, central nervous system depressants, central stimulants, and hallucinogens. In this chapter our intent is to pick up the story from the mid-1970s and to update it. Readers wishing to acquaint themselves with some of the earlier literature might consult our previous article (Grant & Mohns, 1975) or a more recent review by Parsons and Farr (1981).

Our general plan will be to discuss neuropsychological studies grouped under various pharmacological classes of agents. We do this for didactic convenience, while recognizing that in many instances several classes of drugs will be abused concurrently or sequentially (Wesson, Carlin, Adams, & Beschner, 1978). To emphasize this point, one of our final sections will consider specifically the issue of polydrug use, so that the reader might place neuropsychological correlates of multisubstance use in context.

2. Alcohol

Since 1975 there has been a veritable explosion of neuropsychological studies of alcoholics. In the United States this efflorescence has been fueled partially by the establishment of a separate federal funding agency for studies on alcohol (NIAAA). Worldwide, the advent of computerized tomography of the brain has led to increased interest in examining neuroradiological–neuropsychological correlates.

Because of the large number of studies, we will group them in terms of the sorts of research problems with which they primarily deal. These groups of research areas include: specification of the qualitative and quantitative features of neuropsychological change in recently detoxified alcoholics; processes of recovery; prediction of neuropsychological test results from knowledge of drinking history; relationship to aging; relationship to other medical disorders; and relationship to treatment outcome.

2.1. Studies of Recently Detoxified Alcoholics

The bulk of neuropsychological investigations of alcoholics has been conducted with inpatients abstinent 1 to 4 weeks. This is an important point when it comes to drawing inferences about possible long-term effects of alcohol on the brain: While it is true that most studies have taken care not to include any patients who were still in the acute phase of an alcohol withdrawal syndrome, it is nevertheless also becoming evident that even a month of detoxification is probably not ample to achieve maximal neuropsychological recovery from long drinking bouts. Therefore, although the observations to be reviewed below are important in their own right, they should be seen as reflections of the state of an alcoholic's brain during what might be termed a "subacute" phase of recovery from an alcohol-related organic mental disorder. In our view, the extent to which phenomena seen in this subacute phase are predictive of permanent deficits has yet to be established. Permanent deficits can only be confirmed through studies of alcoholics abstinent several years.

2.1.1. Disturbances of Abstracting Ability and Complex Perceptual Motor Functioning

Fitzhugh, Fitzhugh, and Reitan (1960, 1965) were among the first to show that recently detoxified alcoholics who had no obvious neurological disorder evidenced mild disturbances of abstracting ability (as measured by the category test) and of complex perceptual motor skills (as measured by the tactual performance test) in the context of normal psychometric intelligence.

This observation was confirmed by Jones and Parsons (1971, 1972), and the Oklahoma group headed by Parsons has continued in the forefront of detailing neuropsychological change in recently detoxified alcoholics. For example, Klisz and Parsons (1977) demonstrated that alcoholics had difficulties in reasoning as measured by the Levine Hypothesis Testing task. Exploring these reasoning deficits further, the same authors (Klisz & Parsons, 1979) found that alcoholics failed to reach criterion on the Wisconsin Card Sorting task significantly more often than did controls. Further, alcoholics who did reach criterion still scored worse than controls.

The Oklahoma group has also taken the lead in extending to women alcoholics findings that were previously established in male alcoholics. Fabian, Parsons, and Silverstein (1981) reported that women alcoholics performed worse than nonalcoholic controls on a factor analytically derived ability factor that loaded on tests of nonverbal learning and perceptual, spatial, and problem-solving skills. Similar deficits in women alcoholics were found by Nichols-Hochla and Parsons (1982) and Fabian and Parsons (1983).

A number of other groups of investigators have likewise been active in exploring the neuropsychology of recently detoxified alcoholics. At the University of Washington, Donovan and associates (Donovan, Queisser, & O'Leary, 1976) examined 90 alcoholics in their late 40s approximately 1½ weeks after their last drink. These investigators confirmed the presence of nonverbal cognitive disturbance as reflected in depressed Shipley Institute of Living abstraction scale scores. They also affirmed previous observations that such alcoholic groups tend to score in the "external" direction on the group embedded figures test. In another study this Seattle group (O'Leary, Donovan, & Chaney, 1977) compared the functioning of 38 alcoholics abstinent 9–14 days with a matched control group of hospital workers. Neither group had neuropsychiatric diagnoses other than alcoholism. The alcoholics scored worse on the category test, on the WAIS comprehension, arithmetic, picture arrangement, and picture completion subscales, and on both parts of the trail making test, and they had a depressed brain age quotient. The alcoholics were well matched to the controls on verbal IQ, hence education or "native intelligence" differences should not have been playing a part. These investigators also confirmed an increased tendency toward field dependence in the alcoholics and showed that increased externality was significantly correlated with neuropsychological impairment.

A group of Detroit investigators has confirmed perceptual motor and visual perceptual deficits among recently detoxified alcoholics (Ellenberg, Rosenbaum, Goldman, & Whitman, 1980). Eckardt and associates (Eckhardt, Ryback, & Paulter, 1980) found that alcoholics abstinent 2–6 days and 14–31 days were impaired on 12 of 24 neuropsychological tests. The principal impairments were, once again, on category test, tactual performance test, certain WAIS performance scales, as well as global indices of impairment. Interestingly, this same group of investigators (Eckardt, Parker, Noble, Paulter, & Gottschalk, 1979), had earlier found that these alcoholic subjects were statistically indistinguishable from medical controls.

Our own research team has an ongoing prospective investigation examining the interaction of aging, abstinence, and neuromedical risk factors in the prediction of neuropsychological deficit in alcoholics (Grant, Adams, & Reed, 1984). In our recently detoxified group (average duration of abstinence equals 4.4 weeks) we have noted only modest differences between alcoholics and sociodemographically matched controls. Indeed, there were no statistically significant differences on univariate analyses of individual test results. On the other hand, when composite scores were generated, either from factor analysis of extended Halstead-Reitan test results or from blind clinical ratings, some difficulties in learning and problem solving were revealed in recently detoxified alcoholics. Although the differences between the alcoholics and controls were statistically significant on the composite measures, the changes must be regarded as subtle. Our findings were in marked contrast to those of Miller and Orr (1980), who found, on Halstead-Reitan testing, that their recently detoxified alcoholics (mean duration of abstinence, 31.6 days) performed equivalent to a brain-damaged comparison group, and much worse than psychiatric controls. These differences in extent of deficit found between our study and Miller and Orr's illustrate that subject selection plays an important role in results that will be obtained. While our two groups were roughly comparable in age—mean age in Grant et al. (1984) was 42.0; mean age in Miller and Orr (1980) was 46.6—subjects in the Miller and Orr study were selected from among alcoholics referred for neuropsychological assessment; our alcoholics, on the other hand, specifically excluded patients who had positive histories of neurological, developmental, severe medical, or severe psychiatric disturbances. Furthermore, Miller and Orr probably assured that their alcoholics would perform somewhat worse when they matched them to a comparison group that was ten years younger and one year better educated. As we continue this review, we will have more to say about the role of age, education, and neuromedical factors as confounding influences in examining the ethanol–cerebral-dysfunction relationship. For the moment, it should be sufficient to caution the reader that many studies have not taken sufficient care to exclude potentially important sources of neuropsychological variability.

Returning to the main theme, that of evidence for impaired abstracting and perceptual motor abilities in recently detoxified alcoholics, investigators from all over the world have reported results very comparable to those of the U.S. researchers reviewed above. In Canada, Wilkinson and Carlen (1980b) found that 68 alcoholics in their late 40s abstinent for 2 to 4 weeks had impairments (compared to norms) on the category test, the trailmaking test part B, and the tactual performance test of the Halstead-Reitan battery. In the UK, Ron and associates (Ron, Acker, & Lishman, 1980) examined 100 alcoholics who were abstinent 12–120 days (mean, 34 days). These were extremely heavy drinkers (an almost unbelievable 462 g of ethanol per day when drinking "heavily"). Whereas low-intelligence alcoholics were not very different from low-intelligence controls on a series of neuropsychological tests, higher intelligence alcoholics performed worse than higher intelligence controls on abstracting–flexibility of thinking (Wisconsin Card Sort), on WAIS digit symbol, on WAIS block design, and also on tests of verbal memory. The University of Bergen group, employing the Halstead-Reitan procedure to examine Norwegian alcoholics abstinent an average of 33.7 days, also reported deficits on category test and tactual performance test as well as the overall Halstead Impairment Index (Loberg, 1980). Danish (Lee, Moller, Hardt, Haubek, & Jensen, 1979), Swedish (Bergman, Borg, & Holm, 1980), Italian (Miglioli, Buchtel, Campanini, & deRisio, 1979), and Australian (Cala, Jones, Mastaglia, & Wiley, 1978) investigators have also noted varying levels of perceptual and perceptual motor deficits in their recently detoxified alcoholic samples.

2.1.2. Memory in Recently Detoxified Alcoholics

One of the uncommon but dramatic neuropathologic end stages of alcoholism, the Wernicke Korsakoff syndrome, has spurred lively investigations among neuropsychologists interested in memory. In contrast, memory has not received extensive study in recently detoxified and long-term abstinent alcoholics. Much of what has been done can be credited to the Boston VA Psychology group of Nelson Butters and associates. This group found that short-term memory, as reflected in a nonverbal version of the Peterson and Peterson task (with an 18-second delay) was impaired both in short-term (less than 10 years of alcoholic drinking) and long-term (greater than 10 years) alcoholics (Butters, Cermak, Montgomery, & Adinolfi, 1977). These patients were abstinent 3 weeks at time of testing. This finding was confirmed in a second study by Ryan and associates (Ryan, Butters, & Montgomery, 1980). Memory for faces was preserved in another group of alcoholics abstinent one month (Dricker, Butters, Berman, Samuels, & Carey, 1978), but the investigators suggested that these tests might be too easy for the alcoholics. Butters' group noted further that, whereas verbal paired associate learning remained relatively intact among recently detoxified alcoholics, the more unfamiliar and complex symbol digit paired associate learning was impaired (Brandt, Butters, Ryan, & Bayog, 1982; Glosser, Butters, & Kaplan, 1977).

Several other groups of investigators have also considered memory functioning in recently detoxified alcoholics. Loberg (1980) confirmed relative preservation of verbal memory by reporting no changes in logical memory or verbal paired associate learning in his Norwegian sample. Ryan (1980) showed that alcoholics abstinent 4 weeks had some difficulty learning and remembering noun pairs, but could remember as well as controls when properly cued. We have already noted that Ron and his colleagues (Ron et al., 1980) found that both verbal and nonverbal free recall were impaired in relatively intelligent alcoholics abstinent an average of 34 days. Similar findings have been reported by Miglioli et al. (1979). Cutting (1978) showed that alcoholics abstinent 1 month performed worse than medical controls on picture memory (and also on a fluency task involving "s" words). On the other hand, Guthrie and Elliott (1980) found that alcoholics free from drink an average of 14 days were impaired with respect to controls on verbal logical memory and verbal paired associate learning. Mohs and associates (Mohs, Tinklenberg, Roth, & Kopell, 1978) reported that alcoholics hospitalized for 7 weeks showed slower information processing rate, but short-term memory capacity similar to controls on a memory scanning task.

In summary, those studies that have examined memory functioning in alcoholics have shown impairment in short-term and intermediate memory in recently sober patients. Both verbal and nonverbal memory are affected, though the latter apparently somewhat more so. Alcoholics' capacity for learning appears to be preserved, although it takes them longer, and tends to require more effort to accomplish such learning. Retrieval may be impaired in the free recall situation, but proper cueing shows that storage has occurred and that essentially full recall can be achieved.

2.1.3. NP-CT Scan Correlates in Recently Detoxified Alcoholics

In the mid-1970s the increasing availability of brain computerized tomography made it practicable to examine neuropsychological performance among substance abusers in the context of neuroradiological criterion information. Cala et al. (1978) examined 26 inpatient alcoholics in an uncontrolled study. The duration of abstinence is not given, but one assumes these would qualify as recently detoxified alcoholics. Patients received the Wechsler Adult Intelligence Scale, Wechsler Memory Scale, a neurological examination, and brain computerized tomography. Seventy-three percent of these alcoholics were reported to have cerebral atrophy. When CT scans were subjected to a four-point clinical rating procedure, there was a significant correlation between degree of atrophy and age ($r = .49$) and also between atrophy and length of heavy drinking ($r = .53$). On the other hand, correlations between atrophy and neuropsychological performance were generally nonsignificant, the best relationships being achieved for digit symbol, block design, and object assembly. One problem here is how to interpret these correlations—the authors use the Spearman rank order technique—but we noted that 13 of their 26 patients had tied ranks of "4+" on CT impairment.

A number of groups have now confirmed the presence of CT-measured cerebral abnormality in recently detoxified alcoholics. Bergman *et al.* (1980) examined a group of alcoholics abstinent 2 weeks on the average. Their mean age was 44, with a range of 20–65. The alcoholics received both neuropsychological tests and brain CT scanning. Twenty-two percent of the alcoholics were judged to have neuropsychological impairment (versus 9% of controls). One notes, however, that 80% of the alcoholics had an elementary school education or less, versus 54% of the controls. This educational disadvantage in the alcoholics is reflected by their worse performance on IQ (104.2 versus 113.0 for the controls). Although the authors do state that neuropsychological differences between alcoholics and controls persisted after adjustment on level of performance on a synonym task, we believe, for reasons that we describe further in the final section of this chapter, that such covariance procedures are unlikely to control adequately for education or intelligence mismatches.

Bergman *et al.* (1980) partitioned their samples into age decades. Alcoholics had more CT-measured abnormalities in all except for the youngest decade for most indices, but sulcal width was increased even in the youngest alcoholics. The striking cortical atrophy even in the latter subgroup suggests that premorbid influences (or acute effects) might be operating. If this were not so, it would be difficult to explain why the investigators did not observe much change in sulcal abnormality as age increased (except in the very oldest group, age 60–65). One would assume, that if the sulcal deficits were drinking related, then with aging (and therefore more exposure to ethanol) there should be a progressive change in this parameter.

The UK Institute of Psychiatry group (Ron *et al.*, 1980) examined 100 alcoholics whose mean age was 43.5 and who were abstinent 12 to 120 days (mean, 34 days). There were 41 controls aged 40. Alcoholics revealed evidence of cerebral atrophy on a number of CT indices. Because alcoholics were substantially worse than controls on an estimate of premorbid IQ (based on performance on the New Adult Reading Test), the investigators split the alcoholic sample into a more intelligent and less intelligent group. When the low-intelligence alcoholics were compared to controls of similar level of intelligence, there were essentially no differences in neuropsychological test performance between the two groups. The controls outperformed these alcoholics only on a measure of nonverbal memory. The results for the high intelligence group were different. Here, alcoholics did worse than controls on the Wisconsin Card Sort test, digit symbol, block design, and measures of verbal and nonverbal memory. Confirming the earlier results of Cala *et al.* (1978), Ron *et al.* (1980) found no strong relationships between neuropsychological and CT measures; similarly, drinking history was not related to neuropsychological or CT indices when age was controlled statistically.

The Addiction Research Foundation group in Toronto has produced a series of reports concerning neuropsychological, electrophysiological, neuroradiological, and other biological parameters in alcoholics. This group has confirmed CT-scan-measured cortical atrophy in groups of recently detoxified alcoholics (Wilkinson & Carlen, 1980a, 1980b; Wilkinson & Carlen, 1983). Neuropsychologically, they have found alcoholics to be impaired on tests of abstract reasoning (e.g., category test), perceptual motor tasks (e.g., tactual performance test, digit symbol, block design, object assembly, visual reproduction), and memory tests. Although in some analyses this group has found a relationship between CT indices and neuropsychological performance, these significant relationships disappeared when age was adjusted statistically.

Kroll and associates (Kroll, Siegel, O'Neill, & Edwards, 1980) examined 16 alcoholics who were abstinent 2 weeks. There were no controls for neuropsychological tests, but file films were used as controls for CT measures. Eleven of 16 alcoholics were found to be impaired on CT scanning, and most alcoholics exhibited perceptual motor task deficit ranging in extent from mild to severe. The Spearman correlation between neuropsychological performance on the Michigan Battery and CT measured atrophy was .9, but the small sample size makes generalization on the relationship between neuropsychological performance and CT-measured atrophy risky at best.

Most of the studies reviewed so far examined alcoholics in their forties or older. There are two reports concerning younger alcoholics, and their results contradict each other. Lee *et al.* (1979) found that 22 of

37 young (age 30) Danish alcoholics, who had been abstinent approximately 4 weeks, were impaired on neuropsychological tests, and 18 of 37 had CT abnormalities. Again, there were no relationships between CT and NP measures. In contrast, Hill and Mikhael (1979), examining 15 alcoholics abstinent at least 1 month (and up to 6 months), with a history of 14 years of heavy drinking, found no major differences on CT scans between these alcoholics and matched groups of abstinent heroin addicts and non-substance-abusing controls. The average age of the alcoholics was 34.3. Although Hill and Mikhael found no CT abnormalities in their alcoholics, they reported that 12 of 15 were impaired neuropsychologically. It should be noted, however, that the educational level of the controls was 12.7 years versus 11.0 for the alcoholics, and that the controls' IQ was almost 13 points higher.

In summary, neuroradiological investigators of recently detoxified alcoholics have revealed that CT-scan-measured abnormalities are common in such patients. The ventricular dilation and increased sulcal width that have been observed suggest generalized brain atrophy. Interestingly, CT indices generally have been found not to relate significantly to drinking history nor to neuropsychological test performance.

We have used the term "recently detoxified alcoholics" purposefully: Most CT studies have been performed on hospitalized patients several weeks after their last drink. Thus, observed impairments, be they neuropsychological or tomographic, cannot be said with any certainty to reflect permanent brain change. Indeed, the findings of Carlen and associates (Carlen, Wortzman, Holgate, Wilkinson, & Rankin, 1978) and Ron *et al.* (1980) suggest that CT-measured abnormalities might normalize as a function of prolonged abstinence. If this is so, then all the studies reviewed in this section must be viewed as reflecting neurobehavioral disturbance that is part of a slowly reversible detoxification process—or "subacute organic mental disorder," to use the term that we suggested earlier. This brings us to the next important set of investigations in alcoholism: those dealing with the neuropsychology of recovery.

2.2. Recovery with Abstinence

A number of investigators have now explored changes in neuropsychological performance in the recently detoxified alcoholic (i.e., the patient who is in his first 2 months of sobriety). We will comment here only on some of the recent work in this area, and recommend that readers wishing to obtain a review of studies prior to the mid-1970s should consult either our own previous article (Grant & Mohns, 1975) or the more recent excellent reviews of Parsons and Farr (1981) and Ryan and Butters (1982).

There is now general agreement that considerable improvement occurs in the first 3 weeks of abstinence. Most investigators have found that the rate of improvement tends to slow down about the third week, although Eckardt et al. (1979, 1980) reported no recovery after the first week of abstinence.

Some of the most careful work mapping this early phase of recovery has been by Goldman and his associates. The importance of their work lies in the fact that the design of their studies has allowed them to consider the issues of recovery and practice both separately and conjointly. For example, their study of synonym learning examined one group of 11 alcoholics three times (at days 5, 15, and 25), a second group two times (at days 15 and 25), and a third group once only (at day 25). These performances were then compared to those of controls (Sharp, Rosenbaum, Goldman, & Whitman, 1977). These investigators found that alcoholics were impaired on verbal learning ability at 5 days, but that this impairment disappeared by 15 days. Further, they could not demonstrate any practice effect.

Looking for varying patterns of short-term recovery, these investigators noted further that improvement on some tests could be demonstrated as a function of passage of time (without previous exposure to the test), whereas on other tests, prior exposure was required before test improvement could be demonstrated. For example, alcoholics scored better on the digit symbol test at day 21, whether or not they were tested on this task previously. On the other hand, improvement on the trail making test occurred only in those subjects who had taken the test before (Ellenberg et al., 1980; Goldman, 1983).

On the basis of these observations, Goldman has suggested that there may be two kinds of neuropsychological recovery from alcohol toxocity: spontaneous recovery (as illustrated by digit symbol); and experience-dependent recovery (or active learning), as illustrated by the trail making test.

We find some support for the experience-dependent notion. Some of our own recent research findings are consistent with this notion. For example, we have now re-examined 30 long-term abstinent alcoholics 1 to 2 years after their initial testing with an extended Halstead-Reitan Neuropsychological Test Battery (HRB). At initial testing these subjects (age = 38, education = 13) were abstinent continuously a minimum of 18 months. At repeat testing they were abstinent approximately an average of 3.5 years. A group of nonalcoholics closely matched for age and education were also re-examined at a similar interval. An experienced clinician performed blind ratings (i.e., without knowing group membership) of pairs of HRB protocols and rendered a judgment of change on a five-point scale (1 = substantial improvement, 2 = some improvement, 3 = essentially no change, 4 = some worsening, 5 = marked worsening). The clinician was aware that these pairs of protocols reflected retests of the same individual, and therefore he used his best judgment to try not to classify protocols as improved unless they truly appeared to exceed what might be expected from practice. On this basis, the clinician classified 7% of controls as having improved somewhat. Presumably this reflects whatever individual unexplained variability there might be in the subjects, as well as error in clinician rating. At the same time, the long-term abstinent alcoholics were classified as improved in 23% of the cases (one control and no alcoholic rated worse).

There seem to be two possible explanations for this unexpected improvement among long-term abstinent alcoholics. On the one hand, it might be the case that even after 18 months or more of abstinence (which is the time frame at which they were tested initially) the subjects had not yet fully recovered; an alternative possibility, and one that would fit with Goldman's active learning hypothesis, is that the alcoholics were actually underperforming with respect to their "true" potential when they were first tested. In this instance their improvement at second testing might reflect their ability to capitalize on this dormant potential.

Several other studies of recovery during the early phases of abstinence deserve mention. Ayers and associates (Ayers, Templer, Ruff, & Barthlow, 1978) showed improvement on the trail making test from weeks 1–5 independent of practice. Miglioli et al. (1979) retested 15 of their alcoholics at two months (initial testing was at a week and a half). These subjects were impaired on both verbal and nonverbal memory initially, improved to the level of controls on verbal memory at follow-up, but remained impaired on delayed recall of the Rey figure; the alcoholics also showed very little improvement with respect to their own initial performance on the latter test.

Berglund and associates (Berglund, Bliding, Bliding, & Risberg, 1980) tried to correlate changes in cerebral blood flow (CBF; using a xenon inhalation technique) to neuropsychological performance over a period of 1, 3, 5, and 7 weeks of abstinence. These investigators found that digit span, trail making, and choice reaction time all improved steadily over the 7-week period. Koh's blocks (which were only administered at weeks 1 and 7) and rail walking did not improve. In regard to cerebral blood flow, there was a steady decrease in the variability of the measure, and there was a progressive increase over time in the difference between blood flow in the right and left hemispheres (under normal conditions, the right hemisphere registers more CBF than the left). The specific location of CBF increase on the right was in the upper frontal and parietal occipital areas. Berglund *et al.* (1980) also noted a statistically significant association between improvement in trail making part B and increase in the frontal temporal cerebral blood flow bilaterally. Changes in Koh's blocks seemed to be paralleled by increased blood flow in the right frontal region (although, as noted above, Koh's test did not improve significantly from a statistical standpoint).

In summary, studies of recently detoxified alcoholics show substantial improvement in neuropsychological performance and perhaps in regional cerebral blood flow in the first 2 months of abstinence. Tests dependent on verbal learning and memory seemed to show the most marked improvement. There is some evidence that even after 2 months, functions dependent on nonverbal abilities might remain impaired.

Several studies have now examined the question of long-term abstinence (i.e., beyond 2 months). We will turn initially to those studies that have concerned themselves with the first year of abstinence.

We noted already that Carlen *et al.* (1978) performed CT scans twice with eight alcoholics, initially at 2.5 to 11 weeks of abstinence, and the follow-up scans at 33 to 97 weeks. The authors noted CT-measured improvement in several cases and suggested, "Reversible atrophy was noted only in those patients who abstained from alcohol, showed clinical improvement, and had their initial CT scan before demonstrable clinical improvement was complete." On careful rereview of the Carlen *et al.* report, we conclude that although there is some tentative evidence suggesting recovery on CT scan, the authors' assertions must be tempered somewhat. Part of the problem lies in how much change in CT measures really constitute "improvement." For example, changes of 1 mm are probably unreliable, and even changes of 2 mm are suspect. Likely sources of error include slight differences in the plane of tomographic cut, rotation in the plane of cut, and actual measurement error. In other words, we suggest that it is hard to be confident in CT-measured changes that are less than 2 mm in width. Using this more conservative criterion, only two of eight of the Carlen group's cases showed a reduction in ventricular width, and two of eight showed a comparable reduction in sulcal width. Since there was an overlap in one case, it can be said that three cases definitely improved on some dimension. Furthermore, if we apply these stricter criteria, then there is no relationship between CT measured improvement and clinical improvement, since two of the four cases who were rated as being "3+" or better probably had no reliable change in CT.

The same research team (Carlen & Wilkinson, 1980) examined 23 alcoholics 2 to 3 weeks and 3 months after cessation of drinking with neuropsychological, CT, and other biological measures. The authors found that 20% showed clinically significant improvement on WAIS and Wechsler Memory Scale. Because of the way the report is written, it is hard to judge the extent to which CT scans changed over the 2½ month period. While stating that there were no differences in CT change among those alcoholics who remained abstinent versus those who kept on drinking, they suggest that there was a bigger improvement in sulcal width (i.e., narrowing) in younger abstinent alcoholics than for the total group of abstinent alcoholics. A sulcal width change score was also reported to be correlated significantly with age ($r = -.67$, $p < .05$), suggesting greater recoverability in younger subjects.

Ron et al. (1980) also examined a group of 23 alcoholics in the first few months of abstinence and then a year later. Nine of their 23 subjects remained abstinent or were minimal drinkers in the interim. There was a trend, but no statistically significant improvement on CT measures. In a few cases of alcoholics who resumed drinking, the authors suggested (but there are no hard data to back this up) that there might have been worsening on cortical indices in relation to resumed drinking.

While the extent, or for that matter, even the occurrence of CT improvement in longer term abstinent alcoholics remains uncertain, it has become clearer that alcoholics who maintain sobriety for long periods function significantly better neuropsychologically than recently detoxified alcoholics. In some studies such alcoholics performed indistinguishably from age–education-matched controls. Although Kish and associates (Kish, Hagen, Woody, & Harvey, 1980) did not note any difference on WAIS, Graham-Kendall, trail making, and Raven advanced matrices between alcoholics abstinent 21 and 110 days, Leber and associates (Leber, Jenkins, & Parsons, 1981) showed that alcoholics abstinent 3 weeks performed worse on visual–spatial tasks than those abstinent 11 weeks. The longer term abstinent group was indistinguishable from controls. At the same time, the recently detoxified alcoholics were comparable on verbal learning and memory to the longer term abstinent group. Guthrie and Elliott (1980) also found some continuing improvement in 35 alcoholics who were tested at 14 days and 6 months of abstinence.

Our research team has now conducted several studies with longer term abstinent alcoholics. In the initial study (Grant, Adams, & Reed, 1979), we showed that younger alcoholics (mean age = 37) who were abstinent a minimum of 18 months performed equally to age–education-matched controls on an extended HRB augmented by tests of verbal and nonverbal memory. A 1-year followup of some of these same subjects confirmed their normal HRB performance (Adams, Grant, & Reed, 1980). More recently we have completed an investigation of 71 recently detoxified (4.4 weeks) and 65 long-term abstinent (3.7 years) alcoholics and compared their performance to closely matched controls (Grant *et al.*, 1984). The subjects ranged in age from 25 to 59 (mean = 42) and had an average education of 13 years. The alcoholics reported 14 years of drinking in the alcoholic range and hence were broadly comparable to most samples previously examined. We found no differences between the long-term abstinent alcoholics and controls on any neuropsychological test. When the test battery was factor analyzed, four factors emerged: loading on tests that could be conceptualized as general intelligence (factor 1), learning and memory (factor 2), spatial and perceptual skills (factor 3), and speed of information processing-attention (factor 4). Multivariate analysis of variance showed that whereas recently detoxified alcoholics showed deficiencies in factor 2, the long-term abstinent alcoholics had no deficiencies on any of the four factors compared to controls, and outperformed the recently detoxified alcoholics on factor 2. Since we do not know how the subjects who constituted the long-term abstinent alcoholics would have performed in their immediate recovery phase, we cannot state unequivocally that this is evidence of long-term recovery; nevertheless, it is striking that when we compared two groups of alcoholics who differed only, so far as we could tell, on length of time since their last drink, then those abstinent 4 weeks showed some of the deficits that have been reported in many other studies of recently detoxified alcoholics; on the other hand, the long-term abstinent group was essentially indistinguishable from non-substance-abusing controls.

Fabian and Parsons (1983) performed a study whose design was very similar to the one we described above. They compared a group of recently detoxified alcoholic women with another group that had been abstinent for 4 years (there was also a matched non-alcohol-abusing control group). These investigators found that the long-term abstinent alcoholics were statistically indistinguishable from controls, although their performance on some tests was intermediate between the recently detoxified group and the nonalcoholics. Only on digit symbol did the long-term abstinent group perform significantly worse than the controls (but also significantly better than the recent group). Interestingly, Fabian and Parsons also factor-analyzed their neuropsychological test results. A nonverbal–spatial–perceptual motor factor also emerged from their analysis, and they, too, found that whereas recently detoxified alcoholics performed worse on this factor than controls, the long-term abstinent group could not be distinguished statistically from those controls.

Berglund *et al.* (Berglund, Gustafson, Hagberg, Ingvar, Nilsson, Risbert, & Sonneson, 1977) employed a repeated-measures design to examine 53 alcoholics 3.7 years after treatment. They did not observe major improvement in their subjects, although they did perform better on block design and memory tests at follow-up. It was noted that those subjects who drank less in the interim improved more on blocks than those who drank more heavily. O'Leary *et al.* (1977) also performed a test–retest experiment. Their 24 alcoholics improved significantly on trail making from 9 to 14 days of abstinence to 12 to 16 months. Nevertheless, their alcoholics performed more poorly than controls on each testing. Their results, and those of Berglund, would suggest either that interim drinking interfered with maximal recovery, or that some degree of permanent damage was being reflected in follow-up testing.

Butters' group has systematically examined possible changes in memory in alcoholics as a correlate of long-term abstinence. In one study they determined that alcoholics abstinent for 8 months functioned normally in regard to short-term memory (Peterson test), showed some impairment in verbal paired associate learning, and even more striking deficits in symbol digit paired associate learning (Ryan & Butters, 1980). This defect in symbol paired associate learning was confirmed in a subsequent study of a larger group of alcoholics abstinent 8 years (Brandt *et al.*, 1983). Another of their studies explored memory for remote events in a group of alcoholics whose average age was 55 years, who had very prolonged drinking histories (25 years), and who were abstinent an average of 40 weeks (Albert, Butters, & Levin, 1980). They found no defects in long-term memory, with the exception that uncued recall of "Hard events" from the 1970s were remembered less well. While the authors suggest that the explanation for this is that "hard events of the seventies" present a more complex task and therefore, presumably, uncover a subtle deficit, a more plausible explanation would seem to be that as these patients' drinking careers progressed, they became more and more oblivious to events about them. If that were so, then those events that were unnoticed could hardly be remembered later (actually, Albert, Butter, and Levin acknowledge in passing that this might be an explanation for their findings).

In summary, emerging work with alcoholics who have been abstinent for prolonged periods of time is putting much of the original neuropsychological research on alcoholism into better perspective. It has become clear that much previous research (and some present research as well) has concerned itself with what we are now calling the recently detoxified alcoholic (i.e., the person who is in the first 2 months since last drink). It is becoming fairly clear that considerable neuropsychological improvement occurs in the first 3 weeks of sobriety. Until recently, it was generally accepted that by 4 or 5 weeks virtually all recovery that was going to occur had already taken place. Thus, it was commonplace to infer that neuropsychological or CT changes found in such alcoholics probably represented permanent effects.

The work reviewed in this section introduces a note of caution and emphasizes the importance of considering longer term recovery processes. Both neuropsychological and CT studies reviewed here indicate that there might well be subacute organic mental disorder that takes months (perhaps even years) to resolve fully. Further, the time course of recovery of different functions is variable: Abilities dependent on verbal skills seem to restitute most quickly (within 3 weeks in some cases); on the other hand, those dependent on nonverbal learning and memory recover much more slowly. It is too early to tell from the few studies available whether, and in what populations, such recovery becomes complete.

2.3. Prediction of Neuropsychological Test Performance from Drinking History

If alcohol produces brain damage, then it should be possible to predict neuropsychological and neuroradiological outcomes from drinking history. Efforts to establish such a dose effect relationship have been remarkably disappointing. Studies that have shown some modest predictiveness have generally been counterbalanced by an equal number that have shown no prediction at all.

We will consider first those recent studies which have found an association between estimates of drinking and neuropsychological performance. Cutting (1978) divided his recently detoxified alcoholics into "moderate" and "heavy" drinkers in terms of their history. He found that the heavy group performed worse on picture memory and verbal fluency. Blusewicz and associates (Blusewicz, Dustman, Schenkenberg, & Beck, 1977; Blusewicz, Schenkenberg, Dustman, & Beck, 1977) divided their young (early 30s) alcoholics into nonsevere and severe categories. Nonsevere alcoholics did not perform any worse than age-matched controls on the WAIS and elements of the Halstead battery. Severe alcoholics performed worse than nonseveres and controls. One word of caution about this study: The numbers are fairly small ($N = 7$ for nonsevere; $N = 13$ for severe). Bergman and associates (Bergman, Holm, &

Agren, 1981) examined a group of alcoholics in their 50s who had very long drinking histories. They divided this group into a nonsevere category (those who drank on average less than 100 gm ethanol per day and had an average drinking history of 29 years) and a severe category (drank more than 100 gm ethanol per day and had a drinking history of 36 years). The two groups were equivalent in terms of verbal memory and field dependence (both groups were field dependent). The severe alcoholics performed worse on Grassi blocks, the Graham-Kendall, and the trail making test. One potential problem here is an educational mismatch: The severe alcoholics tended to be less well educated than the nonseveres, and this was reflected in higher IQ scores of the nonseveres.

Some of the most extensive and innovative efforts at prediction have been undertaken by Eckardt and associates (Eckardt, Parker, Noble, Feldman, & Gottschalk, 1978; Eckardt et al., 1980). These investigators expressed concern that much previous research on prediction utilized overly simple univariate or multivariate linear statistical models. As an alternative, they evolved a series of nonlinear (quadratic) equations in which neuropsychological test performance in recently detoxified alcoholics was predicted from combinations of demographic and drinking history variables. The difference in their approach from that of standard multiple regression was that predictor variables were expanded to degree 2: This means that, in addition to entering values for the variables themselves, these values were then also squared and various cross products of the predictors were also entered. In this way they were able to achieve very substantial prediction of test performance. The authors suggested that different combinations of recent drinking and remote drinking variables, in association with age and education, uniquely predicted different tests. For example, the equation for predicting digit symbol was different than that predicting TPT time; the equation for predicting block design different from object assembly, and so forth. We will consider efforts to replicate the findings of Eckardt et al. (1980) in the paragraphs that follow.

Before getting to this, we should review some of the work that has not succeeded in developing predictive models of neuropsychological deficits from drinking history. Parsons' group, which was one of the first to suggest such a relationship (Jones & Parsons, 1971; 1972), has, in its more recent work, been unable to definitively establish prediction. For example, Klisz and Parsons (1977) found no difference in performance of longer term versus shorter term alcoholics. Klisz and Parsons (1979) confirmed that duration of alcoholism was unrelated to performance on Wisconsin Card Sort in an examination of 60 alcoholics. Fabian *et al.* (1981) found that years of drinking did not relate to factor 1 neuropsychological ability scores when age was controlled statistically. Nichols-Hochla and Parsons (1982) found that they could predict decreased performance in severe versus nonsevere alcoholics if severity were defined by presence of a history of delirium tremens, fuzzy thinking, hallucinations, and other neuropsychiatric signs. Interestingly, however, the reported amount of drinking by the severes was equivalent to that of nonseveres. This points to the possibility that the severe group was actually one that had some predisposition to neuropsychological impairment, or differential vulnerability to the toxic effects of alcohol.

Our own group (Grant *et al.*, 1979) has examined the question of prediction in considerable detail. In our initial study of younger (mean age 37), recently detoxified (minimum 3 weeks) and longer term abstinent (minimum 18 months) alcoholics, we employed standard multiple regression techniques to try to predict neuropsychological test results from a combination of age, education, medical, nutritional, and drinking history variables. We were able to achieve low order, but significant prediction. But the only variables that contributed substantially to this prediction were age and education.

Since this was one of the studies that Eckardt *et al.* (1980) criticized on the basis of using overly simplistic statistical models, we followed their suggestion and performed a series of nonlinear analyses with an expanded group of subjects (Adams & Grant, 1984). In this study we examined 84 recently detoxified (4 weeks) and 72 long-term abstinent (4 years) alcoholics who were aged 25 to 59 (average, approximately 42). We developed estimates of recent and distant alcohol consumption resembling those of Eckardt and associates. Since our test procedures already included most of the tests used by those investigators, we were able to attempt a replication of their study using methods that were quite directly comparable. Out of the many quadratic equations that were generated in these studies, *none* achieved prediction for our recently detoxified alcoholic group. In the case of the longer term abstinent alcoholics, eight predictions did achieve significance (that is, for eight of our tests we were able to predict performance on the basis of combinations of age, education, and recent and remote drinking history). The disappointing feature was that virtually none of our equations resembled those of Eckardt's group, even when predicting for the very same tests. Further, in attempting to predict tests that had some intrinsic similarity (e.g., tactual performance test time and location), we found that the predictive equations were most dissimilar. These observations suggest to us that in the absence of some theoretical model (i.e., some

notion as to what shape the quadratic surface that is thought to represent the relationship of alcohol and neuropsychology should take), what
one gets is a few significant equations that are unreliable.

Examining the matter further, we then developed linear counterparts of the eight nonlinear equations that were statistically significant.
We found that in six of the eight cases a simpler standard multiple
regression also achieved significant prediction. Interestingly, although
the adjusted R^2 was less for the standard regression, the F ratio was
usually greater. This suggests that we can have somewhat greater confidence that the linear model might replicate. Inspecting further, we discovered that the variables that were significant predictors in the simpler
multiple regression were also those whose squares and cross products
contributed to predictiveness in the quadratic equations. We concluded
that the use of polynomials ran the risk of artifactually inflating the
correlation coefficient through ipsative use of variance. In summary, our
attempts to replicate Eckardt *et al.* (1980) led us to conclude that the use
of polynomials without a theoretical model of the shape of the neuropsychology–drinking relationship was not a useful exercise, since it led
to inappropriate reutilization of predictor terms and capitalization on
chance factors (Adams & Grant, 1984).

As we try to understand why we can or cannot predict neuropsychological test performance from drinking history, several studies of
prediction involving nonalcoholic social drinking samples are of interest. The most extensive exploration of this area has been by Parker
and Noble. In one of their first studies (Parker & Noble, 1977), these
investigators examined 102 bright southern Californians (53% had 16
years of education or more). The subjects were reported not to have had
a drink for the 24 hours preceding testing. Their mean age was approximately 43. Tests included the category test, Shipley Institute of Living,
Wisconsin Card Sort, and tests of verbal learning and memory. The
investigators computed a number of indices of drinking, and attempted
to relate them to test performance. The variable that was modestly successful in predicting some neuropsychological test performance was the
dose of ethanol per drinking occasion. Among light drinkers the correlation between dose per occasion and category score was .23 ($N = 65$;
correlation was adjusted for age but not education). Among heavier
drinkers most test results were related to dose per occasion. The highest
r values were achieved for vocabulary, category errors, and learning of
verbal material. Interpretation of these results is rendered problematic
by the fact that one of the highest correlations of drinking per occasion
was with vocabulary ($r = .41$, age- but *not* education-adjusted). It seems
unlikely that people's vocabulary becomes more impaired as a function

of increased social drinking, since vocabulary is thought to be among the most "resistant" of the neuropsychological ability areas. A more tenable explanation might be that less intelligent people drink more than they should. Parker and Noble acknowledge this as a possible "alternative hypothesis, albeit less likely . . ." (1977, p. 1231). We would argue that in this study, at least, this is the *most* likely explanation.

In a continuation of their studies, Parker and Noble (1980) found that Wisconsin Card Sort trials to criterion and other errors were related both to drinking history and age. In this study it is noteworthy that older subjects were drinking more per occasion, hence the possible independent contributions of recent alcohol consumption and aging are difficult to sort out. In their most comprehensive treatment of this subject, Parker and associates (Parker, Parker, Brody, & Schoenberg, 1982) combined a southern California sample ($N = 102$) with a sample from Detroit ($N = 496$). The well educated southern California group had an average age of 43 years and reported four drinking occasions per week, with an average of 2.5 drinks per occasion. The average age for the Detroit sample was 37.5 years, and the average education was 13. They drank on three occasions per week, 2.5 drinks per occasion. All subjects were men, and their length of abstinence at testing was 24 hours. In their data analyses the authors considered both age and amount drunk per occasion. They concluded that age worsened Shipley abstracting score by about .2 units per year; on the other hand, increasing the amount drunk per occasion by 1 oz. predicted a decrement of 1.57 abstraction units. Another way of putting this is that increasing ethanol consumption by 1 oz. regularly resulted in a "brought foward" aging change in abstracting ability of 3.7 years.

In another study of social drinkers, MacVane and associates (MacVane, Butters, Montgomery, & Farber 1982) were able to show modest associations between current modal quantity/frequency of alcohol consumed and several neuropsychological tests, including Peterson performance at 30-second delay, digit symbol, and certain Wisconsin Card Sort scores. The patterns of relationships differed somewhat depending on whether the correlations were performed with histories from light or heavy social drinkers.

What can we glean from these studies of social drinkers that might be helpful with prediction in alcoholism research? We consider that the important factors may be *recent drinking history* and *length of abstinence at time of testing*. Studies of Parker's and MacVane's groups all specified a 24-hour dry period prior to testing. It would seem to us that this would ensure that the heaviest social drinkers would be in a state of mild withdrawal at the time they were tested. It would also follow that the extent of their difficulty with this mild abstinence syndrome would be proportional to the amount they have been drinking recently. Thus, the parsimonious conclusion from the social drinking studies is that sensitive neuropsychological tests can indeed predict the degree to which heavy social drinkers will experience mild cognitive dysfunction 24 hours after they "went on the wagon." To suggest that these studies are uncovering permanent effects, or even "premature aging," appears to be inappropriate, given that the literature we have already reviewed points to a considerable capacity for recovery even among alcoholics who had been drinking much more heavily than most persons in the social drinking studies. The fact that estimates of lifetime ethanol consumption or maximal ethanol consumption in some distant past generally were not predictive of neuropsychological performance in these social drinking studies also supports the inference that what are being measured here are acute, probably reversible effects. It would be interesting to examine a group of heavy social drinkers who had been given ample time to equilibrate to an alcohol-free state (e.g., had remained abstinent for a month or more).

The most important implication is that studies with alcoholics must bear the recent drinking-recoverable deficit equation in mind when they attempt to predict permanent brain effects from alcohol consumption history. Some of our own recent results illustrate this point. When we tried to predict scores on a factor analytically derived factor thought to reflect ability to learn and problem solve, we found that only one of four drinking history variables had any significant association with performance. This variable was amount drunk in the past year. Years of drinking in the alcoholic range, amount of ethanol consumed in lifetime, and maximum amount of alcohol drunk during time of heaviest drinking were all noncontributory (Grant *et al.*, 1984).

Why is such a commonsense relationship (i.e., the prediction of brain disturbance from amount of past drinking) so difficult to establish? To understand this we must consider confounding sources of variance in neuropsychological test results, which include aging, premorbid, and coexistent medical–neurological conditions, and nutrition. Each of these factors will be discussed below.

2.4. Aging in Relation to Neuropsychological Deficit among Alcoholics

Aging can affect our understanding of neuropsychological change in alcoholism in a number of obvious but important ways. It is evident that the aging process itself, even among healthy adults, is associated with systematic changes in neuropsychological test performance. Indeed, these age-related changes, which involve increases in reaction time, slowing of information processing speed, some reduction in abstracting ability, decrements in perceptual motor integrative tasks, and some difficulties in learning and remembering, bear a superficial similarity to changes that have been ascribed to alcohol abuse: so much so, that some have propounded a "premature aging" theory of the effects of alcohol (for a review of this issue in relation to alcoholism, see Ryan & Butters, 1982; see also Parker et al.'s 1982 discussion of this problem in relation to social drinking).

Several interactions of the effects of alcohol and aging are theoretically possible. Older people have had more time to drink; hence, lifetime alcohol exposure will be related to age. It is also possible that the neurophysiological changes that accompany aging render the brain more sensitive to whatever toxic effects ethanol might possess; hence, quantities of alcohol deemed "safe" in younger folk might be toxic to the aging central nervous system. Older people are also more likely to have accumulated neuromedical risk factors that might have neuropsychological effects; their brains are more likely, one might say, to have suffered the "slings and arrows of outrageous fortune." Examples of such confounding events can include head injury, arteriosclerosis, emphysema with hypoxemia, diabetes, and prescription of medications that have psychoactive properties. Another factor is speed of recovery in relation to age. It is likely that older people would take longer to recover from a drinking bout than younger ones; in that most neuropsychological research with alcoholics has concerned recently detoxified subjects, it is possible that some of these patients performed unusually poorly not because they had amassed permanent brain damage, but because they need more time to recover. Another possibility would be that recovery is less complete in older persons. Finally, from a psychometric standpoint, it can be difficult to compare the performance of younger and older persons in any sample, since deficient performance on some tasks might reflect test-taking style (e.g., a tendency for older people to take fewer risks and to shun mistakes), and what some have called the cohort effect (i.e., older people might be "obsolete" in respect to the tasks demanded by some neuropsychological tests).

Despite all these self-evident reasons to consider aging in alcoholism research, and despite the commonly discussed "premature aging" notion, very few investigations have systematically examined alcohol-associated neuropsychological decline in respect to aging. The studies that have been done can be grouped into two general categories: studies that have compared the performance of younger alcoholics with that of older controls (these studies have generally operated with the premature aging premise); and studies that have looked for possible interactions between alcoholism and aging (such studies might add support either to the premature aging or independent effects hypotheses).

Turning first to comparisons between younger alcoholics and older controls, Blusewicz and associates (Blusewicz, Dustman, Schenkenberg, & Beck, 1977; Blusewicz, Schenkenberg, Dustman, & Beck, 1977) reported that young alcoholics resembled older controls on WAIS and selected Halstead tests. One difficulty in interpreting this study arises from the fact that the younger alcoholics resembled the older controls more on verbal than on performance WAIS items. Also, alcoholics were generally one year less well educated than the young controls. This raises the possibility that alcoholics were inherently less bright than the young controls. Ryan and Butters (1980) found that alcoholics generally performed more poorly on symbol digit paired associate learning than age matched controls, and were essentially indistinguishable from controls a decade or more older. Nichols-Hochla and Parsons (1982) examined 35 women alcoholics, and compared their performance on a number of neuropsychological tests to that of age-matched controls and controls aged 65. They found that alcoholics occupied an intermediate position between the young and old controls on most tests.

An increasing number of studies have attempted to look at the interaction of age and alcohol history. Ellenberg et al. (1980) reported a selective decrease in visual perceptual abilities among older alcoholics (but not younger ones). Klisz and Parsons (1977) found that young alcoholics needed more time but solved as many problems as young controls on the Levine hypothesis-testing task. Old alcoholics, on the other hand, had solved fewer problems than old controls (as well as requiring more time). This suggests a selective progression of deficit among aging alcoholics. Unfortunately, the old alcoholics were almost 2 years less well educated than their control counterparts. Although the authors found no correlation between Levine results and education (a surprise, since virtually every neuropsychological test is influenced by education), we believe, nevertheless, that educational disadvantage cannot be excluded as a cause of observed selective deficits in the older alcoholics. Parker and Noble (1980) found that the amount drunk per occasion had more predictive power on Wisconsin Card Sort "other errors" and "number of trials to criterion" in old alcoholics than young. This finding is interesting, and might well reflect the fact that older, heavy social drinkers achieved less recovery 24 hours after their last drink (which was the time at which Parker and Noble did their testing) than did their younger counterparts. Klisz and Parsons (1979) found that

Wisconsin Card test performance was negatively affected by age in alcoholics but not controls. Again, alcoholics were less well educated.

One of the most extensive studies of alcohol-aging interactions was that of Bergman *et al.* (1980). These investigators examined 130 alcoholics abstinent 2 weeks and compared them on the basis of neuropsychological tests and CT scanning to 195 nonalcoholic controls. They partitioned their samples into five age bins consisting of four decades from age 20 to 59 and one quinquennium, 60 to 65. With regard to CT indices, Bergman *et al.* found progressive ventricular enlargement and vermis atrophy in both alcoholics and nonalcoholics in relation to age. The alcoholics were not distinguishable statistically from controls, although they always showed more ventricular enlargement and vermis atrophy. A third CT measurement, sulcal width, showed greatest differences in the young samples. Mean sulcal width was nine times greater in the 20–29-year-old alcoholics than in their control counterparts. Interestingly, sulcal width did not increase very much as a function of aging, and the differences between controls and alcoholics tended to decrease in the older age groups. Thus, the only "interaction" would be more consistent with the notion that people with bad brains got into heavy drinking younger, rather than that there might be a differential acceleration of brain change in relation to aging among alcoholics. From a neuropsychological standpoint, the alcoholics in the Bergman *et al.* study fared worse than controls at all age levels on the category test. There appeared to be no differential increase in category test errors as a consequence of age in alcoholics. This finding is at variance with the results of several other studies, which we reviewed earlier (see Grant, Reed, & Adams, 1980). As can be seen from Figure 1 (taken from Grant *et al.*, 1980), which summarizes the results of several studies, there is a differential increase in category error as a function of age in recently detoxified alcoholics. While failing to confirm these category results, Bergman *et al.* (1980) did, nevertheless, find that memory for designs errors increased in alcoholics at a greater rate in relation to age than they did for controls.

Our own research has consistently failed to uncover interaction effects on neuropsychological tests in relation to aging and drinking history. In our initial exploration of younger alcoholics (mean age 37, maximum age 46) we performed a two-way analysis of variance, one independent factor being group assignment and the other being age split at age 40 into "younger" and "middle aged" (Grant *et al.*, 1979). Neuropsychological performance was the dependent variable. We found a significant age split effect, but no group effect. These results

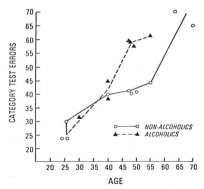

Figure 1. Relationship between age and category test errors in recently detoxified alcoholics and nonalcoholics

suggested that neuropsychological decrements were occurring evenly in
relation to age in all groups. An expansion of this work has recently
been completed with larger samples of recently detoxified and long-term
abstinent alcoholics and controls (Grant *et al.*, 1984). In one set of analyses, we derived factor scores on each of four factors from a factor analysis of our neuropsychological test battery. These factors were described
in an earlier section of this chapter; in brief, factor 1 was a verbal–
intelligence factor, factor 2 a learning and problem-solving factor, factor
3 a spatial abilities factor, and factor 4 a speed-of-information-processing–attentiveness factor. We looked for independent and joint contributions of age and alcoholism to neuropsychological performance by conducting a 3 (groups) × 2 (age under 40/40 and above) multivariate
analysis of variance in which performance as measured by the four
factors constituted dependent variables. We observed significant group
and age main effects. The group main effect was carried by the recently
detoxified alcoholics and was noted on factor 2 (problem solving and
learning). The age main effect was observed in both factors 2 and 4 (the
latter being attention–speed-of-information-processing). *There were no
interactions.* These results suggested to us that both recency of drinking
among alcoholics and aging in all groups of subjects contributed to
neuropsychological decline; but the pattern of deficits was different for
alcoholism and aging. Whereas the recently detoxified alcoholics had
difficulties in learning and problem solving, older subjects in all groups
had difficulties on these tasks *and* were also slower and somewhat less
attentive. A recent study by Brandt and associates (1983) produced similar findings. In this study the authors conducted a series of two-way
analyses of variance with tests such as symbol digit paired associate

learning, embedded figures, and other tests of learning and memory as dependent variables. For most tests there were both alcohol and age main effects, but no interactions.

In summary, the effects of aging on alcoholics, and the interrelationships of the deficits caused by each of these conditions are only now beginning to be studied in detail. Some of the most recent studies, which have been specifically designed to look for age–alcoholism interactions, suggest that the deficits attributable to these two causes may be similar in some ways but different in others.

2.5. Medical and Nutritional Factors in Relation to Neuropsychological Performance in Alcoholics

It is commonly agreed that alcoholism can be associated with a number of health hazards, including increased risk of arteriosclerosis, hypertension, neoplasia, liver cirrhosis, pancreatitis, neurological disease based on nutritional deficiency, and trauma in the form of head injuries. Since all of these conditions can theoretically affect neuropsychological functioning, studies that are interested in isolating the specific effects of ethanol as a toxin must consider these pre-existing and coexisting sources of morbidity. Regrettably, little work has been done on relating these various health problems to neuropsychological performance in alcoholics.

A few studies have recently concerned themselves with the possible effects of liver disease on NP function. In this instance, when we say liver disease, we mean alcoholic cirrhosis that is already established but that has not progressed to end-stage liver failure with hepatic encephalopathy. Smith and Smith (1977) examined two groups of alcoholics abstinent 2 to 4 weeks. One group had evidence of cirrhosis of the liver and the other did not. Cirrhotic patients scored worse on WAIS similarities, digit symbol, block design, and object assembly. Lee *et al.* (1979) examined 37 alcoholics with neuropsychological tests and brain CT scans; 35 of these also had liver biopsies. There was no statistically significant relationship between CT indices or neuropsychological tests and degree of liver change. For example, seven of 35 patients were diagnosed as having alcoholic cirrhosis. Five of these seven (71%) were neuropsychologically impaired. By way of comparison, 61% of the remaining 28 patients were neuropsychologically impaired. Gilberstadt and associates (Gilberstadt, Gilberstadt, Zieve, Buegel, Collier, & Mc-Clain, 1980) examined cirrhotic and noncirrhotic alcoholics. The cirrhotic alcoholics did not have clinical evidence of encephalopathy. In this study the cirrhotics performed worse than other alcoholics on digit symbol, block design, WAIS performance IQ, simple reaction time, and speed of writing words. There were no differences on trailmaking, part B. Interestingly, performance on several neuropsychological tests was significantly related to reduced serum albumin. There was also a relationship between a severity of cirrhosis index and neuropsychological performance, but this relationship disappeared when albumin level was held as a covariate. Gilberstadt's study is perhaps the best designed of these several investigations, and points to a relationship between progressive liver dysfunction and neuropsychological decline. Whether the liver disease is causally implicated in neuropsychological deficit or whether both brain and liver pathology are progressing in parallel remains to be ascertained. Finally, Kroll et al. (1980) reported that CT scans were worse in patients who had liver disease; unfortunately, sufficient documentation is not provided to evaluate this assertion.

Alcoholics can have general nutritional deficiencies, the most extreme forms being represented by beri-beri and Wernicke Korsakoff syndrome in association with thiamin deficiency. Neville and associates (Neville, Eagles, Samson, & Olson, 1968) noted that the mean intake of protein, fat, and carbohydrates in 34 alcoholics was about two-thirds that of normal requirements. Also, 15–30% of these alcoholics had deficient intake in thiamin, niacin, and riboflavin. Wood (1972) observed that as alcohol intake among alcoholics increased, consumption of carbohydrates, proteins, and vitamins as percent of total caloric intake declined. Observations from both of these studies can be thought to represent in part the displacement by alcohol of other foodstuffs—the so-called "empty calorie" contribution of ethanol. Alcohol is also known to reduce absorption of certain essential nutrients, including protein, fat, cyanocobalamin, folate, thiamin, and some amino acids (Israel, Valenzuela, Salazer, & Ugarte, 1969; Lindenbaum & Lieber, 1969; Halsted, Robles, & Mezey, 1972; Roggin, Iber, Kater, & Tabon, 1969). Part of the reason for this impaired absorption may relate to structural changes in the architecture of the small bowel induced by chronic alcohol intake (Lindenbaum & Lieber, 1969).

In view of these observations, it is surprising that few studies have tried to relate general nutritional status of alcoholics to neuropsychological functioning. In our own study of younger alcoholics (Grant et al., 1979), we obtained detailed nutritional histories from our groups of recently and long-term abstinent alcoholics and controls. Thereafter, a dietitian blind to group membership rated nutritional status of all subjects on a six-point scale. Although recently detoxified alcoholics were overrepresented in the less than adequately nourished group, there were not statistically significant differences. Furthermore, when nutritional status was entered into a multiple regression equation that attempted to predict neuropsychological test performance (in combination with other factors such as age, education, drinking history, and medical risk), nutritional status did not contribute significantly to prediction of such performance.

Guthrie and Elliott (1980) divided their recently abstinent alcoholics into a group of 41 malnourished patients and 51 adequately nourished patients. Malnutrition was defined by presence of glossitis, angular stomatitis, reduction in ideal weight, and certain biochemical parameters. Twenty-nine percent of the malnourished alcoholics had moderate to severe neuropsychological impairment, as compared to 10% of the nonmalnourished group. Unfortunately, their statistical design did not take other possible sources of variability in neuropsychological performance into account—for example, education, drinking history, and other medical risk factors, to name a few.

The problem of head injury is also central to understanding neuropsychological deficit in alcoholism. It is a common clinical observation that alcoholics are at risk for head injury. Even though we screened out people with severe head trauma and prolonged unconsciousness, both in our studies of younger alcoholics and in our more recent investigation of a broader age range of these patients (Grant et al., 1979; Grant et al., 1984), we found in both of these studies an overrepresentation of head injury in alcoholic groups as compared to controls. In our second study we performed a series of multiple regressions attempting to predict performance on four factor analytically derived neuropsychological indices from a combination of age, education, head injury risk, medical risk, and several drinking history variables. Three of our four equations achieved significant prediction. In two of these, head injury risk accounted for 3.4–4.9% of the variability in the neuropsychological target in question. Although this is not a large proportion of accountable variance, we should note that alcohol history accounted for 2.9–3.4% of the variance. Thus, even in a group of alcoholics carefully selected to avoid major head injuries with prolonged periods of unconsciousness, repeated episodes of minor head injuries appeared to contribute to neuropsychological outcome. These results may take on added meaning when one considers that many neuropsychological studies did not screen out patients on the basis of head injury. Indeed, several studies actually reported substantial rates of head injury in their sample (e.g., Bergman et al., 1980; Klisz & Parsons, 1977). Several studies attempted to define severity of alcoholism in terms of occurrence of neurological events such as delirium tremens (e.g., Blusewicz, Dustman, Schenkenberg, & Beck, 1977; Nichols-Hochla & Parsons, 1982). It is not very surprising that patients who have exhibited neurological phenomena in the past might also be found to have neuropsychological deficit. A reasonable inference from such results is that persons with neurological vulnerability might respond to alcohol both with delirium tremens and with neuropsychological dysfunction.

In summary, there are some suggestions that factors such as liver disease, history of nonmassive head injury, general medical status, and nutrition might play a role in the prediction of neuropsychological deficit among alcoholics. Future progress in understanding the multivariate causal matrix in the production of neuropsychological deficit in alcoholism must consider all of these factors, in addition to ethanol intake and duration of abstinence, in order to understand the circumstances of production, qualitative and quantitative features, and time course of neuropsychological deficit in this disorder.

2.6. Relationship of Wernicke-Korsakoff Syndrome

Several authors have been interested in the possible continuity of neuropsychological disturbance in alcoholism with the frontal limbic encephalopathy that characterizes Wernicke-Korsakoff Syndrome. Much of this work has been done by Nelson Butters and his associates. These investigators have found some evidence for and against the continuity notion. For example, Butters *et al.* (1977) found Korsakoff patients to be markedly impaired on all memory tasks with respect to alcoholics. Alcoholics in this study were not different from controls on any test except the Peterson nonverbal procedure with an 18-second delay. Butters *et al.* (1977) concluded that this study did not support the continuity theory. On the other hand, Glosser *et al.* (1977) did find nonverbal deficits in alcoholics similar to those of Korsakoff patients, and Kapur and Butters (1977) reported a continuity in visual perceptual deficits. Ryan *et al.* (1980) found that alcoholics were intermediate in their performance between controls and Wernicke-Korsakoff patients. They believed that their findings were consistent with a continuum of impairment notion. Wilkinson and Carlen (1983) performed repeated examinations of amnesic and nonamnesic alcoholics. A discriminant analysis was apparently successful in classifying these subjects with some degree of reliability on the basis of changes in performance in block design and picture completion. The exact meaning of these findings is not evident at this time.

Other investigators have attempted to equate performance of alcoholics to patients with diffuse cerebral dysfunction, right hemisphere pathology, and frontal limbic pathology. These studies are well reviewed by Ryan and Butters (1982). Suffice it to say that there are conflicting data for and against these various hypotheses, but the general weight of this evidence is in favor of a diffuse disorder in alcoholism rather than a selective system disturbance. This inference is given support by a recent study by Goldstein and Shelly (1982). These authors found that impaired alcoholics resembled diffusely brain injured patients more than they did frontal, posterior, left hemisphere, or right hemisphere patients.

2.7. Relationship of Neuropsychological Performance to Treatment Outcome

The exploration of the possible relationship of neuropsychological deficit to treatment outcome is still in its infancy. Several studies have now suggested that alcoholics who showed the least impairment in the early phases of their detoxification tended to show less recidivism at follow-up. For example, Gregson and Taylor (1977) found that performance on a Pattern Cognitive Impairment test administered at 4–6 weeks of abstinence was successful in predicting number of days of abstinence post discharge. About 20% of the variance in days of abstinence was accounted by the neuropsychological test when age, previous job status, religious affiliation, AA membership, and other factors were also examined as predictors. Berglund et al. (1977), who retested a group of 53 alcoholics 3.7 years after treatment, also found that those who were less impaired initially had better behavioral outcome. Guthrie and Elliot (1980) found that lack of neuropsychological impairment at 14 days tended to predict abstinence at 6-month follow-up. Fabian and Parsons (1983) also reported that when they followed a recently detoxified group 1.8 years later, those that stayed sober came from the group that was neuropsychologically indistinguishable from controls at initial testing on six of eight of their factor 1 tests. The nine subjects who resumed drinking in their study were older and more impaired at the outset.

In contrast, Donovan and Kivlahan (1983) found that neuropsychological tests were not powerful predictors of relapse. At the same time, they noted that neuropsychological status initially predicted employability as measured by an "adjusted work week" 9 months later.

Prigatano (1977) raised the question of whether certain treatments might not have effects on neuropsychological function. In one study he found that alcoholics receiving disulfiram (Antabuse) scored worse on category test and speech sounds perception than did milieu-treated alcoholics. This raised the question of possible adverse effects of disulfiram. This concern diminished somewhat when a second group of six alcoholics receiving disulfiram was examined. These subjects did not show the effects noted earlier. In that the first group of disulfiram-treated alcoholics was less well educated than the second, it is possible that the differences between the disulfiram and milieu groups were accountable on this basis.

In summary, there are few studies to relate neuropsychological status of alcoholics to their ultimate treatment outcome. The evidence that has accumulated so far suggests that healthier neuropsychological status and better education predict better outcome of treatment.

3. Drug Abuse Other than Alcohol

In contrast to the large number of neuropsychological investigations of alcohol abuse, there have been relatively few studies examining the long-term brain effects of other substances of abuse. Because our group (Grant & Mohns, 1975) and Parsons and Farr (1981) have already performed fairly extensive reviews of this topic, we will focus in this chapter only on some of the more recent studies or on those earlier reports that have not received sufficient attention previously.

3.1. Opiates

Fields and Fullerton (1974) conducted the first of the recent comprehensive neuropsychological studies of heroin addicts. Using the Halstead-Reitan Battery, they were not able to find any differences between a group of hospitalized young Vietnam-era addicts and medical controls. Hill and Mikhael (1979) examined 15 heroin addicts who were not heavy users of other drugs. They compared their performance with that of 15 alcoholics and 12 non-substance-abusing controls. The subjects were in their late 20s or early 30s. Forty percent of the heroin addicts were found to be impaired on the HRB, with decrements being reported particularly on TPT memory, TPT location, tapping, and the impairment index. Unfortunately, the duration of abstinence of the addicts was not specified in this group, and the impairment index was computed in a nonstandard fashion. CT scans were also performed on all subjects and the heroin addicts were found to have somewhat smaller ventricles and sulci than the other two groups. The reasons for this were not clear, but it was at least evident that there were no CT signs of atrophy in the heroin addicts. In a continuation of this work, Hill and associates (Hill, Reyes, Mikhael, & Ayre, 1978) examined 70 heroin addicts in their late 20s and early 30s. The subjects were divided into current users (either use within the last 2 months, or on methadone), and remitted addicts (abstinent more than 2 months and not receiving methadone). Current users scored significantly worse than controls on the category test and Raven's matrices. Remitted addicts were not statistically different than controls. When the two groups of heroin addicts were combined, and their performance compared to that of 23 early-30s alcoholics and 14 controls, then the alcoholics scored significantly worse than the heroin addicts, and they, in turn, scored significantly worse than controls on category, TPT total time, and TPT location.

Rounsaville and associates (Rounsaville, Novelly, & Kleber, 1980) examined 72 opiate addicts on a selected battery of neuropsychological tests, and compared their performance to that of 60 epileptics. Both groups were in their late 20s. Using published norms on the various tests administered, the investigators concluded that 79% of the heroin addicts were impaired with 53% rated as having "moderate to severe" impairment. In comparison, 86% of the epileptics were impaired, 53%, moderate to severe. Because 43% of these addicts had abused other drugs (29% abused 4–6 different drugs), the investigators examined other correlates of impairment. Alcohol abuse history and abuse of cocaine were both related to increased impairment in this study. In addition, impairment was also related to lowered education level and history of hyperactivity. A substantial difficulty in interpreting this study is posed by the fact that almost half of these subjects were really polydrug abusers; additionally, 15 of 72 were still using heroin or other opiates at time of testing, while 37 of 72 were on methadone. This means that only 20 of 72 addicts were drug free at time of testing. It does not seem surprising that such a group of "wet" drug addicts would have impairments on neuropsychological tests. This study, consequently, tells us very little about possible long-term effects of opiates on the brain.

In a continuation of their work, Rounsaville and associates (Rounsaville, Jones, Novelly, & Kleber, 1982) reported comparing the performance of this same sample of heroin users to 29 CETA employees. Group differences were found only on measures of motor speed (finger tapping), and here, the heroin users performed better than the CETA controls. Overall, both groups functioned in a range that can be described as mildly impaired. Both samples may well have come from a population with serious educational disadvantages and/or long-standing neuropsychological deficits. Nevertheless, the failure to find group differences suggests that heroin abuse may not add significantly to already existing deficits. In this study the authors report that at 6-month follow-up their 59 addicts showed some improvement on block design, picture arrangement, digit symbol, and fluency ("H" words). Since a comparison group was not tested twice, it is difficult to know what was practice effect and what represented real improvement. Furthermore, the authors reported no relationship of intercurrent drug use to follow-up neuropsychological status. (Readers should also consult our Section 3.7 on polydrug abuse for some additional data on neuropsychological impairment in relation to opiates emerging from the Collaborative Neuropsychological Study of Polydrug Users.)

3.2. Cannabis

Readers are again referred to our previous review (Grant & Mohns, 1975) for some of the earlier neuropsychological work on marijuana use. In brief, the earlier controlled investigations, for example, Mendelson and Meyer (1972), Grant and associates (Grant, Rochford, Fleming, & Stunkard, 1973), Rubin and Comitas (1975), and the Greek marijuana studies (Miras & Fink, 1972) found no evidence that various gorups of marijuana users who were otherwise healthy had any evidence of cerebral disturbance. In contrast, clinical case reports and some clinical studies (e.g., Campbell, Evans, Thompson, & Williams, 1971; Kolansky & Moore, 1971, 1972; Soueif, 1971) suggested that a number of cognitive impairments were associated with heavy cannabis use. The uncontrolled nature of these studies, coupled with the probability that clinical and nutritional disturbances not directly related to cannabis might be contributing to neuropsychological impairment, have generally caused these latter studies to be given less weight.

More recent findings have tended to bear out the conclusions of the earlier controlled investigations. Perhaps the largest and most important study was conducted by Satz and associates (Satz, Fletcher, & Sutker, 1976). They compared the performance of very heavy marijuana-using Costa Ricans (average, 9 cigarettes per day × 17 years!) to that of closely matched, nonusing natives. An extensive battery of neuropsychological tests was employed, including tactual performance test, a modified WAIS, the Benton visual retention test, speech sounds perception, tapping, word learning, facial recognition, and tests of visual and verbal memory. No significant differences emerged on any tests.

Our own group (Rochford, Grant, & LaVigne, 1977) performed a replication of our earlier study with medical students who were users and nonusers of marijuana. We confirmed that there were no significant neuropsychological differences between these two very well educated groups. Similarly, Carlin and Trupin (1977), also studying a highly educated sample of 10 users contrasted with an equal number of nonusers, found no HRB or WAIS differences between these groups.

These consistently negative results make it seem likely that even rather heavy marijuana users are generally not adversely affected from a neuropsychological standpoint under the sorts of conditions of use that prevail in North America and Europe. It remains possible, of course, that there might be circumstances or populations that might be vulnerable to cannabis neurotoxicity. Indeed, the work of Heath and associates (Heath, Fitzjarrell, Fontana, & Garcy, 1980) with rhesus monkeys, which found changes in the synaptic clefts of limbic structures, changes in the rough endoplasmic reticulum, and the presence of nuclear inclusion bodies, suggests that there might at least be species-specific vulnerabilities to cannabis. Thus, though from a clinical standpoint neuropsychologists could reasonably assume that neuropsychological impairment in their patients is not caused by marijuana use, one supposes that there might be circumstances under which cannabis might be considered to be a factor, in combination with other conditions, in explaining observed impairment.

3.3. Phencyclidine (PCP)

Phencyclidine (PCP, "angel dust") was regarded in the 1960s as a relatively undesirable contaminant among afficionados of hallucinogenic drugs; but in the 1970s the agent acquired followers in its own right. Because phencyclidine produces dramatic changes in orientation and memory, and can cause dissociation, paranoia, violence, and catatonic phenomena, clinicians have naturally been concerned that its chronic use may be associated with permanent brain damage. Unfortunately, neuropsychological sequelae of PCP abuse have not been well studied to date, although some early and current data exist that describe isolated neuropsychological functions in relation to PCP ingestion. For example, Davies and Beech (1960) reported a disruption of attention span, and Davies (1961) observed a slowing of tapping (pure motor) speed following phencyclidine administration. Luby and associates (Luby, Cohen, Rosenbaum, Gottlieb, & Kelly, 1959) described changes in cognition that left phencyclidine-toxic persons with deficits similar to those seen in schizophrenics: Disturbances were noted in abstracting, attention, perception, and motor performance. Domino (1964) speculated that PCP had its greatest effect on the sensory cortex. Cohen (1977) and Fauman and Fauman (1977) reported the presence of organic brain syndrome among even "sober" PCP abusers, which Cohen characterized as consisting of memory gaps, disorientation, visual disturbances, and speech difficulties. Luisada (1978) reported that PCP abusers experienced episodes of anterograde amnesia following consumption of this agent. Yesavage and Freman (1978) tested the ability of hospitalized PCP abusers to estimate correctly the passage of time (30 seconds) and found a relationship between oral (as opposed to insufflated) ingestion and greater impairment on this time-sense measure. Since PCP users generally are polydrug users, two studies have attempted specifically to explore NP effects in this context. For example, Ware (1978) examined eight polydrug users who were also regular PCP users, and compared their performance on an extensive battery of neuropsychological tests to that of polydrug users who did not use PCP. She found differences on performance IQ, especially in block design and object assembly, with the PCP users scoring worse. Unfortunately, the PCP users also scored worse on VIQ; in particular, on information and comprehension. In addition, PCP users scored worse on speech perception. These results raise the strong possibility that these PCP users were less intelligent in the first place, hence few conclusions can be drawn about the unique neuropsychological effects of this agent. This inference is strengthened by Ware's own finding that her PCP users and polydrug users were equivalent on a clinical rating system.

Carlin and associates (Carlin, Grant, Adams, & Reed, 1979) compared a group of polydrug users who consumed large amounts of PCP with polydrug users inexperienced in taking PCP and with controls. The Halstead-Reitan Battery was used, and results were reported in terms of analysis of individual test results and clinical ratings in six ability areas as well as a global clinical rating. Six of 12 PCP users, five of 12 PCP naive polydrug users, and none of the controls were rated as impaired clinically. The general pattern of impairment, involving mild disturbances in abstracting ability and complex perceptual–motor skills, was comparable for the two drug-using groups.

In summary, the neuropsychological correlates of heavy PCP abuse are still unknown. Given what is known about the pharmacological and clinical effects of this anesthetic drug, this should be a high-priority area for neuropsychological research in substance abuse.

3.4. Sedative-Hypnotics

This is another group of drugs that, despite considerable suggestive clinical and research evidence linking them with long-term neuropsychological impairment, have received relatively little systematic scrutiny.

In work preliminary to the Collaborative Polydrug Studies, Adams and associates (Adams, Rennick, Schooff, & Keegan, 1975) found substantial impairment among a sedative-hypnotic-abusing group on elements of the Lafayette Repeatable Battery. Several weeks of detoxification were associated with improvement in a number of ability areas, suggesting that some of the initially observed deficit was reversible. Judd and Grant (1975) compared sedative-hypnotic abusers to a group of medical patients and neurological patients. The sedative-hypnotic group achieved scores intermediate between neurological patients and medical patients, with particular disturbances being manifest on the category test and tactual performance test. These subjects were abstinent approximately 3 weeks at time of testing.

Bergman *et al.* (1980) examined 55 barbiturate abusers approximately 18 days after admission. Their performance was compared to that of controls, who were matched to drug users on the Synonyms test but had a somewhat higher IQ (control IQ, 109 vs. 100 for drug users). The sedative-hypnotic abusers were impaired on tests of reasoning, verbal and nonverbal learning, and speed–flexibility (trailmaking).

The findings of these three studies, coupled with those emerging from the Collaborative Neuropsychological Study of Polydrug Users (review in Section 3.7 of this chapter), strongly suggest that abuse of the sedative-hypnotic drugs is associated with neuropsychological impairment whose pattern is somewhat similar to that seen among some alcoholics. Again, this is another important area for future neuropsychological research.

3.5. Volatile Substances

Volatile substances have increasingly been a subject for health concern as their use in industry and various products for the home gradually has placed more and more people at risk either to low-level background exposure or to accidental massive contamination. The volatile substances in question range from simple organic chemicals such as formaldehyde and acetaldehyde to more complex substances such as acrolein, toluene, n-hexane, methylbutylketone, acrylamide, and triorthocresyl phosphate. Photochemical smog and seepage from formylurea insulation are two examples of sources of low-level exposure to aldehydes among the general population. In industrial settings, workers can suffer either increased levels of chronic exposure or occasional, accidental, and massive exposure to such substances, since they are used in the manufacture of a rich array of organochemicals and synthetic materials (Committee on Aldehydes, NRC, 1981).

Regarding neurological and neuropsychological effects, one distinguishing feature of the volatiles is the tendency of several of the substances to produce marked peripheral neuropathy that is sometimes irreversible. Peripheral neuropathy has been reported in association with exposure to n-hexane (Gonzalez & Downey, 1972; Goto, Matsumura, Inoue, Murai, Shida, Santa, & Keiroiwa, 1974; Herskowitz, Ishii, & Schaumburg, 1971; Korobkin, Ashbury, Sumner, & Neilsen, 1975; Towfighi, Gonatas, Pleasure, Cooper, & McCree, 1976), methylbutylketone (Mallov, 1976; Prockop, Alt, & Tison, 1974), triorthocresyl phosphate, and acrylamide (Prockop et al., 1974). Interestingly, toluene, which is the desired active ingredient in most paints and glues that substance abusers sniff, appears not to cause peripheral neuropathy, although it might have other central nervous system effects, as will be reviewed in subsequent discussions.

The first report that recreational sniffing of toluene might be associated with CNS disturbance appears to be that of Grabski (1961). His patient presented with nystagmus, titubation, and equivocal Babinski sign after 7 years of sniffing. The same patient was followed up by Knox and Nelson (1966). By this time he had a 14-year history of toluene inhalation and was 33 years of age. He was found to have ataxia, tremulousness, snout and Babinski reflexes, with evidence of cerebral atrophy on pneumoencephalography (PEG). The PEG showed enlarged ventricles, widened sulci, especially in the frontal lobes, and cerebellar atrophy.

Massengale and associates (Massengale, Glaser, LeLievre, Dodds, & Klock, 1963) and Dodds and Santostefano (1964) reported on their experience with a group of preadolescent boys who had sniffed glue for 3–42 months with an average of 82 occasions of use. These boys were given tests of visual motor coordination, spatial perception, and the continuous performance task. There were no differences between 12 glue sniffers and 21 age-matched nonsniffers, and no correlation between the amount of glue used and neuropsychological change. Barman and associates (Barman, Sigel, Beedle, & Larson, 1964) examined 8 boys under 16 years of age who were referrals from juvenile hall. They performed two testings, the first being 3 hours after arrest (since the boys were caught sniffing, they were presumably still somewhat intoxicated at this time), and the second test 1 week later. These boys' Bender Gestalt performance was clearly impaired initially and improved somewhat 1 week later. The authors do not specify whether this performance was in the normal range at retest.

Berry and associates (Berry, Heaton, & Kirby, 1977) compared 37 polydrug users who had also inhaled volatiles on an average of 7,000 occasions with 11 polydrug users who were not users of volatiles. Subjects were abstinent about 72 hours at time of testing. The Halstead-Reitan Battery and WAIS were used as neuropsychological tests. Both groups were approximately 60% Mexican-American, their age was approximately 18, and their level of education was approximately grade 9. Both groups of subjects showed increased levels of neuropsychological impairment on a variety of measures, but the volatile-using polydrug users performed worse than the non-volatile-using sample.

Tsushima and Towne (1977) examined a group of polydrug users who were also habitual glue sniffers, and compared their performance on neuropsychological tests to those of a group with "mild" involvement with drugs. Although the glue sniffers were clearly more impaired on NP tests in this study, it should be noted that these subjects were also at an educational disadvantage, and also scored 21 points lower in Peabody IQ. This suggests that there might have been intellectual differences between the glue sniffers and the others from the very start. The authors attempted to relate number of drugs abused to neuropsychological change; they found some modest relationships between performance on the Stroop test and this index.

Korman and associates (Korman, Matthews, & Lovitt, 1981) used a similar design, comparing polydrug users experienced and inexperienced with inhalants. Although this was a large study (61 inhalant users and 41 other polydrug users) and the authors found that the volatile users performed worse on neuropsychological tests, the results are difficult to interpret. This is primarily because inhalant users had a verbal IQ of 75 and the polydrug users, 85. Part of the problem might be cultural disadvantage (subjects were Mexican-Americans). Even given this, however, it is clear that the volatile substance abusers were probably substantially less intelligent than the other drug users. This being the case, one would expect a wide range of neuropsychological differences that might have something to do with choice of drug (i.e., it might be the case that glue sniffers are less intelligent in the first place), but that might have very little to do with the actual effects of these drugs.

Several studies have now reported chronic neurological effects in painters exposed to paint fumes over prolonged periods of time. Mallov (1976) noted that spray painters were at increased risk to develop peripheral neuropathy. The ingredient thought to be responsible for this is methylbutylketone. Arlien-Soborg and associates (Arlien-Soborg, Bruhn, Gyldensted, & Melgaard, 1979) described a "chronic painter's syndrome" in a group of 70 house painters who had an average exposure of 27 years. Most were examined about a month after their last exposure, and were referred because of various neurological complaints. The volatile substances in the paint appear to be various turpentine substitutes, which were reported to be a mix of aliphatic and aromatic hydrocarbons. Fifty of the subjects had neuropsychological tests, 38 had CT scans, and 12 had pneumoencephalograms. Thirty-nine of the 50 who underwent neuropsychological testing were reported to have slight to moderate NP abnormalities, with memory most frequently impaired and construction dyspraxia next. Thirty-one of these 50 were also reported to have evidence of cerebral atrophy either on CT scanning or pneumoencephalography. Sulcal enlargements were seen, but not ventricular atrophy. EEGs were also obtained in 46 of the 50 patients. Slight abnormalities were found in 7 of 46, and moderate abnormalities in 2 of 46. Bruhn and associates (Bruhn, Arlien-Soborg, Gyldensted, & Christensen, 1981) performed a 2-year follow-up on 26 of these patients. None had any interim exposure. Of these 26 patients, 4 were reported to be somewhat improved and 2 somewhat worse on the basis of neuropsychological testing and CT scans. In most instances there were no intertest differences. Thus, it appears that the painter's syndrome might reflect permanent brain damage.

Several articles have noted adverse effects in glue sniffers who have switched from toluene-containing substances to others containing n-hexane. Towfighi *et al.* (1976) reported on two cases of men who had 10-year histories of sniffing non-hexane-containing glue with no apparent major ill effect. Within 1 to 2 years after switching to a different brand of glue (which was later found to contain n-hexane), these individuals developed signs of peripheral neuropathy—loss of sensation, especially in the legs, instability, and foot drop. Memory loss was reported. One year after discontinuing the use of n-hexane, one individual reported full recovery and another, partial recovery from the peripheral neuropathy. Korobkin (Korobkin *et al.*, 1975) also noted that a 29-year-old man who suffered from peripheral neuropathy in association with sniffing n-hexane containing cement, had some neurological improvement after switching to a compound free of n-hexane. Prockop *et al.* (1975) reported on seven young men (age range 17–22) who developed weakness and peripheral neuropathy that got progressively worse for 8 weeks after cessation of sniffing. Two of them developed total paralysis, including bulbar palsy; one died. Several others reported a change in the odor of the lacquer thinner they were sniffing some time prior to developing neurological symptoms. The solvent responsible for this catastrophic set of events was not clarified.

In summary, the volatile substances are of importance to neuropsychologists for two reasons: First, a minority of addicts select them as drugs of abuse; second, and of greater importance, is that many persons are at risk of exposure either through occupation or through slow increases in low background activity of these substances. From what is known so far, it appears that toluene itself has a fairly high margin of safety—although there is evidence that very long-term and heavy exposure does produce cerebral atrophy and neuropsychological deficit. Some of the other volatiles appear to have predilections for the peripheral nervous system—n-hexane and methylbutylketone have been particularly implicated.

3.6. Hallucinogens, Central Stimulants, and Miscellaneous Substances

Central stimulants and hallucinogens are two other classes of substances of major abuse. We are not aware of any recent neuropsychological studies of cocaine, methylphenidate, dextroamphetamine, or related central stimulants. In our previous review (Grant & Mohns, 1975) we noted that there have been some case reports of stroke and neuroradiological studies suggesting vasculitis in relation to central stimulant administration, both in animals and in man. These reports remain few in number, and their clinical significance is unclear. The Collaborative Neuropsychological Study of Polydrug Users (see below) also failed to uncover any relationship between reported central stimulant abuse and neuropsychological disturbance.

The situation is somewhat similar for hallucinogens. Although there are a large group of these substances, some related structurally to serotonin (e.g., LSD, psilocybin, dimethyltriptamine) and others related to norepinephrine (mescaline, dimethoxymethyl amphetamine or STP, methylene dioxyamphetamine or MDA, trimethoxy amphetamine or TMA, nutmeg, paramethoxy amphetamine or PMA), only LSD has received any scrutiny at all from neuropsychologists. The studies are all dated, and were considered in our previous review (Grant & Mohns, 1975). The results were mixed, and the general inference we drew then was that there was no systematic information implicating LSD in neuropsychological disturbance. In view of the fact that the Collaborative Neuropsychological Study of Polydrug Users (see below) also failed to uncover any relationship between reported hallucinogen use and deficit, we must conclude that these drugs, if they are toxic in the long-term, must exert only rather subtle effects and perhaps only in certain predisposed individuals.

Two substances that are becoming more commonly abused at the time of this writing are amyl nitrite and related inhalable nitrates, and phenylpropanolamine, an agent used to suppress appetite in over-the-counter diet pills. The nitrites and nitrates, although originally developed as vasodilators for the treatment of angina (some products are also marketed as room deodorizers), have been popular, particularly in the homosexual community, as enhancers of the orgasmic experience. Users report that inhalation of amyl nitrite just at the point of ejaculation inevitability greatly increases orgasmic pleasure. At the same time some male homosexuals have found amyl nitrite useful as a relaxer of the anal sphincter. To our knowledge, there have been no clinical reports of neurological toxicity from these substances. There have been no systematic neuropsychological studies of chronic use of amyl nitrite, but the popularity of these drugs clearly warrants their further examination.

With regard to phenylpropanolamine, the abuse potential of this drug is only now coming to be recognized. This agent was initially used to alleviate symptoms of the common cold; more recently it has been marketed as a modestly effective anorexigenic. The street community is only now becoming fully conversant with possible interesting effects of phenylpropanolamine. One would expect, on the basis of experience with central stimulants, that the primary psychological effects would be emotional and ideational—ranging from improved mood to excitement, elation, anxiety, agitation, paranoia, and violence. We noted earlier that there have been case reports of vascular catastrophes and possibly angiopathy in relation to central stimulant use. These studies need further confirmation. It is possible that similar effects might be seen for phenylpropanolamine.

3.7. Polydrug Use

We began this section on abuse of substances other than alcohol with the caution that heavy use of virtually any of these drugs (perhaps with the exception of cannabis) is associated with exposure to, and probably abuse of, several other agents as well. Therefore, in the present context it is virtually impossible to study the brain effects of any one of the many substances of abuse in isolation. One must accept the fact that most drug-dependent individuals will be multisubstance abusers. For this reason our group has been involved for several years in the study of "polydrug" users. The term polydrug may be imprecise, but it does serve the function of stressing the multidrug use history that characterizes these populations.

Our preliminary studies (Grant et al., 1976) found that about 50% of polydrug users who were detoxified for an average of 60 days in a residential treatment center exhibited mild neuropsychological impairment, with deficits in certain WAIS performance items as well as on the category and tactual performance elements of the Halstead-Reitan Battery. Our second study (Grant & Judd, 1976) administered both the Halstead-Reitan Battery and the EEG to a group of 66 polydrug abusers abstinent for 3 weeks. Forty-five percent had neuropsychological impairment and 43% had EEG abnormalities. EEG and neuropsychological ratings agreed 75% of the time. Five months later, 30 of the original 66 subjects were followed, and 8 (27%) were thought to have neuropsychological deficit. Thus there was apparently a substantial improvement over a period of 5 months in relation to reduced drug taking; at the same time an inordinate number of rather young subjects (the average age was 26) manifested neuropsychological deficits. In this study there was an association between greater use of sedative-hypnotics and alcohol and NP deficit. Our colleagues in Detroit (Adams *et al.*, 1975), who studied a group of polydrug users whose modal drug class consisted of the central nervous system depressants, also found a high rate of neuropsychological impairment in these subjects on the basis of the Lafayette Repeatable Battery.

As a result of these preliminary findings, the National Institute on Drug Abuse sponsored the eight-center Collaborative Neuropsychological Study of Polydrug Users. This project examined 161 recently detoxified polydrug users (3 weeks abstinent) and followed them up at 3 months (Grant, Adams, Carlin, Rennick, Judd, & Schooff, 1978; Grant, Adams, Carlin, Rennick, Judd, Schoof, & Reed, 1978). There were two comparison groups: psychiatric inpatients and day treatment patients (to match on the "psychopathology" dimension); and non-substance-abusing, nonpsychiatric controls. All subjects were carefully interviewed concerning their substance use and their medical and developmental histories. Neuropsychological examinations were with the Halstead-Reitan Battery (HRB). Analytic techniques included usual comparisons of test results and summary indices; in addition, selected tests were factor analyzed, and all test protocols were rated by a clinician blind to group membership. Similar procedures were employed at the 3-month follow-up with all groups. It was found that 37% of polydrug users were impaired (compared to 26% of psychiatric patients and 8% of controls). Multivariate analysis of variance established an association between amount of sedative-hypnotic and opiate use and impairment. Patterns of impairment in psychiatric patients differed from those found in polydrug users (polydrug users had more generalized deficits), and there was an association between diagnosis of schizophrenia and neuropsychological impairment in the psychiatric patients (but not in polydrug users who had only 3 diagnosed schizophrenics). Medical "risk" factors accounted for a small proportion (20%) of the variance in the Halstead Impairment Index.

At 3-month follow-up, 25% of the polydrug users who were initially rated as impaired, improved. At the same time, a substantial proportion (34% of all retested subjects) had persisting mild impairments. In some instances these impairments could be related to intercurrent drug use; in others it appeared that premorbid risk factors might be contributing. This still left the possibility that among some of the remaining impaired subjects, either permanent or only slowly reversible drug-related impairment was being observed. Further analyses of these data suggested that there might be two types of impaired polydrug user. One type was the "street-wise" abuser whose impairment tended to be associated with heavy use of all drugs, opiates in particular. This group also reported most of the intravenous substance abuse in the sample. A second group of impaired polydrug users consisted of subjects with a more conventional or "straight" lifestyle orientation. These subjects tended to be less experienced with a broad range of drugs of abuse, but were particularly heavy consumers of sedative hypnotics (Carlin, Stauss, Adams, & Grant, 1978; Carlin, Stauss, Grant, & Adams, 1980).

In summary, the Collaborative Neuropsychological Study of Polydrug Users, which attempted to consider possible central nervous system effects of multisubstance abuse in context, provides suggestions that multiple drug abuse is associated with neuropsychological impairment. Some of that impairment might well reflect the drift of persons with mild, perhaps subclinical deficits into careers of drug abuse; other impairments might be attributable to slowly resolving substance organic mental disorders; still others might reflect permanent damage. There appear to be at least two different classes of impaired polydrug abusers—"street-wise" heavy abusers of all drugs and opiates in particular, and a smaller group of heavy abusers of sedative-hypnotic drugs, who have more conventional lifestyles and tend to be less experienced with many other agents. Long-term follow-up of these samples is needed to clarify some of the causal links. In particular, studies of children and adolescents at risk through periods of vulnerability to substance abuse are sorely needed.

4. Summary

Of all the substances of abuse, alcohol has received the most extensive neuropsychological scrutiny. Yet even here, the prevalence, quantitative and qualitative features, natural history, and causal matrix of neuropsychological impairment remain far from clear. It is evident that recently detoxified alcoholics manifest disturbances in abstracting ability, complex perceptual–motor skills, and learning even though they may no longer have obvious abstinence phenomena or intellectual deterioration. Computerized tomography confirms a disproportionate prevalence of atrophy in such groups. At the same time, newer research demonstrates that there may be considerable room for improvement among such groups, neuropsychologically and perhaps even tomographically. We have suggested that recently detoxified alcoholics (i.e., those abstinent 1–2 months) might be considered to be at risk for a subacute organic mental disorder that is very slowly reversible. Evidence for such an inference comes from a few longitudinal studies that have followed alcoholics over their first year of recovery, and from comparisons of long-term detoxified alcoholics with those abstinent a month or so. This is an important area for continued inquiry. Present evidence suggests to us that *permanent* deficit might inappropriately have been inferred from studies of alcoholics who were suffering from subacute (but partially reversible) disorder.

The other findings emerging from alcohol research suggest that although both alcoholism and aging are related to neuropsychological deficit, the qualitative features of such deficit are more consistent with independent decrements than a premature aging model. Further, nutritional status and neuromedical risk factors appear to be contributing to the resultant picture in ways that have not yet been clearly delineated. Treatment outcome appears to be weakly but significantly related to neuropsychological status early in the detoxification process: Generally speaking, unimpaired persons are more successful at abstaining than those who are impaired.

Turning to other drugs of abuse, research findings are much more preliminary. Nevertheless, it appears that one-third to one-half of polydrug users will exhibit neuropsychological deficit in the first weeks after they have ceased active drug taking. Many will recover 3–6 months later, but the prevalence of impairment will remain higher than that of matched controls. There is an association between impairment and abuse of opiates and sedative–hypnotic drugs, and the populations that such impaired persons represent might be different: These are "streetwise," heavy multidrug users, who show opiate-related impairment, and "straight" drug users with less extensive multiple drug use, who manifest sedative-hypnotic-related deficits.

Phencyclidine might possibly cause long-term cerebral dysfunction, but studies are too few and are complicated by polydrug effects. Of the volatiles, toluene probably produces permanent cerebral atrophy after very lengthy exposure; n-hexane and methylbutylketone are more toxic, and cause peripheral neuropathy.

There is no compelling evidence that use of cannabis, hallucinogens, or central stimulants is associated with neuropsychological impairment. Finally, we know of no NP studies of abusers of amyl nitrite or phenylpropanolamine.

Our review of these studies has highlighted for us the numerous methodological pitfalls in human neuropsychological research. Two of these, subject selection and inappropriate use of covariance, deserve special emphasis.

Alcohol and polydrug-abusing subjects are likely to be different than the general population on a number of important dimensions that might have neuropsychological implications. Some of these dimensions include subtle neuropsychological deficits that antedate drug or alcohol use, educational disparities, and exposure to neuromedical risk. Each of these factors can independently, and in combination with the others, yield a picture of neuropsychological deficit that might inappropriately be causally linked in a "dose effect" relationship to abuse of various agents. Although it might be impossible ever to fully avoid specification artifact, since randomized designs are for all practical purposes impossible in this kind of work, it should nevertheless be possible, through careful matching, to minimize the influence of factors other than substance abuse.

A good case in point is the influence of education, which is known to have an important effect on neuropsychological performance. Many studies with alcoholics and drug abusers report slight and, in a number of instances, fairly significant educational disadvantage in drug-abusing groups when compared to controls. Some investigators have attempted to adjust for the inequality by using analysis of covariance. What tends to be forgotten here is that covariance procedures *almost invariably underadjust,* thereby lulling the unwary investigator into assuming that he has "taken care of" the initial educational mismatch (Elashoff, 1969). Although it is possible to correct for attenuation due to unreliability of the variables in question (Campbell & Boruch, 1975), such reliability-corrected covariance has not been reported in the studies that we reviewed. Even such corrections, of course, would remain imperfect, since many of the assumptions underlying covariance are violated in studies of natural groups of the type that are found in research on the neuropsychology of substance abuse. Some key assumptions are that assignments to groups have been made at random, that the covariate (in our example, education) is independent of group assignment, and that the outcomes (in our case, neuropsychological test scores) are distributed normally.

All this is another way of emphasizing that there are no generally effective, after-the-fact statistical exercises that will correct for initial sample mismatch on key variables that are known to influence neuropsychological performance. Since randomization is not practical, careful case-by-case matching seems to be the next best strategy. Whether or not we gradually approach some understanding of the causal role of substances of abuse in permanent brain impairment will depend on future investigators' efforts to specify accurately sources of variability in neuropsychological performance that have confounded our ability to clarify the drug-abuse–brain-function relationship.

ACKNOWLEDGMENT

The authors extend a heartfelt thanks to Debi Taylor for a superb job with the preparation of this manuscript.

5. References

Adams, K. M., & Grant, I. Failure of nonlinear models of drinking history variables to predict neuropsychological performance in alcoholics. *American Journal of Psychiatry*, 1984, *141*, 663–667.

Adams, K., Grant, I., & Reed, R. Neuropsychology in alcoholic men in their late thirties: One year follow-up. *American Journal of Psychiatry*, 1980, *137*, 928–931.

Adams, K. M., Rennick, P. M., Schooff, K. G., & Keegan, J. F. Neuropsychological measurement of drug effects: Polydrug research. *Journal of Psychedelic Drugs*, 1975, *7*, 151–160.

Albert, M. S., Butters, N., & Brandt, J. Memory for remote events in alcoholics. *Journal of Studies on Alcohol*, 1980, *41*, 1071–1081.

Albert, M. S., Butters, N., & Levin, J. Memory for Remote Events in Chronic Alcoholics and Alcoholic Korsakoff's Patients. In H. Begleiter (Ed.), *Biological effects of alcohol*. New York: Plenum Press, 1980.

Arlien-Soborg, P., Brunn, P., Gyldensted, C., & Melgaard, B. Chronic painters' syndrome: Chronic toxic encephalopathy in house painters. *Acta Neurologica Scandinavica*, 1979, *60*, 149–156.

Ayers, J. L., Templer, D. I., Ruff, C. F., & Barthlow, B. A. Trailmaking test improvement in abstinent alcoholics. *Journal of Studies on Alcohol*, 1978, *39*, 1627–1629.

Barman, M. L., Sigel, N. B., Beedle, D. B., & Larson, R. K. Acute and chronic effects of glue sniffing. *California Medicine*, 1964, *100*, 19–22.

Berglund, M., Bliding, G., Bliding, A., & Risberg, J. Reversibility of cerebral dysfunction in alcoholism during the first weeks of abstinence: A regional cerebral bloodflow study. *Acta Psychiatrica Scandinavica*, 1980, Supp. *286*, 119–127.

Berglund, M., Gustafson, L., Hagberg, B., Ingvar, D. H., Nilsson, L., Risbert, J., & Sonneson, B. Cerebral dysfunction in alcoholism and presenile dementia. A comparison of two groups with similar reduction of the cerebral bloodflow. *Acta Psychiatrica Scandinavica*, 1977, *55*, 391–398.

Bergman, H., Borg, S., & Holm, L. Neuropsychological impairment and the exclusive abuse of sedatives or hypnotics. *American Journal of Psychiatry*, 1980, *137*, 215–217.

Bergman, H., Holm, L., & Agren, G. Neuropsychological impairment and a test of the predisposition hypothesis with regard to field dependence in alcoholics. *Journal of Studies on Alcohol*, 1981, *42*, 25–33.

Berry, G. J., Heaton, R. K., & Kirby, M. W. Neuropsychological deficits of chronic inhalant abusers. In B. Rumach & A. Temple (Eds.), *Management of the poisoned patient*. Princeton, N.J.: Science Press, 1977, 9–31.

Blusewicz, M. J., Dustman, R. E., Schenkenberg, T., & Beck, E. C. Neuropsychological correlates of chronic alcoholism and aging. *Journal of Nervous and Mental Disease*, 1977, 165, 348–355.

Blusewicz, M. J., Schenkenberg, T., Dustman, R. E., & Beck, E. C. WAIS performance in young normal, young alcoholic, and elderly normal groups: An evaluation of the organicity and mental aging indices. *Journal of Clinical Psychology*, 1977, 33, 1149–1153.

Brandt, J., Butters, N., Ryan, C., & Bayog, R. Cognitive loss and recovery in chronic alcohol abusers. *Archives of General Psychiatry* in press.

Bruhn, P., Arlien-Soborg, P., Gyldensted, C., & Christensen, E. L. Prognosis in chronic toxic encephalopathy. *Acta Neurologica Scandinavica*, 1981, 64, 259–272.

Butters, N., Cermak, L. S., Montgomery, B. A., & Adinolfi, A. Some comparisons of the memory and visuoperceptive deficits of chronic alcoholics and patients with Korsakoff's disease. *Alcoholism: Clinical and Experimental Research*, 1977, 1, 245–257.

Cala, L. A., Jones, B., Mastaglia, F. L., & Wiley, B. Brain atrophy and intellectual impairment in heavy drinkers: A clinical, psychometric and computerized tomography study. *Australia/New Zealand Journal of Medicine*, 1978, 8, 147–153.

Campbell, A. M. G., Evans, M., Thompson, J. L. G., & Williams, M. J. Cerebral atrophy in young cannabis smokers. *Lancet, 1971, 2*, 1420.

Campbell, D. P., & Boruch, R. F. Making the case for randomized assignment to treatments by considering the alternatives: Six ways in which quasi-experimental evaluations in compensatory education tend to underestimate effects. In C. A. Bennett & A. A. Lumsdaine (Eds.), *Evaluation and experience: Some critical issues in assessing social programs*. New York: Academic Press, 1975.

Carlen, P. L., & Wilkinson, D. A. Alcoholic brain damage and reversible deficits. *Acta Psychiatrica Scandinavica*, 1980, 62, 103–118.

Carlen, P. L., Wortzman, G., Holgate, R. C., Wilkinson, D. A., & Rankin, J. G. Reversible cerebral atrophy in recently abstinent chronic alcoholics measured by computed tomography scans. *Science, 1978, 200*, 1076–1078.

Carlin, A. S., & Trupin, E. W. The effect of long-term chronic marijuana use on neuropsychological functioning. *International Journal of the Addictions*, 1977, 12, 617–624.

Carlin, A., Grant, I., Adams, K., & Reed, R. Is phencyclidine (PCP) abuse associated with organic mental impairment? *American Journal of Drug and Alcohol Abuse*, 1979, 6, 273–281.

Carlin, A. S., Stauss, F. F., Adams, K. M., & Grant, I. The prediction of neuropsychological impairment in polydrug abusers. *Addictive Behaviors*, 1978, 3, 5–12.

Carlin, A. S., Stauss, F. F., Grant, I., & Adams, K. M. Drug abuse style, drug use type, and neuropsychological deficit in polydrug users. *Addictive Behaviors*, 1980, 5, 229–234.

Cohen, S. Angel dust. *Journal of the American Medical Association, 1977, 283*, 515–516.

Committee on Aldehydes, National Research Council. *Formaldehyde and other aldehydes*. Washington, D.C.: National Academy Press, 1981.

Cutting, J. Specific psychological deficits in alcoholism. *British Journal of Psychiatry*, 1978, 133, 119–122.

Davies, B. M. Oral sernyl in obsessive states. *Journal of Mental Science, 1961, 109* 1–14.

Davies, B. M., & Beech, H. R. The effect of 1-arylcyclohexamine (sernyl) on twelve normal volunteers. *Journal of Mental Science, 1960, 106*, 912–924.

Dodds, J., & Santostefano, S. A comparison of the cognitive functioning of glue sniffers and nonsniffers. *Journal of Pediatrics*, 1964, *64*, 565–570.

Domino, E. F. Neurobiology of phencyclidine (sernyl), a drug with an unusual spectrum of pharmacological activity. *Neurobiology*, 1964, *6*, 303–347.

Donovan, D., & Kivlahan, D. Neuropsychological impairment of alcoholics before and after treatment: Implications for treatment outcome. Paper presented at the annual meeting of the International Neuropsychology Society, Mexico City, 1983.

Donovan, D., Queisser, H., & O'Leary, M. R. Group embedded figures as a predictor of cognitive impairment among alcoholics. *International Journal of the Addictions*, 1976, *11*, 725–739.

Dricker, J., Butters, N., Berman, G., Samuels, I., & Carey, S. The recognition and encoding of faces by alcoholic Korsakoff's and right hemisphere patients. *Neuropsychologia*, 1978, *16*, 683–695.

Eckardt, M. J., Ryback, R. S., & Paulter, C. P. Neuropsychological deficits in alcoholic men in their mid-thirties. *American Journal of Psychiatry*, 1980, *137*, 932–936.

Eckardt, M. J., Parker, E. S., Noble, E. P., Feldman, D. J., & Gottschalk, L. A. Relationship between neuropsychological performance and alcohol consumption in alcoholics. *Biological Psychiatry*, 1978, *13*, 551–565.

Eckardt, M. J., Parker, E. S., Noble, E. P., Paulter, C. P., & Gottschalk, L. A. Changes in neuropsychological performance during treatment for alcoholism. *Biological Psychiatry*, 1979, *14*, 943–954.

Elashoff, J. D. Analysis of covariance: A delicate instrument. *American Education Research Journal*, 1969, *3*, 383–401.

Ellenberg, L., Rosenbaum, G., Goldman, M. S., & Whitman, R. D. Recoverability of psychological functioning following alcohol abuse: Lateralization effects. *Journal of Consulting and Clinical Psychology*, 1980, *48*, 503–510.

Fabian, M. S., & Parsons, O. A. Differential improvements of cognitive functions in recovering alcoholic women. *Journal of Abnormal Psychology*, 1983, *92*, 81–95.

Fabian, M. S., Parsons, O. A., & Silverstein, J. A. Impaired perceptual cognitive functioning in women alcoholics. *Journal of Studies on Alcohol*, 1981, *42*, 217–229.

Fauman, M. A., & Fauman, B. J. The differential diagnosis of organic base psychiatric disturbance in the emergency department. *Journal of the American College of Emergency Room Physicians*, 1977, *6*, 315–323.

Fields, F. R. J., & Fullerton, J. R. The influence of heroin addiction on neuropsychological functioning. *Veterans Administration Newsletter for Research in Mental Health and Behavioral Science*, 1974, *16*, 20–25.

Fitzhugh, L. C., Fitzhugh, K. B., & Reitan, R. M. Adaptive abilities and intellectual functioning of hospitalized alcoholics. *Quarterly Journal of Studies on Alcohol*, 1960, *21*, 414–423.

Fitzhugh, L. C., Fitzhugh, K. B., & Reitan, R. M. Adaptive abilities and intellectual functioning in hospitalized alcoholics: Further considerations. *Quarterly Journal of Studies on Alcohol*, 1965, *26*, 402–411.

Gilberstadt, S. J., Gilberstadt, H., Zieve, L., Buegel, B., Collier, P. O., & McClain, C. J. Psychomotor performance defects in cirrhotic patients without overt encephalopathy. *Archives of Internal Medicine*, 1980, *140*, 519–521.

Glosser, G., Butters, N., & Kaplan, E. Visuoperceptive processes in brain damaged patients on the digit symbol substitution test. *International Journal of Neuroscience*, 1977, *7*, 59–66.

Goldman, M. Reversibility of psychological deficits in alcoholics: The interaction of aging with alcohol. Paper presented at the Eleventh Annual Meeting of the International Neuropsychology Society, Mexico City, 1983.

Goldstein, G., & Shelly, C. A multivariate neuropsychological approach to brain lesion localization in alcoholism. *Addictive Behaviors*, 1982, *7*, 165–175.

Gonzales, E., & Downey, J. Polyneuropathy in a glue sniffer. *Archives of Physical Medicine and Rehabilitation*, July 1972, pp. 333–337.

Goto, I., Matsumura, M., Inoue, N., Murai, Y., Shida, K., Santa, T., & Keiroiwa, Y. Toxic polyneuropathy due to glue sniffing. *Journal of Neurology, Neurosurgery, and Psychiatry*, 1974, *37*, 848–853.

Grabski, D. A. Toluene sniffing producing cerebellar degeneration. *American Journal of Psychiatry*, 1961, *118*, 461–462.

Grant, I., & Judd, L. Neuropsychological and EEG disturbances in polydrug users. *American Journal of Psychiatry*, 1976, *133*, 1039–1042.

Grant, I., & Mohns, L. Chronic cerebral effects of alcohol and drug abuse. *International Journal of the Addictions*, 1975, 883–920.

Grant, I., Adams, K., & Reed, R. Normal neuropsychological abilities of alcoholic men in their late thirties. *American Journal of Psychiatry*, 1979, *136*, 1263–1269.

Grant, I., Adams, K., & Reed, R. Aging, abstinence and medical risk factors in the prediction of neuropsychological deficit among long-term alcoholics. *Archives of General Psychiatry*, 1984, *41*, 710–718.

Grant, I., Reed, R., & Adams, K. Natural history of alcohol and drug related brain disorder: Implications for neuropsychological research. *Journal of Clinical Neuropsychology*, 1980, *2*, 321–331.

Grant, I., Mohns, L., Miller, M., & Reitan, R. A neuropsychological study of polydrug users. *Archives of General Psychiatry*, 1976, *33*, 973–978.

Grant, I., Rochford, J., Fleming, T., & Stunkard, A. A neuropsychological assessment of the effects of moderate marijuana use. *Journal of Nervous and Mental Disease*, 1973, *156*, 278–280.

Grant, I., Adams, K., Carlin, A., Rennick, P., Judd, L., & Schooff, K. The collaborative neuropsychological study of polydrug users. *Archives of General Psychiatry*, 1978, *35*, 1063–1074.

Grant, I., Adams, K. M., Carlin, A. S., Rennick, P. M., Judd, L. L., Schooff, K., & Reed, R. Neuropsychological effects of polydrug abuse. In D. R. Wesson, A. S. Carlin, K. M. Adams, & G. Beschner (Eds.), *Polydrug abuse: The results of a national collaborative study*. New York: Academic Press, 1978.

Gregson, R. A. M., & Taylor, M. A. Prediction of relapse in men alcoholics. *Journal of Studies on Alcohol*, 1977, *38*, 1749–1760.

Guthrie, A., & Elliott, W. A. The nature and reversibility of cerebral impairment in alcoholism. *Journal of Studies on Alcohol*, 1980, *41*, 147–155.

Halsted, C. H., Robles, E. A., & Mezey, E. Intestinal malabsorption induced by feeding a folate deficient diet and ethanol to alcoholic patients. *American Journal of Clinical Nutrition*, 1972, *25*, 449.

Heath, R. G., Fitzjarrell, A. T., Fontana, C. J., & Garcy, R. E. Cannabis Sativa: Effects on brain function and ultrastructure in Rhesus monkeys. *Biological Psychiatry*, 1980, *15*, 657–690.

Herskowitz, A., Ishii, N., & Schaumburg, H. N-Hexane neuropathy. *The New England Journal of Medicine*, July 8, 1971, pp. 82–85.

Hill, S. Y., & Mikhael, M. A. Computerized transaxial tomographic and neuropsychological evaluations in chronic alcoholics and heroin abusers. *American Journal of Psychiatry*, 1979, *136*, 598–602.

Hill, S. Y., Reyes, R. B., Mikhael, M., & Ayre, F. *A comparison of alcoholics and heroin abusers: Computerized transaxial tomography and neuropsychological functioning.* Paper presented at the National Council on Alcoholism meetings, St. Louis, 1978.

Israel, Y., Valenzuela, J. E., Salazer, I., & Ugarte, G. Alcohol and amino acid transport in the human small intestine. *Journal of Nutrition,* 1969, 222–224.

Jones, B., & Parsons, O. A. Impaired abstracting ability in chronic alcoholics. *Archives of General Psychiatry,* 1971, *24,* 71–75.

Jones. B., & Parsons, O. A. Specific versus generalized deficits of abstracting ability in chronic alcoholics. *Archives of General Psychiatry,* 1972, *26,* 380–384.

Judd, L. L., & Grant, I. Brain dysfunction in chronic sedative users. *Journal of Psychedelic Drugs,* 1975, *7,* 143–149.

Kapur, N., & Butters, N. Visuoperceptive deficits in long-term alcoholics and alcoholics with Korsakoff's psychosis. *Journal of Studies on Alcohol,* 1977, *38,* 2025–2035.

Kish, G. B., Hagen, J. M., Woody, M. M., & Harvey, H. L. Alcoholic's recovery from cerebral impairment as a function of duration of abstinence. *Journal of Clinical Psychology,* 1980, *36,* 584–589.

Klisz, D. K., & Parsons, O. A. Hypothesis testing in younger and older alcoholics. *Journal of Studies on Alcohol,* 1977, *38,* 1718–1729.

Klisz, D. K., & Parsons, O. A. Cognitive functioning in alcoholics: The role of subject attrition. *Journal of Abnormal Psychology,* 1979, *88,* 268–276.

Knox, J. W., & Nelson, J. R. Permanent encephalopathy from toluene inhalation. *New England Journal of Medicine,* 1966, *275,* 1494–1496.

Kolansky, H., & Moore, W. T. Effects of marijuana on adolescents and young adults. *Journal of the American Medical Association,* 1971, *216,* 486–492.

Kolansky, H., & Moore, W. T. Toxic effects of chronic marijuana use. *Journal of the American Medical Association,* 1972, *222,* 35–41.

Korman, M., Matthews, R. W., & Lovitt, R. Neuropsychological effects of abuse of inhalants. *Perceptual and Motor Skills,* 1981, *53,* 547–553.

Korobkin, R., Asbury, A., Sumner, A., & Nielsen, S. Glue-sniffing neuropathy. *Archives of Neurology,* 1975, *32,* 158–162.

Kroll, P., Siegel, R., O'Neill, B., & Edwards, R. P. Cerebral cortical atrophy in alcoholic men. *Journal of Clinical Psychiatry,* 1980, *36,* 584–589.

Leber, W. R., Jenkins, R. L., & Parsons, O. A. Recovery of visuo-spatial learning and memory in chronic alcoholics. *Journal of Clinical Psychology,* 1981, *37,* 192–197.

Lee, K., Moller, L., Hardt, F., Haubek, A., & Jensen, E. Alcohol-induced brain damage and liver damage in young males. *The Lancet,* October 13, 1979, pp. 759–761.

Lindenbaum, J., & Lieber, C. S. Alcohol induced malabsorption of vitamin B12 in man. *Nature,* 1969, *224,* 806.

Loberg, T. Alcohol misuse and neuropsychological deficits in men. *Journal of Studies on Alcohol,* 1980, *41,* 119–128.

Luby, E. D., Cohen, B. D., Rosenbaum, G., Gottlieb, J. S., & Kelly, R. A study of a new schizophrenomimetic drug—sernyl. *Archives of Neurology and Psychiatry,* 1959, *81,* 363–369.

Luisada, P. V. The phencyclidine psychosis: Phenomenology and treatment. In R. C. Petersen, (Ed.), *Phencyclidine Abuse: An Appraisal. N.I.D.A. Research Monograph Series,* 1978, *21,* 241–253.

MacVane, J., Butters, N., Montgomery, K., & Farber, J. Cognitive functioning in men social drinkers: A replication study. *Journal of Studies on Alcohol,* 1982, *43,* 81–95.

Mallov, J. MBK neuropathy among spray painters. *Journal of the American Medical Association,* 1976, *235,* 1455–1457.

Massengale, O., Glaser, H., LeLievre, R., Dodds, J., & Klock, M. Physical and psychologic factors in glue sniffing. *The New England Journal of Medicine*, 1963, *269*, 1340–1344.

Mendelson, J. H., & Meyer, R. E. Behavioral and biological concomitants of chronic marijuana smoking by heavy and casual users. In *Technical papers of the first report of the National Commission on Marijuana and Drug Abuse*. Washington, D.C.: U.S. Government Printing Office, 1972, pp. 69–246.

Miglioli, M., Buchtel, H. A., Campanini, T., & deRisio, C. Cerebral hemispheric lateralization of cognitive deficits due to alcoholism. *Journal of Nervous and Mental Disease*, 1979, *167*, 212–217.

Miller, W. R., & Orr, J. Nature and sequence of neuropsychological deficits in alcoholics. *Journal of Studies on Alcohol*, 1980, *41*, 325–337.

Miras, C. N., & Fink, M. Investigation of very heavy, very long-term cannabis users: Greece. In *Technical papers of the National Commission on Marijuana and Drug Abuse*. Washington, D.C.: U.S. Government Printing Office, 1972, pp. 53–55.

Mohs, R. C., Tinklenberg, J. R., Roth, W. T., & Kopell, B. S. Slowing of short-term memory scanning in alcoholics. *Journal of Studies on Alcohol*, 1978, *39*, 1908–1915.

Neville, J. N., Eagles, J. A., Samson, G., & Olson, R. E. Nutritional status of alcoholics. *American Journal of Clinical Nutrition*, 1968, *2*, 1329–1340.

Nichols-Hochla, N. A., & Parsons, O. A. Premature aging in female alcoholics: A neuropsychological study. *Journal of Nervous and Mental Disease*, 1982, *170*, 241–245.

O'Leary, M. R., Donovan, M. A., & Chaney, E. F. The relationship of perceptual field orientation to measures of cognitive functioning and current adaptive abilities in alcoholics and nonalcoholics. *Journal of Nervous and Mental Disease*, 1977, *165*, 275–282.

Parker, E. S., & Noble, E. P. Alcohol consumption and cognitive functioning in social drinkers. *Journal of Studies on Alcohol*, 1977, *38*, 1224–1232.

Parker, E. S., & Noble, E. P. Alcohol and the aging process in social drinkers. *Journal of Studies on Alcohol*, 1980, *41*, 170–178.

Parker, E. S. Parker, D., Brody, J. A., & Schoenberg, R. Cognitive patterns resembling premature aging in male social drinkers. *Alcoholism: Clinical and Experimental Research*, 1982, *6*, 46–52.

Parsons, O. A., & Farr, S. D. The neuropsychology of alcohol and drug use. In S. B. Filskov & T. J. Boll (Eds.), *Handbook of clinical neuropsychology*. New York: Wiley, 1981.

Prigatano, G. Neuropsychological functioning in recidivist alcoholics treated with disulfiram. *Alcoholism: Clinical and Experimental Research*, 1977, *1*, 81–86.

Prockop, L., Alt, M., & Tison, J. "Huffer's" neuropathy. *Journal of the American Medical Association*, 1974, *229*, 1083–1084.

Rochford, J., Grant, I., & LaVigne, G. Medical students and drugs: Further neuropsychological and use pattern considerations. *International Journal of the Addictions*, 1977, *12*, 1057–1065.

Roggin, G. M., Iber, F. L., Kater, R. M. H., & Tabon, F. Malabsorption in the chronic alcoholic. *Johns Hopkins Medical Journal*, 1969, *125*, 321–330.

Ron, M. A. Brain damage in chronic alcoholism: A neuropathological, neuroradiological and psychological review. *Psychological Medicine*, 1977, *7*, 103–112.

Ron, M. A., Acker, W., & Lishman, W. A. Morphological abnormalities in the brains of chronic alcoholics: A clinical, psychological and computerized axial tomographic study. *Acta Psychiatrica Scandinavica*, 1980, *62*, 41–46.

Rounsaville, B. J., Novelly, R. A., & Kleber, H. D. Neuropsychological impairment in opiate addicts: Risk factors. *Annals of New York Academy of Science*, 1980, *362*, 79–90.

Rounsaville, B. J., Jones, C., Novelly, R. A., & Kleber, H. Neuropsychological functioning in opiate addicts. *Journal of Nervous and Mental Disease*, 1982, *170*, 209–216.

Rubin, V., & Comitas, L. *Ganja in Jamaica: A medical and anthropological study of chronic marijuana use.* The Hague: Mouton, 1975.

Ryan, C. Learning and memory deficits in alcoholics. *Journal of Studies on Alcohol,* 1980, *41,* 437–447.

Ryan, C. Alcoholism and premature aging: A neuropsychological perspective. *Alcoholism: Clinical and experimental research,* 1982, *6,* 22–30.

Ryan, C., & Butters, N. Learning and memory impairments in young and old alcoholics: Evidence for the premature aging hypothesis. *Alcoholism: Clinical and Experimental Research,* 1980, *4,* 288.

Ryan, C., & Butters, N. Cognitive effects in alcohol abuse. In B. Kissin & H. Begleiter (Eds.), *Biology of alcoholism: Vol. 6—Biological pathologenesis of alcoholism.* New York: Plenum Press, 1982.

Ryan, C., Butters, N., & Montgomery, K. Memory deficits in chronic alcoholics: Continuities between the "intact" alcoholic and the alcoholic Korsakoff's patient. In H. Begleiter (Ed.), *Biological Effects of Alcohol.* New York: Plenum Press, 1980.

Satz, P., Fletcher, J. M., & Sutker, L. S. Neuropsychological, intellectual, and personality correlates of chronic marijuana use in native Costa Ricans. *Annals of the New York Academy of Sciences,* 1976, 266–306.

Sharp, J. R., Rosenbaum, G., Goldman, M. S., & Whitman, R. D. Recoverability of psychological functioning following alcohol abuse: Acquisition of meaningful synonyms. *Journal of Consulting and Clinical Psychology,* 1977, *45,* 1023–1028.

Smith, H. H., & Smith, L. S. WAIS functioning of cirrhotic and noncirrhotic alcoholics. *Journal of Clinical Psychology,* 1977, *33,* 309–313.

Soueif, M. I. The use of cannabis in Egypt: A behavioral study. *Bulletin of Narcotics,* 1971, *23,* 17–28.

Towfighi, J., Gonatas, N. K., Pleasure, D., Cooper, H. S., & McCree, L. Glue sniffers' neuropathy. *Neurology,* 1976, *26,* 238–243.

Tsushima, W. T., & Towne, W. S. Effects of paint sniffing on neuropsychological test performance. *Journal of Abnormal Psychology,* 1977, *86,* 402–407.

Ware, L. A. Neuropsychological functioning in users and nonusers of phencyclidine. *Dissertation Abstracts,* Southern Illinois University at Carbondale, 1978, p. 124.

Wesson, D., Carlin, A., Adams, K., & Beschner, G. (Eds.), *Polydrug abuse: The results of a national collaborative study.* New York: Academic Press, 1978.

Wilkinson, D. A., & Carlen, P. L. Relationship of neuropsychological test performance to brain morphology in amnesic and nonamnesic chronic alcoholics. *Acta Psychiatrica Scandinavica,* 1980, Suup. 286, 89–101. (a)

Wilkinson, D. A., & Carlen, P. L. Neuropsychological and neurological assessment of alcoholism: Discrimination between groups. *Journal of Studies on Alcohol,* 1980, *41,* 129–139. (b)

Wilkinson, D. A., & Carlen, P. L. *Recoverability in recently abstinent alcoholics: Results of repeated neuropsychological, EEG, CT scan and neurological examination.* Paper presented at the Annual Meeting of the International Neuropsychology Society, Mexico City, 1983.

Wood, M. B. A dietary study of alcoholism. *Food and Nutrition; Notes-Review,* 1972, *29,* 33–41.

Yesavage, J. A., & Freman, A. M. Acute Phencyclidine (PCP) intoxication: Psychopathology and prognosis. *Journal of Clinical Psychiatry,* 1978, *39,* 664–666.

12

Family Adaptation to Substance Abuse

EDWARD KAUFMAN

1. General Introduction

At the present time, it is commonly accepted that substance abuse has a profound effect on the family, generally leading to greater psychopathology in all or most individual members as well as to impairment in family system functioning. Several other major issues remain controversial. The first of these deals with which specific types of family dysfunction lead to substance abuse, particularly to abuse of different drugs and alcohol. A second major problem in the field is the relative lack of a clear demonstration of the efficacy of family therapy, and particularly of its superiority over other forms of treatment for substance abuse. Our current knowledge of these areas will be a major emphasis of this chapter. Lastly, another area worthy of exploration is the extent to which family pathology precedes and leads to substance abuse as opposed to the extent to which it is a consequence of substance abuse.

Two separate literatures on the family and alcohol and the family and drug abuse have evolved. There have been few attempts to synthesize these into a single cohesive theory. Until recently, most studies on the alcoholic family have focused on a male alcoholic in his 40s and his overinvolved spouse (Kaufman & Pattison, 1981). Most studies of drug abusers have focused on addicts in their teens and 20s and their parents

EDWARD KAUFMAN • Department of Psychiatry and Human Behavior, University of California Irvine, California College of Medicine, 101 City Drive South, Orange, California 92668.

(Kaufman, 1974, 1980). All too frequently and yet unknowingly, these studies have focused on the same family. This is because families wherein a parent is an alcoholic frequently produce both drug-abusing and alcoholic children (Ziegler-Driscoll, 1977). In addition, the incidence of drug abuse by alcoholics and of alcohol by primary drug abusers is increasing (Kaufman, 1981). This leads to further fusion of the two types of families. The present focus on the family and substance abuse is a three-generational one where we look at the entire family system and its relation to substance abuse. This chapter will provide both an historical perspective and a review of current knowledge in the above areas as they relate separately to alcohol and drug abuse and an integration of these findings. Further implications for research and treatment will also be discussed.

2. Historical Perspectives

A historical perspective is presented because many early findings are still quite valid while others have been subject to criticism in more recent times.

2.1. Alcohol and the Family

Most early clinical studies focused on the male alcoholic and his nonalcoholic wife. It was implied that the wife was neurotic and chose an alcoholic husband, and later that the wife became neurotic because of her husband's alcoholism. Perhaps even more misogynistic was the view that the wife "drove her husband to drink." But current empirical research reveals that the fable of the noxious wife is just that (Kogan & Jackson, 1965). There is no validity to several earlier typologies of "typical wives of alcoholics" (Edwards, Harvey, & Whitehead, 1973; Kaufman & Pattison, 1981; Paolino, McCrady, Diamond, & Longabaugh, 1976). The same problem obtains in the few studies of men who marry women alcoholics, which often indicate significant psychopathology in these men, but there is probably no specific type of man who marries an alcoholic woman (Busch, Kormendy, & Feuerlein, 1973; Rimmer, 1974).

A more fruitful approach has been the study of marital interactional dynamics, role perceptions, and marital patterns of expectations and sanctions about the use of alcohol. Alcoholic couples demonstrate neurotic interactional behavior similar to that of other neurotic couples; both alcoholic and neurotic marriages differ from healthy ones. Thus, alcoholic marriages may not be unique, but similar to other neurotic marriages in which alcoholism is part of the neurotic interaction (Becker & Miller, 1976; Billings, Kessler, Gomberg, & Weiner, 1979; Drewery & Rae, 1969; Hanson, Sands, & Sheldon, 1968).

2.1.1. Alcoholism and Children in the Family System

Much of the literature on alcoholic families has focused primarily on the marital partners' parental neglect, the roles and functions of children in the family and the consequences of alcoholism for the children; Cork (1969) called them "the forgotten children." Children are often the most severely victimized members of an alcoholic family. They have growth and developmental problems (Chafetz, Blane, & Hill, 1971; El-Guebaly & Orford, 1977). Further, these children are often subject to gross neglect and abuse. Teenage children are not immune to these adverse consequences, even though they are often considered less vulnerable—perhaps a misperception. Just as significant are the long-term adverse consequences for personality patterning and identity formation, and the possible development of dysfunctional attitudes toward alcohol.

2.2. Drug Abuse and the Family

The early literature concerning the family and drug abuse focused on the symbiotic tie between mothers and their drug-addicted sons and the absence or uninvolvement of fathers. Fort (1954) noted that such mothers were "overprotective, controlling, and indulgent," and that "They were willing to do anything for their sons, except let them alone." In a comparative study conducted by Attardo (1965) of the mothers of drug addicts, schizophrenics, and normal adolescents, the mother's symbiotic need for the child was found to be highest in the mothers of drug abusers. Fort (1954) noted, in a group comprised mainly of ghetto addicts, the "frequent virtual absence of a father figure." However, studies of middle-class families have noted the presence of a "strong" father (Alexander & Dibb, 1975). But Kirschenbaum, Leonoff, and Maliano (1974) noted that the father's position as strong leader of the family seemed to be a fiction—one needed and nourished by the mother as the real head of the family. Schwartzman (1975) also noted that fathers were either "straw man" authoritarian figures or distant but clearly "secondary" to the mother in terms of power.

Several authors have noted that drug use is essential to maintaining an interactional family equilibrium that resolves a disorganization of the family system that existed prior to drug taking (Noone & Reddig, 1976)—for example, the addict or preaddict is scapegoated so that the parents do not work through conflicts between themselves. The "addictor" in the system may be the parent(s) or the spouse, as noted by Pearson and Little (1975). Wellisch, Gay, and McEntea (1970) noted that one partner, usually the male, is supported or taken care of by the other, and so becomes an "easy rider" throughout the relationship. In this author's own early work in this field (Kaufman, 1974), it was noted that fathers of heroin addicts were described as cold, distant, sadistic, competitive with their sons, and seductive of their daughters. Mothers were perceived as distant or overly seductive.

Alexander and Dibb (1975) have stated that "a minority of opiate addicts maintain close emotional and financial relationships with their parents," although this statement was based on only one group of caucasian methadone patients in British Columbia. However, Noone and Reddig (1976) found that a majority of drug abusers and addicts maintain close ties with their families of origin years after they have apparently left home. Alexander and Dibb (1975) noted that when the father was present, he was dominant in 11 of the 18 families studied.

Stanton (1971), in a study of 85 addicts, was one of the first observers to quantify the heroin addict's frequent involvement with family; noting that of addicts with living parents, 82% saw their mothers and 58% saw their fathers at least weekly; 66% either lived with their parents or saw their mothers daily. In 1966, Vaillant (1966) reported 72% of (active) addicts still lived with their mothers at age 22, and 47% continued to live with a female blood relative after age 30. Interestingly, he also noted that of the 30 abstinent addicts in his follow-up study, virtually all were living independently from their parents.

Reilly's (1976) clinical studies noted that families of drug abusers are often dull and lifeless and only become alive when mobilized to deal with the crisis of drug abuse. Reilly reported, additionally, that communication is most frequently negative and that there is no appropriate praise for good behavior. There is a lack of consistent limit setting by parents so that defiance may be punished or rewarded at different times. Reilly's observations continue to be considered quite valid for many types of drug-abusing families.

3. Current Research and Knowledge

3.1. Alcohol and the Family

Recent research has moved away from a focus on the marital partners toward a consideration of the family system, the families of origin, the consequent life-style of children from alcoholic families, and the kin structures of the extended family system (Kaufman & Pattison, 1981). The first conclusion from experimental observations of family systems is that alcohol use in a family is not just an individual matter. The use of alcohol and the consequential behavior are dynamically related to events in the family system. Thus, the use of alcohol is purposeful, adaptive, homeostatic, and meaningful. The problem of alcoholism is not just the consequences of drinking *per se*, but, more important, includes the system functions that drinking fulfills in the dynamics of the family system (Davis, Berenson, Steinglass, & Davis, 1974). Thus we may properly consider alcoholism a family systems problem.

Olson has studied 300 "chemically dependent families" with the Family Adaptability and Cohesion Evaluation Scales (FACES; Olson, Bell, & Porter, 1978) and has found more families to be "chaotically disengaged" than normal controls; that is, few rules, no leadership, and extreme unrelatedness among members (Olson, 1983). He noted little difference in extremes in the "enmeshed" dimension. "Enmeshed" in this context refers to the greater degree of intrapersonal reactivity and togetherness so often described in alcoholic families.

Kaufman and Pattison (1981) have classified the alcoholic and his/her family into four subtypes that have important implications for treatment: (1) Functional, (2) Neurotic Enmeshed, (3) Disintegrated, and (4) Absent. These systems may stabilize at one of these levels or may be progressive toward greater deterioration of family structure. These four family subtypes are based on over 20 years of experience and observation. They have not yet been quantitatively studied, and they represent hypothetical family types based, as indicated, on long-term clinical observations. These are described in the following discussion.

3.1.1. The Functional Family System
(The Family with an Alcoholic Member)

Functional family systems are apparently stable and happy. The parents maintain a loving relationship with a relatively good sexual adjustment. They are successful as parents and their children are well adjusted and have good relationships with each other and their peers. *Drinking by the alcoholic spouse does not evolve as a result of family stresses,* but primarily in response to social strains or personal neurotic conflict. Excessive drinking often occurs outside the home, in binges, at parties, or at bedtime. It is usually alcoholics in the early phases of alcoholism who have functional family systems, as such systems may deteriorate with progressive alcoholism.

These families have learned to function with a minimum of overt conflict. Often they are sensitive to social standards and values and are responsive to social norms and social authorities (ministers, physicians, supervisors). Although such families function well, they are usually not psychologically oriented. They are functional, but not insightful. They are responsive to external change, but resistant to internal change.

Because of lack of obvious conflict and generally good family functioning, the focus of family concern is likely to be on the alcoholic member. The emotional balance of the family is likely to be positive, with a desire to retain and rehabilitate the alcoholic member. Thus, families can usually be engaged in exploration of family rules and roles. Cognitive modes of interaction are usually acceptable, but more uncovering and emotional interactions may be resisted. The exploration of family response to the drinking member will uncover dysfunctional role behavior. New roles and rules for response can be defined, practiced, and reviewed.

In general, short-term family therapy aimed at rule definition and role restructure may suffice. Sometimes such families may move into more psychodynamic family therapy. However, if abstinence is achieved with a new stable family equilibrium, the therapist should be content.

3.1.2. The Neurotic Enmeshed Family System (The Alcoholic Family)

In Neurotic Enmeshed families, drinking behavior interrupts normal family tasks, causes conflict, shifts roles, and demands adjustive and adaptive responses from family members. Alcoholism creates physical problems, including sexual dysfunction and debilitating cardiac, hepatic, and neurological disease, which in turn produce further marital conflicts and role alignments.

A converse dynamic also occurs in that marital and family styles, rules, and conflict may evoke, support, and maintain alcoholism as a symptom of family system dysfunction or as a coping mechanism to deal with family anxiety. The anxiety increases when the alcoholic drinks, which intensifies present behavior. Drinking to relieve anxiety, and family anxiety in response to drinking, spiral into a crisis (Bowen, 1976).

Stresses in any single family member affect the entire family with urgency and immediacy. Communication is often not direct, but through a third party. Conflicts are triangulated (projected) onto another family member (Bowen, 1976). Everyone in the family feels guilty and responsible for each other, but particularly for the alcoholic and his drinking.

Such alcoholic marriages are often highly competitive. Each partner sees himself as giving in to the other. The alcoholic repeatedly tries to control and avoid responsibility through passive-dependent techniques. The spouse tries to control by being forceful, active, blunt and dominating, or by suffering. Neither ever clearly become dominant, but the fight continues indefinitely (Hersen, Miller, & Eisler, 1973). Fighting frequently occurs as the spouses blame each other for the family's problem. This dual projection blinds the couple to their respective roles in creating problems. They may fight endlessly about who is to blame, and they may readily duplicate this position in therapy with the hope that the therapist will judge right and wrong.

The alcoholic in this family system may relinquish his role as a parent. Other roles involving household chores and maintenance are abandoned and turned over to others. The role of the breadwinner is the last to go, and this stage may have to be reached before treatment is sought.

As nonalcoholic members take over management of the family, the alcoholic is relegated to the status of a child, which perpetuates drinking. Coalitions occur between the nonalcoholic spouse and children or in-laws, which tend to distance the alcoholic further. The alcoholic parent is prone to abuse his/her children through neglect. The nonalcoholic spouse may also neglect children in order to direct attention to the alcoholic.

The therapeutic approach to these families is much more difficult and prolonged than that for functional family systems. Educational and behavioral methods as outlined above may provide some initial relief, but are not likely to have much impact on the enmeshed neurotic relationships. Thus, these families require intensive restructuring family therapy. Reenactment (actualization) of family conflicts is very important. These families will resist therapeutic work by attempting to recount the past rather than deal with the present. Because of the mutual enmeshment of the family members, it is usually necessary to work hard on marking boundaries and defining personal roles. These families are often emotionally explosive. Hence the therapist will probably have to engage in frequent joining maneuvers to keep family members involved and to keep the emotional tensions within workable limits. Initially, the therapist will need to be active in defining roles, rules, and tasks for family members until enough disengagement and individual autonomy have developed for the family to operate as an effective system. With these families, disengagement and unmeshing can be assisted by getting family members involved with external support groups such as A.A., Al-Anon, and Alateen, or with church and community group activities.

Although initial hospitalization or detoxification may achieve temporary abstinence, the alcoholic is highly vulnerable to relapse. Therefore, longer-term family therapy is required to develop a nonneurotic family system.

3.1.3. *The Disintegrated Family System (Family Separation and Isolation)*

In the disintegrated system, there is a hisotry of reasonable vocational function and family life. Usually there has been a progressive deterioration, with loss of job, loss of self-respect, family instability, inability to function in the family, and finally separation from the family. Frequently the family has become destitute and has turned to in-laws or welfare for support. The alcoholic is now totally alienated from his/her family.

This type of alcoholic usually presents him/herself at hospitals or clinics without any family and frequently without any recent family contact. Usually he/she requires immediate physical support in terms of room, food, and clothes. He/she is usually without the resources needed to return immediately to independent functioning. The use of family interventions might seem irrelevant in such a case. However, many of these marriages and families have fallen apart only after severe alcoholismic behavior. Further, there is often only "pseudoindividuation" of the alcoholic from marital, family, and kinship ties. These families usually cannot and will not reconstitute during the early phases of rehabilitation. Thus, the early stages of treatment should focus primarily on the individual alcoholic. However, potential ties to spouse, family, kin, and friends should be explored early in treatment, and some contact should be initiated. There should be neither explicit nor implicit assumptions that such familial ties will be fully reconstituted. When abstinence and personal stability have been achieved over several months, more substantive family explorations can be initiated to reestablish parental roles and family and kinship relationships—still without reconstitution. These family definitional sessions can then serve as the springboard for either appropriate redefinition of separate roles or for a reconstituted family structure. In either case, it is important for both the alcoholic and his/her family to renegotiate new roles and relationships on the basis of his/her identity as a rehabilitated alcoholic. Some families may not desire a reunion but can achieve healthy separation. Families who do desire a reunion must establish a new base for family relationships.

Should family reconciliation occur, it is desirable to continue family sessions to monitor and stabilize new family roles. When reconciliation is not achieved, it is important to negotiate acceptable roles that leave all parties with reasonable self-respect and respect for each other. Otherwise, the alcoholic is likely to be victimized by continuing conflict with the separated family.

3.1.4. The Absent Family System (The Long-Term Isolated Alcoholic)

Although the Absent system may be a final stage, more often this type of alcoholism syndrome is marked by total loss of family of origin early in the drinking career. Such persons usually have never married, or have had brief, fleeting relationships. They may have relatives or in-laws with whom they maintain perfunctory contact. These persons rarely have close friendships, and they have minimal social or vocational skills. Their significant others are boarding home operators or "bottle gang" buddies. For these alcoholics the problem is not rehabilitation, for the vast majority have never acquired major coping skills. They have little ability to form effective social relationships, and they do best in partially institutionalized social support systems. There are two major family interventions to consider. The first is the elaboration of still extant friend and kin contacts. Often these social relationships can be re-vitalized and can provide meaningful social support. Second, in some of the younger alcoholics of this type, there can be a positive response to intense socialization in peer groups, such as AA, church fellowships, and recreational and vocational rehabilitation. These alcoholics, who are exceptions to the rule, can develop positive new self-help skills and can become able to engage in satisfactory marriage and family life.

It would be helpful to know the current prevalence of these four family types among all alcoholics. However, such a study would require widespread epidemiological sampling, as almost no single treatment program observes all types of families and most observe only one. Un-fortunately, the failure to sample and observe all subtypes has often led to overgeneralizations about the families of alcoholics.

3.2. Drug Abuse and the Family

3.2.1. Heroin and the Family

This author's own studies of the *effects* of drug abuse on the family include a four-year study of 75 families and 78 heroin addicts in New York City and Los Angeles County (Kaufman, 1981).

Of the 75 families, 88% of the mothers were emotionally enmeshed with their drug-abusing children, mainly sons, to the extent that their happiness and emotional pain were totally dependent on the behavior of and closeness with these children. Although 43% of the fathers were absent or emotionally disengaged from the drug abuser and the entire family, 41% of the fathers were enmeshed with the drug abuser, as well as with the total family. The relatively high percentage of enmeshed fathers in this study includes fathers who were overinvolved with female heroin addicts and fathers who stayed enmeshed in Italian and Jewish families, which were highly represented in this sample compared to other studies of families of drug abusers.

The role of siblings of drug abusers has also been consistently overlooked. Siblings tend to fall equally into two basic categories: very good and very bad. The "bad" group is composed of fellow drug abusers whose drug use is inextricably fused with that of the "identified" or "index patient" (I.P.). The "good" group includes children with parental family roles who assume an authoritarian role when the father is disengaged and/or are themselves highly successful vocationally. Some of these successful siblings had individuated from the family, but many were still enmeshed. Hendin, Pollinger, Ulman, and Carr (1981) noted that drug-free siblings were frequently the "good ones" even before drug use began, and that this dichotomy contributes to the difficulties of the I.P. Another small group of "good" siblings were quite passive and not involved with substance abuse. Some of these develop "anger in" disorders such as depression and headaches (Hendin et al., 1981). Enmeshed drug-abusing siblings provide drugs for each other, inject drugs into one another, set the other up to be arrested, or even pimp or prostitute themselves for one another. At times, a large family may show sibling relationships of all the above types. Many successful older siblings were quite prominent in their fields and, in these cases, the patient siblings withdrew from any vocational achievement rather than compete in a seemingly no-win situation. In a few cases, drug abusers were themselves parental children who had no way of asking for relief of responsibility, except through drugs. More commonly, they were the youngest child, and their drug abuse maintained their role as the family baby. They were frequently the child who got the most attention, and whose drug abuse kept him/her from ever abandoning the parental nest, serving an important function for parents who need to have children around to be concerned with and/or to prevent the parents from experiencing strong feelings of boredom in their spousal relationship.

The extent of what was considered pathologic enmeshment varied quantitatively. In an extreme case, a mother who was frequently psychotic with repeated psychiatric hospitalizations was symbiotically tied to her drug-abusing son. Early in their family therapy, the I.P. left residential treatment. His mother was able to refuse to take him back at the expense of triggering an overt psychotic episode, which was resolved with the help of her private psychiatrist. After the mother emerged from her psychosis, she poignantly told her son, "I will not hold you to me anymore." Enmeshed mothers tend to think, act and feel for the drug-abusing child. Several mothers regularly ingested prescribed minor tranquilizers or narcotics that were shared overtly or covertly with their sons. Many mothers suffered an agitated depression whenever their son or daughter "acted out" in destructive ways. Mothers who took prescription tranquilizers or abused alcohol frequently, increased their intake whenever the drug abuser "acted out" (Kaufman, 1981).

My own studies of female addicts are just a beginning. The relationship between mothers and daughters tends to be extremely hostile, competitive, and at times chaotic (Kaufman, 1981). Half the mother–daughter relationships in this study were severely enmeshed. When her mother suicided, a daughter also made a serious suicide attempt. Several other mothers threatened or attempted suicide. One father suicided after his wife ordered him out of the house for his brutality.

A third of the fathers were alcoholic; however, all but four of these had abandoned their families or died from alcoholism and had not been a part of treatment. Most studies of the fathers of drug abusers find the incidence of parental alcoholism to be about 50% (Ziegler-Driscoll, 1977). The one father in my study who had himself been a heroin addict raised five of his six children as heroin addicts (Kaufman, 1981). Frequently, in totally enmeshed families, both parents collaborate with the addict to keep him or her infantilized under the guise of protecting him/her from arrest or other dangers. A pattern of father–son brutality was quite common, although it was seen in fathers who were enmeshed as well as disengaged. With disengaged fathers, brutality was frequently their only contact with their children, which pushes the I.P. into coalitions with the mother against the father. Physical brutality was common between Italian fathers and their sons. However, this was a multigenerational problem that was a part of enmeshed intimacy. Additionally, it is much easier for these fathers to hit a child once or twice than to enforce discipline over hours and days (Kaufman, 1981).

3.2.2. Nonnarcotic Adolescent Substance Abuse and the Family

The patterns of family response to adolescent drug use develop particularly early in the adolescent's drug abuse career. These problems may start upon entry into junior high school, and school problems may be the first warning sign. Parents are also lied to about needs for funds that are diverted to drugs, and when this is discovered, they may be coerced by threats of violence from debtors, commission of crimes, and protection of their child from incarceration. Adolescents continue to drink and use drugs at home after they have been prohibited from doing so, even with legal enforcement. They lie about their substance abuse and destructive behavior, and the lies themselves frequently become a major concern to parents. They also engage parents in frequent power struggles about whether they are high or have used drugs. Parents frequently become totally preoccupied with the adolescent's behavior and their inability to contain it, causing a great deal of parental anguish and suffering (Hendin *et al.*, 1981). Their concern for their adolescent child may create distance between the spouses that may function as a needed buffer or may painfully create sexual and emotional withdrawal.

The family system frequently revolves around the drug abuser as a scapegoat upon whom all intrafamilial problems are focused. At other times the adolescent's difficulties keep conflictual parents together or are an attempt to reunite separated parents. Guilt is a frequent currency of manipulation and may be induced by the drug abuser to coerce the family into continued financial and emotional support of drug use, or guilt may be employed by parents to curb individuation. Many mothers have severe depression, anxiety, or psychosomatic symptoms that are blamed on the I.P., thereby reinforcing the pattern of guilt and mutual manipulation. Mother's drug and alcohol abuse and suicide attempts are also blamed on the drug abuser.

Physical expressions of love and affection (by the parent) are either absent or used to deny and obliterate individuation or conflict. Anger about interpersonal conflicts is not expressed directly, unless it erupts into explosive violence. Anger about drug use and denial of it is expressed quite frequently and is almost always counterproductive. All joy has disappeared in these families, as lives are totally taken up with the sufferings and entanglements of having a substance-abusing child. However, in many cases the joylessness preceded the addiction (Reilly, 1976).

Adolescent drug abuse is also frequently an expression of defiance that actually leads to infantilization and continuing intense family ties, albeit conflictual ones. These adolescents frequently undergo pseudoindividuations through institutionalizations, runaways with crises that result in brief reunions with parents, and scrapes with authorities that result in incarcerations or parental bail-out. Thus, though drug-abusing adolescents may be hundreds of miles from home, their behavior continues to deeply affect and be affected by parental ties.

3.3. Family Factors That Prevent the Abuse of Both Drugs and Alcohol

As yet, no specific patterns have been uncovered that differentiate the family that produces a drug problem from one that produces an alcohol problem (Kaufman, 1980). A healthy family system, however, can prevent substance abuse even in the face of heavy peer pressure to use and abuse drugs (Blum, 1972). As warm and mutual family ties diminish, the adolescent becomes more vulnerable to peer pressure. The key to healthy family functioning is the family's ability to flexibly adapt to different stresses with varied but effective coping mechanisms. Thus, extreme closeness is necessary when children are small, but as they enter adolescence, the family must permit them to become autonomous without excessive control and guilt. The healthy family requires a balance in the following processes: assertiveness, control, discipline, negotiation, roles, rules, and system feedback (Olson, Sprenkle, & Russell, 1979). Not only should a family be able to adapt to expected stresses in the life cycle, but to unanticipated stresses such as physical illness, accidents, job loss, relocations, deaths of family members, divorce, inclusion of new members (including step families), and external catastrophes.

A critical way the family can avoid substance abuse is to eliminate those family factors that are known to heavily predispose to the abuse of drugs and alcohol. The most common finding in the families of substance abusers is parents who are themselves substance abusers, specifically alcoholic fathers and prescription-drug-abusing mothers. Parental abuse of drugs and alcohol is a much more important determinant of adolescent abuse than parental attitude toward the child's drug and alcohol use (Kandel, Kessler, & Margulies, 1978). Even parents who use minor tranquilizers in prescribed doses have a greater incidence of drug- and alcohol-abusing adolescents (Kandel et al., 1978). Alcohol- and drug-abusing siblings statistically predispose other siblings to substance abuse, although sibling substance abuse may also spare over sibs through the family dynamics described previously. Parental mental illness, divorce, separation, and frequent moves also predispose to substance abuse (Gibbs, 1982). Another predisposing factor is physical birth trauma (Gibbs, 1982), perhaps because it leads to diminished coping skills and the need for drugs as compensation.

In general, a traditional family structure insulates the adolescent from drug abuse (Blum, 1972). That is, in a family where there are (1) greater degrees of parental control, (2) a high premium on achieving and high expectations of children by parents, and (3) structured, shared parent/child activities, there is a lowered likelihood of substance abuse (Blum, 1972). However, if any of these three orientations are overdone and excessive, they may themselves lead to or perpetuate substance abuse. The key to determining if these traditional values are overdone is the adolescent's response. If the child vigorously and repeatedly defies parental controls, is overwhelmed by expectations, and avoids activities, then the parents' escalating but ineffective demands may lead to drug abuse or secondarily may be associated with it (Brooks, Lukoff, & Whiteman, 1978).

Studies on the family that differentiate the effects of the female drug abuser and alcoholic from the male have only recently been performed (Kaufman & Pattison, 1981). My observations of female addicts who are spouses of male addicts are that these women are generally quite passive. They have frequently been introduced to drugs by their male partners, who feel a strong need to control their spouse's heroin use. Male addict partners buy drugs for their spouses, inject them, and beat them if they buy heroin from someone else, discouraging and prohibiting any autonomy, including that associated with drug use. There is also a significant minority of assertive female addicts, particularly those who have come to addiction on their own rather than having been led to it by a male. Binion (1980) has studied a group of women heroin users in Detroit. She noted (surprisingly) that female addicts saw their fathers as "helpful loving people," although they "slightly favored their mothers."

More recent research has focused on the female alcoholic than on the female heavy drug abuser. Female alcoholics tend to marry male alcoholics, but male alcoholics rarely marry female alcoholics (Janzen, 1977). Individuals tend to choose spouses with equal levels of ego strength and self-awareness, but with opposite ways of feeling stress. We see frequently that opposites attract in male–female relationships, with each person seeing him/herself as "giving in" to the other. The one who "gives in" the most becomes "deselfed" and is vulnerable to a drinking problem (Bowen, 1974). If it is the wife, she begins drinking during the day to help her through her chores, hiding it from the husband to be ready for ideal togetherness when he returns—until she passes out several times and the problem is recognized.

Women's drinking tends to be more hidden and more triggered by specific life events than that of their male counterparts (Winokur, Reich, Rimmer, & Pitts, 1970). Women alcoholics are more likely to have affective disorders, and men to be sociopathic. Thus males are less accepting of their wives' drinking and tend to leave alcoholic wives more easily. Women tend to start heavy drinking later than men, when their personal inadequacy and lack of self-realization lead to an awareness that the promise of youth cannot be fulfilled. The "empty nest" etiology of alcoholism focuses mainly on women.

4. Specific Treatment Issues

4.1. The Relationship of Family Treatment to Cessation of Substance Abuse

Family therapy is difficult when a nonidentified patient is substance dysfunctional. On the other hand, many families in which one or more members continue to use substances desperately need family intervention and will not accept a therapy that expects them to cease substance abuse as a precondition for beginning therapy. My basic premise is that if the substance abuser's intake is so severe that he or she is unable to attend sessions without being under the influence and/or if functioning is severely impaired, then the first priority is interrupting the pattern of abuse, at least temporarily. Thus my first goal is to persuade the family to pull together to initiate detoxification or at least some measure to achieve temporary abstinence. Generally, this is best done in a hospital and, if the abuse pattern is severe, I will require this in the first session or very early in the therapy. If the substance abuse is only moderately severe or intermittent, such as binge alcoholism or stimulant abuse, then I would offer the family alternatives as to how to initiate this temporary substance-free state. These include social detoxification centers, regular attendance at Alcoholics Anonymous, and Antabuse for alcoholics or detoxification and methadone maintenance for heroin addicts. Antabuse should generally not be given to a family member for daily distribution, as this tends to reinforce the family's being locked into the alcoholic's condition of either drinking or not drinking. Benzodiazepines are discouraged because they tend to become a part of the problem rather than a solution. In general, treatment begins in a much more effective way if the patient is totally immersed in a 28-day residential treatment program that includes detoxification, individual and group therapy, AA and/or Narcotics Anonymous, and family therapy. This type of short-term residential program can be a most effective introduction to therapy. It seems to be more effective with alcoholics than with drug abusers, perhaps because of the effectiveness of AA as aftercare. For many drug-dependent patients, insistence on long-term residential treatment is the only alternative. Most families will not accept this until other methods have failed. In order to accomplish this a therapist must maintain long-term ties with the family, even through multiple treatment failures.

Bowen (1974) and Berenson (1979) have developed treatment approaches for working with families while a member continues to drink problematically. I know of no successful programs for working with the families of presently addicted drug abusers, though some strategic family therapists may theorize that this could be done by first changing family structures. Berenson (1979) offers a series of steps and chores for the nonalcoholic spouse that I have found clinically useful, particularly when the drinking member refuses to enter or return to treatment:

Step 1. Calm down the system and provide clarity (e.g., by clarifying problems and solutions).

Step 2. Create a support system for family members so that the emotional intensity isn't all within the session or in the relationship with the alcoholic. In Al-Anon the group and/or the sponsor may provide emotional support and calm down the situation.

Steps 1 and 2 may also be facilitated by having the spouses and significant others of the alcoholics participate together in a group in addition to or in place of their Al-Anon participation. This group provides support to the spouse, and frequently, after several months, the alcoholic may join the spouse in a couples group. In both this group and in individual therapy the spouse can be given three choices (Berenson, 1979): (a) keep doing exactly what you are doing, (b) detach or emotionally distance from the alcoholic, or (c) separate or physically distance. When the client does not change, it is labelled an overt choice (a). When clients do not choose (b) or (c), the therapist can point out that they are in effect choosing (a). Spouses are helped not to criticize drinking, to accept it, to live with the alcoholic, and to be responsible for their own reactivity regarding drinking.

For the spouse, all three choices may seem impossible to carry out. The problem is resolved by choosing one of three courses of action and following through, or by experiencing the helplessness and powerlessness of these situations being repeated and clarified. As part of the initial contract that I make with a couple, where one member is abusing substances, I request that the spouse commit him/herself to continue in the spouse group even if the I.P. drops out, generally because of resuming heavy substance abuse.

It should be emphasized that whenever we maintain therapy with a "wet" alcoholic or drug-abusing system, we have the responsibility of not maintaining the illusion that families are resolving problems, while in fact they are really reinforcing them.

Another method for dealing with treatment-resistant, drinking alcoholics is "The Intervention" developed at the Johnson Institute in Minnesota and the Freedom Institute in New York (Freedom Institute, 1980). In this technique, the family (excluding the I.P.) and significant network members, including employer, fellow employees, friends, and neighbors, are coached to confront the alcoholic with concern, but without hostility, about the destructiveness of his/her drinking and behavior. They agree in advance about what treatment is necessary and then insist on it.

4.2. The Need for Involvement of the Total Family

The concept of the family of the substance abuser as a multigenerational system necessitates that the entire family be involved in treatment. In my experience, family therapy limited to any dyad is most difficult. The mother-addicted son dyad is an almost impossible one to treat if no other family member is involved. Someone else must be involved if treatment is to succeed. This may be a drug-free sibling, grandparent, aunt, uncle, cousin, or even the mother's lover or ex-spouse. If there is absolutely no one else available, surrogate family members in multiple family therapy provide leverage to facilitate restructuring maneuvers (Kaufman, 1980).

Therapy limited to alcoholics and their spouses has been a traditional family approach to alcoholism. All too often, this has excluded the crucial parents or progeny of the alcoholic. Children are not just victims of alcoholic families. By their reciprocating involvement in these systems, they contribute to the problem and are a necessary part of the solution, regardless of age.

Treatment which focuses on drug addicts and their spouses has been less effective than with alcoholic couples. This had led Stanton and Todd (1982) to suggest that family treatment of male narcotic addicts begin with their parents, and that the addict-spouse couple should not be worked with until the addict's parents are brought to where they can "release" him to his spouse. Phoenix House has found so much difficulty with addicted couples that they insist on separate residential treatment sites for such couples. At the Awakening Family in California, we have met with some success with treating addicted couples in the same program. Success is enhanced by insisting on couple therapy throughout the duration of their stay in the program. Another essential aspect of treating such couples when they have children is focusing on their function as a parenting team. The addict has usually neglected his/her function as a parent, and therapy that involves progeny has the distinct advantage of developing parenting skills *in vivo*. Thus, I strongly advise the utilization of a four-generational approach involving grandparents, parents, spouse, and children. When most other family problems are resolved, a couples group can be very helpful, but it should never be the sole modality of family therapy, as it is in so many treatment programs.

4.3. The Need for an Integrated Approach That Utilizes Techniques Other than Family Therapy Alone

My family approach with drug addicts and abusers would be relatively useless without methadone, propoxyphene or clonidine detoxification, methadone and LAAM maintenance, Naltrexone, residential therapeutic communities, hospital detoxification, Narcotics Anonymous, specialized vocational rehabilitation, individual and group psychotherapy, particularly the techniques of ex-addicts, and so on. Similarly with alcoholics, I would feel powerless without AA, Al-Anon, Al-Ateen, Antabuse, specialized alcohol residential treatment, relaxation techniques, or medication to treat underlying affective and psychotic disorders. Thus, I use an integrated, multidisciplinary approach that meets each patient's needs and problems.

4.4. The Use of Multiple Family Therapy Groups

Multiple family therapy groups (MFT) are composed of from 3 to 15 multigenerational families and function quite well in residential settings where all families of program members are invited to participate. Families share experiences and offer help by acting as helpful extended families to other families and to identified patients. The group is seated in a large circle with several cotherapists seated at equal distances from each other to permit observation of the entire group. The group approach will vary from a directive educational one to restructuring techniques. Groups are frequently held for 3 hours with three or four families worked with intensively while the other families identify strongly with most conflicts, as well as serve as facilitators and supporters of positive change.

I feel that MFT is a very helpful and at times necessary adjunct to family therapy. It is particularly helpful, as stated earlier, with single-parent families. MFT offers relief from the loneliness and pain of having a substance abuser in the family. It forms a new supportive network, a family of families that offers an alternative to a network of drug abuse facilitators and substance abusers. It is relatively easy for parents to offer conflict-free parenting to the children of others and then apply these techniques to themselves (Kaufman & Kaufmann, 1977).

In MFT, members can make changes through group pressure and support that would be otherwise impossible. One of the most striking examples of this is "closing the back door" (i.e., helping parents to refrain from taking their children out of treatment and back home), which is a prime goal of any MFT in a residential setting. Without MFT, many substance abusers will persuade their families to take them back to protect them from "unfairness" and "harsh discipline" in much the same way as they have manipulated their families to protect them from jail, overdoses, and creditors. To work this problem through in individual family therapy requires interruption of the patterns of guilt and severe enmeshment, and considerable restructuring. This is done much more easily in MFT by group support and insight. A mother will say, "Yes, I thought I was protecting my child from jail and suicide by giving him money for drugs. Now I know that was what was killing him. To take him back before he was graduated would also be killing him, and I must not kill my child by taking him back now" (Kaufman, 1980).

4.5. Specific Family Therapy Techniques

The two most necessary ingredients for successful family therapy with substance abusers are (1) a knowledge of substance abuse and its effects on the family, and (2) an experienced family therapist who has established a workable system of family therapy. This system may be behavioral, communications, structural, strategic, psychodynamic, or a personal synthesis of any combination of these. When a therapist has accomplished both of these, any workable system of family therapy will be successful. For further specific application of these techniques, the reader is referred to Family Therapy of Drug and Alcohol Abuse (Kaufman & Kaufmann, 1979).

4.6. Demonstration of the Efficacy of Family Therapy with Substance Abuse

Those of us who use family therapy know that it works. However, this knowledge is not sufficient proof of effectiveness. One problem is that all family therapy is not the same; thus it is difficult to generalize success or failure from one program or individual to another or to the field in general. Another problem is that the entire field of family therapy is just entering a phase of scientific evaluation that, fortunately or unfortunately, will result in as rigorous an evaluation as individual therapy has had in its longer history. Certainly, similar questions can be raised about the efficacy of any system of psychotherapy presently in use.

Numerous papers have been published on the success or failure of

362 | EDWARD KAUFMAN

family therapy dealing with a single case (Dinaburg, Glick, & Feigen-baum, 1975). These can be dismissed with the same ease as the clini-cian's own subjective perceptions. That single case studies continue to be published emphasizes the relatively pristine state of family therapy evaluation. I will not attempt to thoroughly review the evaluation liter-ature on family therapy of substance abusers. Stanton (1979) and Janzen (1977) have already provided us with excellent reviews of drug abuse and alcoholism, respectively. However, I will discuss several examples of family therapy evaluation in this field.

In drug abuse, Silver, Panepinto, Arnon, and Swaine (1975) de-scribe a methadone program for pregnant addicts and their addicted spouses that included 40% of the women becoming drug free in treat-ment and the male employment rate increasing from 10% to 55%. Both rates are much higher than those achieved by traditional methadone programs without family treatment. The problem with this study, as with most evaluations of family approaches to drug abuse, is the lack of follow-up data or control groups.

Ziegler-Driscoll (1977) has reported a study conducted at Eagleville that found, on 4–6 month follow-up, no difference between treatment groups with family therapy and those without. However, the therapists were new to family therapy and the supervisors new to substance abuse. As the therapists became more experienced, their results im-proved. Stanton and Todd's (1982) evaluation of family therapy with heroin addicts on methadone is perhaps the most outstanding study of hard-core drug addicts. These researchers compared paid family thera-py, unpaid family therapy, paid family movie "treatment," and non-family treatment. The results of a 1-year, posttreatment follow-up were that the two family-therapy treatments produced much better outcomes in abstinence from drugs than did nonfamily treatments. The nonfamily treatment and movie groups did not differ from each other.

Hendricks (1971) found at 1-year follow-up that narcotic addicts who had received 5½ months of MFT were twice as likely to remain in continuous therapy than were addicts not responding to MFT. Ka-ufmann's work has shown that adolescent addicts with MFT have half the recidivism rate of clients without it (Kaufman & Kaufmann, 1977). Stanton (1979) noted that of 68 studies of the efficacy of the family therapy of drug abuse, only 14 quantify their outcome, and only 6 of these provide comparative data with other forms of treatment or control groups.

Although there has been more detailed evaluation of the family therapy of alcoholism than of drug abuse, Janzen (1977) stated that "it is not possible to show that family treatment is as good or better than other forms of treatment for alcoholism." However, he also stated that such treatment has advantages to both the family and the alcoholic that other treatments do not offer. Despite their methodological shortcomings, all the studies he cited reported positive results. Meeks and Kelly (1970) reported the success of family therapy with five couples, but no comparison-group couples were included. These researchers described abstinence in two alcoholics and "improved drinking patterns" in the other three, but with no objective measures of family functioning. Cadogan (1973) compared marital group therapy in 20 couples with 20 other couples on a waiting list. After 6 months of therapy, nine couples in the treatment group, but only two couples in the control group, were abstinent. However, again there was no follow-up or use of objective, externally validated measures.

Steinglass (1979), utilizing a comprehensive battery of evaluative instruments with alcoholic families before treatment and at 6-month follow-up, found that five of nine alcoholics were drinking less at follow-up. Overall positive changes in psychiatric symptomatology were minimal. However, when the results of the two therapists were analyzed, the directive, forceful therapist was found to much more successful than the passive one. Steinglass also proposes that brief, intense family therapy programs may shift rigid systems, but may not provide sufficient time for beneficial shifts to be permanently incorporated.

5. Summary

We see that a great deal of work has been done toward describing typical family systems and symptomatology associated with both alcohol and drug abuse. We may not yet specifically know the extent to which the family pathology precedes or is a consequence of substance abuse. We do not know with much certainty which family dysfunctions are specific to substance abuse, nor do we know which types of family dysfunction are associated with the abuse of different drugs or alcohol. Although more treatment evaluation is being conducted than previously, we do not as yet know that family treatment is superior to other forms of treatment for substance abuse. However, we have learned a great deal about the many types of disordered interactions and systems that occur in most substance-abusing families. We have also seen the development of many new family therapy techniques and strategies that can be applied to substance abusers. Multiple family therapy, for example, is one form of treatment that appears to be useful in treating substance abuse families.

Although we cannot claim that family therapy alone is superior to

364 EDWARD KAUFMAN

other treatment methods, we certainly know that it is a valuable, if not essential, component of treatment, and that family system concepts have great value in conceptualizing the particular problems of the substance-abusing family. When the type of family and/or system involvement is tailored to the individual family and its own specific needs, then at least some knowledge of the utilization of the family becomes a necessity for understanding and treating the substance abuser.

6. References

Alexander, B. K., & Dibb, G. S. Opiate addicts and their parents. *Family Process*, 1975, *14*, 499–514.

Attardo, N. Psychodynamic factors in the mother–child relationship in adolescent drug addiction: A comparison of mothers of schizophrenics and mothers of normal adolescent sons. *Psychotherapeutic Psychosomatic*, 1965, *13*, 249–255.

Becker, J. V., & Miller, P. M. Verbal and nonverbal marital interaction pattern of alcoholic and nonalcoholic couples during drinking and nondrinking sessions. *Quarterly Journal of Studies on Alcohol*, 1976, *37*, 1616–1624.

Berenson, D. The therapist's relationship with couples with an alcoholic member. In E. Kaufman & P. Kaufmann (Eds.), *Family therapy of drug and alcohol abuse*. New York: Gardner Press, 1979.

Billings, A. G., Kessler, M., Gomberg, C. A., & Weiner, S. Marital conflict resolution of alcoholic and nonalcoholic couples during drinking and nondrinking sessions. *Quarterly Journal of Studies on Alcohol*, 1979, *40*, 183–195.

Binion, V. J. A descriptive comparison of the family of origin of women heroin users and non-users. In National Institute on Drug Abuse, *Addicted Women: Family dynamics, self perceptions and support systems* (DHEW Pub. No. [ADM]80-762). Rockville, Md.: the Institute, 1980.

Blum, R. H. *Horatio Alger's children*. San Francisco: Josey Bass, 1972.

Bowen, M. Alcoholism as viewed through family systems theory and family psychotherapy. *Annals of the New York Academy of Sciences*, 1974, *233*, 115–122.

Bowen, M. Theory in the practice of psychotherapy. In P. J. Guerin, Jr. (Ed.), *Family therapy: Theory and practice*. New York: Gardner Press, 1976.

Brooks, J. E., Lukoff, I. F., & Whiteman, M. Family socialization and adolescent personality and their association with adolescent use of marijuana. *Journal of Genetic Psychology*, 1978, *133*, 261–271.

Busch, H., Kormendy, E., & Feuerlein, W. Partners of female alcoholics. *British Journal of Addiction*, 1973, *68*, 179–184.

Cadogan, D. A. Marital group therapy in the treatment of alcoholism. *Quarterly Journal of Studies on Alcohol*, 1973, *34*, 1184–1194.

Chafetz, M. E., Blane, H. T., & Hill, M. J. Children of alcoholics: Observations in child guidance clinic. *Quarterly Journal of Studies on Alcohol*, 1971, *32*, 687–698.

Cork, R. M. *The forgotten children*. Toronto: Addiction Research Foundation, 1969.

Davis, D. I., Berenson, D., Steinglass, P., & Davis, S. The adaptive consequences of drinking. *Psychiatry*, 1974, *37*, 209–215.

Dinaburg, D., Glick, I. D., & Feigenbaum, E. Marital therapy of women alcoholics. *Quarterly Journal of Studies on Alcohol*, 1975, *36*, 1245–1257.

Drewery, J., & Rae, J. B. A group comparison of alcoholic and nonalcoholic marriages using the interpersonal and perception techniques. *British Journal of Psychiatry*, 1969, *115*, 287–300.

Edwards, P., Harvey, C., & Whitehead, P. C. Wives of alcoholics: A critical review and analysis. *Quarterly Journal of Studies on Alcohol*, 1973, *34*, 112–132.

El-Guebaly, N., & Orford, D. R. The offspring of alcoholics: A critical review. *American Journal Psychiatry*, 1977, *134*, 357–365.

Fort, J. P. Heroin addiction among young men. *Psychiatry*, 1954, *17*, 251–259.

Freedom Institute. *Intervention: A Process through which family and friends are able to confront the chronically dependent in a controlled and caring environment.* New York: A Freedom Institute Publication, 1980.

Gibbs, J. T. Psychosocial factors related to substance abuse among delinquent families. *American Journal of Orthopsychiatry*, 1982, *52*, 261–271.

Hanson, P. G., Sands, P. M., & Sheldon, R. B. Patterns of communication in alcoholic marital couples. *Quarterly Journal of Psychiatry*, 1968, *42*, 538–547.

Hendin, H., Pollinger, A., Ulman, R., & Carr, A. C. Adolescent marijuana abusers and their families. *NIDA Research Monograph 40*, September 1981, pp. 17–25.

Hendricks, W. J. Use of multifamily counseling groups in treatment of male narcotic addicts. *International Journal of Group Psychotherapy*, 1971, *22*, 34–90.

Hersen, M., Miller, P. M., & Eisler, R. M. Interactions between alcoholics and their wives: A descriptive analysis of verbal and nonverbal behavior. *Quarterly Journal of Studies on Alcohol*, 1973, *34*, 516–520.

Janzen, C. Families in the treatment of alcoholism. *Quarterly Journal of Studies on Alcohol*, 1977, *38*, 114–130.

Kandel, D. B., Kessler, R., & Margulies, R. Antecedents of adolescent initiation into stages of drug use: A developmental analysis. *Journal of Youth and Adolescence*, 1978, *7*(1), 13–40.

Kaufman, E. The psychodynamics of opiate dependence: A new look. *American Journal of Drug and Alcohol Abuse*, 1974, *1*, 349–370.

Kaufman, E. Myth and reality in the family patterns and treatment of substance abusers. *American Journal of Alcohol Abuse*, 1980, *7*(3,4), 257–279.

Kaufman, E. Family structures of narcotic addicts. *International Journal of Addictions*, 1982, *16*(2), 106–108.

Kaufman, E., & Kaufmann, P. Multiple family therapy: A new direction in the treatment of drug abusers. *American Journal of Drug and Alcohol Abuse*, 1977, *4*, 467–478.

Kaufman, E., & Kaufmann, P. Family therapy of substance abusers. In *Family therapy of drug and alcohol abuse*. New York: Gardner Press, 1979.

Kaufman, E., & Pattison, E. M. Differential methods of family therapy in the treatment of alcoholism. *Quarterly Journal of Studies on Alcohol*, 1981, *42*, 951–971.

Kirschenbaum, M., Leonoff, G., & Maliano, A. Characteristic patterns in drug abuse families. *Family Therapy*, 1974, *1*, 43–62.

Kogan, J. L., & Jackson, J. K. Alcoholism: The fable of the noxious wife. *Mental Hygiene*, 1965, *49*, 428–437.

Meeks, D. E., & Kelly, C. Family therapy with the families of recovered alcoholics. *Quarterly Journal of Studies on Alcohol*, 1970, *31*, 399–413.

Noone, R. J., & Reddig, R. L. Case studies on the family treatment of drug abuse. *Family Process*, 1976, *15*, 325–332.

Olson, D. H. *300 chemically dependent families.* Mimeograph, 1983.

Olson, D. H., Bell, R., & Porter, J. *FACES: Family adaptability and cohesion evaluation scales, family social sciences.* St. Paul: University of Minnesota, 1978.

Olson, D. H., Sprenkle, D. H., & Russell, C. S. Circumplex model of marital and family systems: 1. Cohesion and adaptability dimensions, family types, and clinical applications. *Family Process,* 1979, *18,* 3–28.

Paolino, T. J., Jr., McCrady, B. S., Diamond, S., & Longabaugh, R. Psychological disturbances in spouses of alcoholics: An empirical assessment. *Quarterly Journal of Studies on Alcohol,* 1976, *37,* 1600–1608.

Pearson, M. M., & Little, R. B. Treatment of drug addiction: Private practice experience with 84 addicts. *American Journal of Psychiatry,* 1975, *122,* 164–169.

Reilly, D. M. Family factors in the etiology and treatment of youthful drug abuse. *Family Therapy,* 1976, *2,* 149–171.

Rimmer, J. Psychiatric illness in husbands of alcoholics. *Quarterly Journal of Studies on Alcohol,* 1974, *35,* 281–283.

Schwartzman, J. The addict, abstinence and the family. *American Journal of Psychiatry,* 1975, *132,* 154–157.

Silver, F. C., Panepinto, W. C., Arnon, D., & Swaine, W. T. A family approach in treating the pregnant addict. In E. Senay (Ed.), *Developments in the field of drug abuse.* Cambridge, Mass.: Shenkman, 1975.

Stanton, M. D. *Some outcome results and aspects of structural family therapy with drug addicts.* Paper presented at the *National Drug Abuse Conference,* May 5–6, 1971, in San Francisco, California.

Stanton, M. D. Family treatment approaches to drug abuse problems: A review. *Family Process,* 1979, *18,* 251–280.

Stanton, M. D., & Todd, T. C. *The family therapy of drug abuse and addiction.* New York: Guilford Press, 1982.

Steinglass, P. An experimental treatment program for alcoholic couples. *Quarterly Journal of Studies on Alcohol,* 1979, *40,* 159–182.

Vaillant, G. A 12 year followup of New York narcotic addicts. *Archives of General Psychiatry,* 1966, *15,* 599–609.

Wellisch, D. K., Gay, G. R., & McEntea, R. The easy rider syndrome: A pattern of hetero- and homosexual relationships in a heroin addict population. *Family Process,* 1970, *9,* 425–430.

Winokur, G., Reich, T., Rimmer, J., & Pitts, F. N. Alcoholism III: Diagnosis and familial psychiatric illness in 259 alcoholic probands. *Archives of General Psychiatry,* 1970, *23,* 104–111.

Ziegler-Driscoll, G. Family research study at eagleville hospital and rehabilitation center. *Family Process,* 1977, *61,* 175–189.

13

Conjoint Treatment of Dual Disorders

Patricia Ann Harrison, Jodi A. Martin, Vicente B. Tuason, and Norman G. Hoffmann

1. Introduction

Traditional, distinct mod͟ ͟atment for psychiatric illness and substance abuse are no l͟ ͟e to respond to the clinical reality of coexisting disorde͟ ͟Sax, Deal, & Ostreicher, 1980). The body of ev͟ ͟ of dual disorders is increasing rapidly (Alt͟ ͟81; McLellan, Druley, & Carson, 197͟ ͟Othmer, Bingham, & Rice, 1982)͟ ͟ance abuse rehabilitation concep͟ ͟ion and that more comprehensiv͟ ͟pond successfully to patients wi͟ ͟severe emotional dysfunction (͟ ͟ne, 1980; McLellan, MacGahan,͟ ͟1980). Others emphasize that th͟ ͟wide array of problem areas an͟ ͟gram will

Patricia Ann Harrison, Jodi A. ͟ ͟ ͟epartment of Psychiatry, Saint Paul-Ramsey Medi͟ ͟ ͟ Vicente B. Tuason • Department of Psychiatry, ͟ ͟ter, Saint Paul, Minnesota 55101; Psychiatric Services, South ͟ ͟ervices, Albuquerque, New Mexico 87102.

be determined not only by the severity of the patients' disorders but by the training and experience of the staff (Fine, 1980; Zosa, 1978).

Like colleagues throughout the country, mental health and substance abuse professionals at St. Paul-Ramsey Medical Center became increasingly aware of dual problems among patients they serve. An active commitment to programming for individuals with dual disorders was initiated in 1978, with funding from state grants that supported a dual aftercare program for 2 years. The initial emphasis was on screening and assessment of chemical dependency patients for other, undiagnosed psychiatric conditions.

Although the outpatient clinic proved useful in maintaining moderately ill, dual-disorder patients, a need for inpatient programming became apparent. For a 6-month period, an intensive dual treatment program was provided to a small number of psychiatric inpatients. The program was phased out after encountering several obstacles involving length of stay, location, and cost. However, the greater psychopathology and dysfunction of the patients involved in this program documented the need for a continuum of care, including primary inpatient treatment.

The dual program was placed under the immediate supervision of the alcohol and drug abuse programs within the department of psychiatry. Its services, however, were distinct from traditional chemical dependency programming, and were not limited to a centrally located population. Dual treatment was available to any chemical dependency inpatients and outpatients who had other psychiatric disorders; to psychiatric inpatients or outpatients with substance abuse problems; and to any community referral who met dual diagnostic criteria.

Treatment for dual disorder clients has undergone several evolutionary phases, but the commitment to serve the dual population has not faltered. Program structure will continue to change in response to patient needs and staff experience. What has surfaced over time is the complexity of treatment issues and the demands for flexibility.

Often the traditional responses of one subspecialty are not sufficient to meet the needs of patients who cross professional boundaries that are perhaps more distinct and meaningful to the practitioners than to the patients in turmoil. Thus, needs for individual assessment and case management outweigh the limitations of rigid programmatic considerations. Even the term "dual disorder" has proved inadequate to describe patient status; the clientele in reality is multiproblem. Dual program staff is challenged not only by the multiproblem population, but also by the necessity for professionals from all backgrounds to question preconceptions regarding effective treatment approaches.

2. Population Profile

St. Paul-Ramsey Medical Center is a major metropolitan medical center as well as a teaching and research institution. Its psychiatric and chemical-dependency treatment units serve many government assistance clients, most of whom are ineligible to obtain treatment elsewhere than in state hospitals. Since the move toward deinstitutionalization of psychiatric patients, more former state hospital patients are admitted to St. Paul-Ramsey Medical Center (Tuason, Fair-Riedesel, & Hoffmann, 1982). Generally, substance abuse and psychiatric patients exhibit greater chronicity and psychosocial impairment than patients served in other area hospitals and free-standing treatment centers. A summary of characteristics for 207 patients admitted to the program from late 1981 through early 1983 reveals the severity of patient disorders. (This profile is not necessarily representative of area dual patients in general, nor of St. Paul-Ramsey Medical Center patients over past years, as the population has evolved over time.)

All patients presented coexisting substance use and psychiatric disorders diagnosed in accordance with DSM-III criteria. Tables 1 and 2 detail the diagnostic breakdown of dual disorders. Axis I psychiatric disorders were diagnosed for 84% of the patients, and some had multiple diagnoses; 50% of the patients had personality disorders, and multiple diagnoses were given in many cases.

Affective disorders were diagnosed for more than one-third of the population. Almost one-fourth met criteria for schizophrenia or other psychotic disorders. Adjustment disorders, anxiety disorders, impulse control disorders, and personality disorders completed the diagnostic profile of psychiatric illness.

Almost half of the patients who met diagnostic criteria for alcohol abuse or dependency did not report problems with other drugs. Patients abusing only alcohol numbered 49%, while 42% abused alcohol and one or more other drugs. Only 9% reported drug abuse or dependency with no alcohol problems (see Table 2).

Case Illustration*

Matt was a 32-year-old single male with diagnoses of schizoaffective disorder and chronic alcoholism. He was troubled by sexuality issues, and when he experienced intense guilt, he felt compelled to drink or to slash his arms. He had

---

*To protect patient privacy, all case illustrations are composites; the illustrations do, however, depict actual patient characteristics and events.

more than 15 admissions for chemical dependency treatment and multiple psychiatric hospitalizations before entering the dual program.

Two-thirds of the patients were male. More than half were under age 30, and one-third were in their 30s. More than two-thirds were unemployed at time of intake. Of the total, 47% had had three or more admissions to inpatient psychiatric units, and 62% had at least one psychiatric hospitalization; 23% had entered chemical-dependency treatment at least three times, and 51% at least once. A suicide attempt was reported by 50%, and one-third of these patients had a history of three or more attempts. During their dual program involvement, 46% of the patients were taking psychotropic medication.

Preliminary data suggest that the majority of female patients reported being victims of physical abuse and sexual abuse; a smaller but significant proportion of males had also been abuse victims. More than one-third of the males and one-quarter of the females acknowledged having been physically abusive to others. Only a small proportion of either sex acknowledged or were referred for perpetrating sexual abuse, but this is a serious offense, and likely to be underreported.

Table 1
DSM-III Psychiatric Diagnoses of Dual-Disorder Patients
(N = 207)

Axis I	
Major affective disorder	29%
Schizophrenia and other psychotic disorders	24%
Adjustment disorder	19%
Dysthymia	9%
Anxiety disorder	7%
Impulse control disorder	6%
Other	5%
Axis II—Personality Disorders	
Dependent	30%
Antisocial	17%
Avoidant	14%
Borderline	11%
Passive/aggressive	9%
Compulsive	7%
Narcissistic	7%
Histrionic	6%
Schizoid	3%
Schizotypal	1%
Atypical, mixed or other	10%

Table 2
DSM-III Substance Use Diagnoses
of Dual-Disorder Patients (N = 207)

Alcohol only	49%
Alcohol/polydrug	22%
Alcohol/cannabis	17%
Alcohol/single other drug	. 3%
Cannabis only	3%
Single substance other than alcohol or cannabis	3%
Polydrug (no alcohol)	3%

Case Illustration

Pamela was a 19-year-old single female with a schizoid personality disorder and severe alcoholism dating back to age 12. She periodically abused tranquilizers and barbiturates in an attempt to combat the social anxiety she experienced. She had been coerced into an incestuous relationship with an older brother beginning at age 9. She had been raped on two occasions when she had been drinking. All her siblings had chemical-abuse problems. Both parents had been hospitalized on several occasions for unspecified mental illness. Pamela had dropped out of school in the sixth grade and, more often than not, lived on the street. She was admitted frequently to detoxification centers and the hospital with her blood-alcohol levels in excess of 0.40, and had often overdosed on medication. She had been referred for treatment on many occasions, and was even committed, but eloped from all programs.

Most patients identified moderate, severe, or extreme psychosocial stressors over the preceding year, according to DSM-III, Axis IV. More than one-third reported marital separation or divorce, physical or sexual abuse, including rape and incest, attempts at self-injury, or major physical illness or death of family members or significant others. Some of the environmental stress experienced during the previous year may have been in part a consequence of substance abuse or psychiatric illness or both, and in some cases stress may have precipitated the onset or exacerbation of substance abuse or psychiatric illness. Dual patient psychosocial functioning was well below average, with one-half meeting DSM-III, Axis V criteria for grossly impaired functioning.

Case Illustration

Bill, a 21-year-old, had recently lost his job when found smoking marijuana. Weeks earlier, his stepmother had left his father. His own mother had abandoned the family when he was 10. He was brought to the emergency room by the police after being arrested for carrying a gun into a supermarket. He told the police that his brothers blamed him for the stepmother's departure and that they planned to kill him. He was hospitalized on the psychiatric unit. Assessment revealed a paranoid disorder and a long history of hallucinogen and cannabis abuse.

For many of the dual patients, "recovery" must be defined with cautious realism in view of their limitations. Some cannot be expected to achieve optimum recovery goals: permanent abstinence from alcohol and other drugs, a future free of psychiatric hospitalization, a return to the work force, and stable interpersonal relationships. Rather, recovery for some patients must be measured in terms of less frequent and less destructive encounters with chemicals, increased intervals of abstinence and satisfying sobriety, compliance with medication maintenance and fewer hospitalizations, less injurious behavior, and increased social responsibility and self-esteem. Unfortunately, there is little documentation to date of the long-term efficacy of the conjoint treatment of dual disorders undertaken at St. Paul-Ramsey Medical Center, although follow-up efforts are underway.

Data are available on length of attendance, and suggest some success in engaging the population in treatment. Thirty-one percent of the admitted patients (1982 sample) attended only as inpatients. Much of the attrition at discharge occurred for pragmatic reasons: Patients who lived at a great distance or who were referred to residential settings having their own programming were not expected to return. Of those who participated as outpatients, 38% attended 1 to 5 group-therapy sessions, 38% attended 6 to 20 sessions, and 24% attended 20 to 100 sessions.

3. Program Philosophy

Without general documentation of dual treatment efficacy, staff remain committed to their efforts on faith and anecdotal case reports rather than empirical evidence. However, the severe impairment and dysfunction noted in this relatively young population solidly supports the need for early intervention, thorough assessment, and adequate response to all patient needs.

Case Illustration

Kathleen had a borderline personality disorder and alcohol and mixed substance dependence. She had been hospitalized in psychiatric units or substance-abuse treatment centers for much of the past 8 years. She was 22, a high school dropout, and had worked only sporadically. She reported that she had been raped by her father when she was 13, and had herself sexually abused a younger brother.

Dual treatment staff are aware of the severity of the illnesses that bring patients to the program. They know, too, of the hopelessness that can surround psychiatric illness and chemical dependency. One of their major functions, then, is to instill optimism. Staff is mindful that recovery from dual disorders requires a major recuperative phase and reinforces patient progress (even painfully slow and uneven progress) to reduce the guilt often associated with lengthy program involvement. Patients are encouraged to realistically examine their expectations of themselves and to set short-term goals so they can benefit from recognition of their own progress.

Case Illustration

Lucy, 36, was diagnosed as having bipolar affective disorder and alcohol dependence. She had no prior treatment for either. She identified her major problems as being too perfectionistic with regard to expectations of her children's behavior and her housecleaning. She believed her behavior was extremely compulsive. Lucy needed guidance to separate her problems from her children's problems (which were considerable), and to take responsibility only for that which she could control. The family met for several sessions and the adolescent son was referred for individual therapy. Lucy worked hard toward goals of assertiveness and raising her self-esteem.

Interpersonal respect is the hallmark of the program. Patients have experienced disrespect, not only from the unknowing public, but also from careless professionals. Staff is committed to modeling interpersonal respect, and does not practice or permit devaluation of any patient.

Treatment is geared to respect the individual patient's pace and boundaries. Staff do not attack the patient's defenses. They may model other responses and allow the patient to lower defenses when it feels safe, or they may help the patient identify defenses and learn alternative, more satisfying ways to cope. Self-disclosure about personal issues is not demanded during group sessions. Individual counseling is available as needed for patients who choose not to or who are unable to divulge certain information to a group. This information is often of a sexual nature and may be shrouded by shame.

A patient contract is employed to specify expectations of participants. The contract details expectations with respect to attendance, compliance with medication regimens and physician orders, appropriate ingroup behavior, abstinence from alcohol and drugs, the necessity of reporting use of chemicals by self or other group members, the mandate against verbal and physical abuse, and the explicit requirements of confidentiality.

Confidentiality is explained as the obligation of patients to keep private whatever they hear in group, so that all members may feel safe to speak freely. It is explained that this obligation also binds staff members, with certain exceptions. Staff members do not agree to keep secrets from each other or from clinical supervisors, because of the team approach, but explain that team discussions of patients take place privately. The legal limitations of confidentiality are also discussed, including notifying others of a patient's statement of intent to harm them, and the mandate to report abuse of children or vulnerable adults.

Staff also communicates directly and candidly with patients about the interpretation of diagnoses. The limitations of assessment procedures and treatment are explained to the patients, who are encouraged not to equate an evolving diagnosis with self. Patients are encouraged to explore their own experience of their disorders rather than to remain focused on the diagnostic language.

The dual program does not adhere to the AA model for substance abuse treatment. The AA model may be particularly effective for late-stage alcoholism and drug abuse, particularly when the chemical use is not reactive and has truly developed a life of its own. However, it may be severely limited for patients in early stages of substance abuse who have not experienced clear "loss of control." It may also be limited in its flexibility to accommodate the complex clinical picture presented by many dual patients. In some cases, it is necessary to explore substance use as an amelioration of psychiatric distress and a behavioral maladaptation rather than to interpret the use in the traditional disease model. Presentation of a multifactorial etiology of chemical dependency may be more appropriate for dual patients than the disease concept. Alcoholics Anonymous philosophy is offered in traditional treatment as a recovery model, but the dual program allows for the development of alternative models.

Case Illustration

Caroline had a difficult time believing she had substance abuse problems. Although her use patterns were neither frequent nor heavy, her drinking tended to coincide with the onset of mania (see Chapter 4 in this volume), for which she was hospitalized on three occasions. Staff worked with Caroline not to convince her she was an "alcoholic," but to help her accept her affective disorder and the risks of using alcohol and drugs.

4. Program Description

4.1. Overview

The dual program is designed to offer patients the opportunity to explore realistically the impact of dual disorders upon individual lives and to develop alternatives to chemical use or symptomatic behavior. The counseling groups are of varying intensity and frequency; patients are assigned to groups according to assessment of individual needs and coping strengths. The lists below detail the populations most commonly found in each group, although there are exceptions.

Discovery groups meet three times a week for 90 minutes. Most of the patients in these groups have major psychiatric disorders, which typically require maintenance medication for adequate functioning. Typical diagnoses include:

- Organic mental disorders (drug-induced psychosis or substance delusional disorders)
- Schizophrenic disorders
- Paranoid disorders
- Other psychotic disorders (atypical psychosis)
- Major affective disorders (most often mania or major depression with psychotic features)
- Borderline personality disorder, schizotypal personality disorder, or other personality disorders where brief psychotic states may occur

The Horizons group meets twice a week for 90 minutes. The majority of the patients have nonpsychotic psychiatric conditions. Typical diagnoses include:

- Affective disorders largely without psychotic features
- Anxiety disorders
- Impulse control disorders
- Personality disorders not involving psychotic states
- Psychosexual disorders
- Severe adjustment disorders, which limit responsiveness to traditional chemical dependency programs

Horizons group also includes former Discovery group patients who have come to a point in their recovery where they can benefit from a somewhat less intensive treatment experience.

Transitions group meets once a week in the evening for 2 hours. Most patients have had prior involvement in either Horizons or Discovery groups, and for many of them Transitions functions as an aftercare group. For others, participation in Horizons or Discovery group is interrupted by employment or school before they have completed their therapeutic goals, and these patients may also be transferred to the Transitions group. On rare occasions, a patient who is employed, attending school, or unable for some other reason to participate in Horizons or Discovery group, may begin dual program involvement in Transitions.

The dual treatment program also offers educational workshops. The workshops cover topics frequently of concern to group members, so that repeated discussion of the same topics need not disrupt the counseling group process. Occasionally, when the counseling groups are full, a patient will attend workshops while on a waiting list.

One sequence of workshops is designed to cover topics of particular interest to Discovery group members, although other dual patients also participate. The knowledge base shared among patients seems to decrease tension and strengthen the program milieu. The topics covered include:

- Cognitive therapy (an examination of attitudes, beliefs, and feelings, and differentiating feelings from thoughts)
- Symptom patterns (in behavioral terms) of major mental illness
- Common reactions to stress and grief
- Coping skills in dealing with life events
- Functional areas of the brain affected by different types of mental illness
- The role of medications in the treatment of mental illness

Another series of workshops is designed for both Discovery and Horizons members. The topics include:

- Group process, rules, interpersonal skills
- Communication skills, especially assertiveness
- Sexuality as an expression of self
- Family roles and communication patterns

Depending upon the stage of involvement in the dual program, a patient may be attending from one to five sessions each week. In addition, patients in primary chemical-dependency treatment attend a traditional program format during the remaining day or evening hours, and psychiatric inpatients attend scheduled groups on their residential units. Inpatients routinely continue with the same dual group as outpatients following their discharge from the medical center. Usually 40 to 50 patients are involved in the program at any one time. Up to one-fourth of these may be inpatients in primary chemical dependency treatment. Several more may be hospitalized on a psychiatric ward. The majority, however, are outpatients, some of whom began their involvement as inpatients and some of whom were direct outpatient admissions.

4.2. Staffing

The multidisciplinary team and departmental rapport are the cardinal features of the dual treatment program. The staffing team must be capable of functioning in close proximity and of interfacing and coordinating their respective activities. It is necessary for all staff to appreciate their own and other professionals' competencies and limitations, and to confer with one another on case management. This close working relationship requires appreciable initial training and continuing staff development.

Staffing patterns reflect the multiple needs of the patient population, as well as the innovative nature of programming designed for this population. The multidisciplinary staff includes a psychiatrist, a psychologist, certified chemical dependency counselors, and nurses (on the inpatient unit). Plans are underway to add a social worker to the team.

Each discipline contributes its special expertise (psychological assessment, psychiatric diagnosis and medication monitoring, and nursing care plans, among others), but beyond that, staff roles are intentionally flexible, and staff are encouraged to develop their own unique styles and contributions. Staff providing direct counseling services are clinically supervised by a psychologist who has additional training in substance abuse. Staff recognize that patients often arouse strong countertransference reactions, and frequent consultation is encouraged as an aspect of staff self-care. Major case management decisions, especially those concerning disciplinary actions, are made in team consultation meetings.

In reality, the dual program draws on hospitalwide support, especially from the other psychiatric services. Ancillary services provided include biofeedback, crisis intervention during nonprogram hours, neuropsychological assessment, and interpreters for hearing-impaired persons.

4.3. Referral

Referrals to dual treatment come from a variety of sources: inpatient and outpatient chemical dependency and psychiatric programs within the hospital and in the community; the emergency room; medical units; community welfare workers; detoxification centers; community corrections officers; halfway houses; board-and-care homes, and state hospitals; and family and self-referrals. The wide variety and array of referral sources necessitates a flexible and individually responsive intake and assessment procedure. Procedures have been devised to accommodate differing needs of diverse referral sources while establishing that patients meet inclusion criteria.

Frequently, it is to the patient's advantage to have involvement with multiple care providers. Staff are open to coordinating care both with referral sources and with other services beneficial to the patient.

Case Illustration

Linda was the young mother of a 10-year-old daughter. Linda was hearing-impaired and involved with the Mental Health Hearing Impaired program at the medical center. She was on government assistance and had a community social worker and financial worker. During the course of Linda's dual treatment for paranoid schizophrenia and alcohol dependence, her daughter was referred to the community mental health center for assessment. Periodically, the dual counselor coordinated meetings of all the professionals involved with the care of Linda and her daughter.

4.4. Assessment

Assessment of clients referred to the dual program draws upon psychiatric, psychological, and substance abuse expertise. Diagnostic evaluation is an essential element of the assessment and is completed prior to program admittance to ensure that referrals meet diagnostic inclusion criteria. However, the goal of the assessment process is to generate a comprehensive description or understanding of the client's experience of his or her disorders, self-concept, coping strengths, perceived obstacles to recovery, and treatment needs. This comprehensive assessment is an ongoing process and guides both therapeutic intervention and case management over the course of treatment.

When staff are confident that a client meets inclusion criteria and that they have an adequate understanding at least of the client's immediate needs, several treatment options are available. The most frequently employed options are: hospitalization on chemical-dependency or psychiatric units, with dual-program involvement as soon as the patient is symptomatically stable; or outpatient dual-program participation, possibly supplemented by chemical dependency or psychiatric services.

Although the majority of patients are appropriate for group participation, there are several factors that may indicate postponement of group involvement. Moderate levels of some deviant behavior might be tolerated in a group setting, but extreme levels of the following dictate a delay in initiating group participation: lack of impulse control; hypomanic behavior; uncontrollable drowsiness associated with psychotropic medication; extreme agitation; loose associations and ideas of reference; extreme paranoia; delusions. Not only the well-being of the individual patient, but the well-being of the group as a whole is evaluated with respect to any existing symptoms. Current knowledge of the "health" of groups available to the patient, and awareness of how much deviance each group could tolerate along any dimension, are essential elements of "good fit."

Case Illustration

Joe, a young male, self-referred, requested participation in the dual program after hearing of it from a fellow resident in the halfway house where he was living after completing inpatient chemical dependency treatment for alcohol and cannabis dependence. He had 6 weeks of abstinence. His fear of losing control—he had a serious history of explosive physical abuse of female family members—was vivid and demanded response. He also appeared to be significantly depressed. His mother was still hospitalized following a recent overdose, and Joe seemed to pose considerable risk of suicide. He was offered psychiatric hospitalization for his own safety as well as that of others, and he accepted. Joe met criteria for an episode of major depression and intermittent explosive behavior. Staff consulted and agreed that he would not be appropriate for group involvement until some issues regarding strengthened behavior controls could be addressed in individual therapy.

4.5. Orientation

A pre-entry interview is an essential part of program procedure. The purpose of the orientation session is twofold: to allay patient anxieties about the dual program in general, and group experiences in particular, and to engage the patient in a commitment to treatment.

The essential dynamic of the pre-entry session is the foundation for the establishment of trust and a therapeutic relationship. Patients need several things from the counselor. They want to know that they are respected and accepted—with their disorders, their boundaries, their limitations, and their strengths. They want to know that the counselor believes that they can recover, and that the program can assist them. Patients need underpinnings for hope.

4.6. Group Therapy

Patients entering dual groups are aware that all other members have dual diagnoses and problems not unlike their own. Dual-group patients require a milieu that guarantees freedom of speech and emotional (as well as physical) safety. Talking about mental illness or psychotic experiences is still one of the great taboos of our society, especially among the patient's own family. Patients often complain that family members "walk on eggshells" in their presence and tiptoe around the very topics that are uppermost in the patient's consciousness. The end result is that the patients have often internalized the message that it is not okay to talk about the most frightening things that they are experiencing right now. One function of dual disorder groups is to allow—even encourage—patients to ventilate their fears and other feelings about mental illness and drug-induced experiences. Upon sharing similar experiences in group, patients are able to see experiences as symptoms of an illness, not an integral part of their self-definition. Sharing the experiences with others builds a bond of mutual friendship, understanding, and acceptance, a step from self-absorption to other-involvement. Patients often become acutely attuned to powerful emotions accompanying and resulting from such experiences, and cathartic sessions often have a powerful unifying effect on a group.

Case Illustration

During one group session, patients spontaneously began to disclose some of their beliefs when they had been delusional. Rusty told other members that when he joined the group he could not understand why people were not fighting over the chair next to him to "sit at the right hand of the Father." Greg chimed in that what got him referred to the program was signing in at the community mental health center as J. R. Ewing. Richard told the group he had put his head in his lap and cried as he watched on television the hostages home from Iran, at a White House dinner. Only then did he realize that he was not one of the 52 honored guests. After members shared the experiences that now seemed amusing, they were also able to share those that were still painful.

A here-and-now focus is emphasized and group process maintained. However, because patients are often in immediate crisis with overriding concerns, counselors do, at times, resort to individual counseling within the group setting. A dual disorder clientele is highly volatile, and sometimes unfortunately unpredictable, but the counselor attempts to remain fairly consistent in response to patient behavior from day to day. As patients learn what to expect, their anxiety lessens; as anxiety lessens, there is often a reduction in the psychiatric symptomatology that can impede group performance.

Patients are taught, however, that anxiety is a normal part of growth, and that without anxiety psychotherapeutic change cannot be accomplished. Anxiety poses a special threat to substance abuse patients, and this is addressed realistically since patients are likely to turn to chemicals to alleviate anxiety. For many, self-medication has been the primary coping mechanism to combat uncomfortable feelings. Patients are provided with information regarding alternate coping mechanisms. In group, they are encouraged to discuss anticipated anxiety-provoking events that threaten their abstinence. Group time may be devoted to behavioral rehearsal and role-play situations to build patients' confidence that they can confront stressful situations without drinking or using drugs.

Case Illustration

Michael was most anxious about giving in to drink at family events, yet he wanted to attend a cousin's wedding over the weekend. He gave the group an idea of how the pressure was applied by his father, brothers, other relatives, and friends. When the group had the idea, they began to offer, coerce, kid, cajole, and shame him into "having just one," and "having a little fun." Michael's confidence appeared to grow as he discovered that the assertive replies he was preparing were quite effective. He returned to the group the next week, still sober, and quite pleased with himself.

As with any group of recovering individuals, helping patients identify, acknowledge, express, and accept their feelings is an essential ingredient to the psychotherapeutic process. Emphasis is also placed on freedom of choice for behavioral options. Dual disorder patients frequently attribute much of their behavior to external (or internal) forces beyond their control, and assuming responsibility for behavior is an essential step toward recovery. Patients often conceptualize both their feelings and their thoughts as automatic and autonomous. Cognitive behavioral techniques are employed frequently by counselors. Consistent errors in thinking and patterns of cognitive distortion are readily assimilated by patients, and patients learn to assist one another in recognizing cognitive errors.

4.7. Relapse and Crisis Intervention

Sometimes a patient's psychiatric symptoms worsen. This may occur during a group session, or outside program involvement. During a group session, the situation may be handled therapeutically, but on occasion the client must be escorted from group by one therapist, leaving the other to attend to the needs of the group.

Case Illustration

Jerry had begun to show signs of decompensation earlier in the week, although he denied substance use or discontinuing his medication. Shortly after group began, something another patient said triggered his delusional structure, and he became threatening. The other patient then became very defensive and they stood face to face, voices raised. One counselor encouraged the second patient to sit down and not to react to symptomatic behavior. The other counselor told Jerry firmly that he would need to leave the group for a time. He became somewhat calmer and was escorted to the emergency room, where he was evaluated and admitted to a psychiatric unit.

Should patients decompensate during time outside program involvement, they may come to the emergency room to be admitted to psychiatry. In these cases, dual program staff are notified of the admission, and arrangements are made with the attending psychiatrist for readmission to the dual program as soon as possible.

Particularly when patients seek admission on their own, they are supported for their responsibility in recognizing their symptoms and need for assistance, and taking steps to respond to their situation. Some psychiatrists have expressed surprise and pleasure at the patient's awareness, initiative, and responsibility regarding signs of decompensation.

Case Illustration

Wayne had a history of becoming paranoid, drinking to combat his discomfort, and subsequently becoming extremely delusional and requiring hospitalization. One day he showed up on psychiatry and told his doctor that he believed he was breaking through his medication level because he was aware of his early symptoms. The doctor supported his observations and commended his responsibility. He was hospitalized only briefly while his dosage was adjusted.

No matter what the circumstances for readmission, patients are encouraged to view the setback as temporary and as only part of the recovery process. Patients tend to take an "all is lost" view of a temporary setback. For this reason, return to program participation is encouraged and supported at the earliest opportunity.

Relapses to substance abuse can occur with or without an increase in psychiatric symptomatology. Although not grounds for automatic discharge, such behavior is treated as serious. An attempt is made to explore the precipitating factors to such use, so that the event can be turned to therapeutic benefit. A second such relapse, however, almost always results in referral to primary chemical-dependency treatment as a requirement for further program participation. Sometimes the chemical use leads to such severe disruption of patient functioning that continued program participation is not feasible, and referral options are then discussed with the patient. Relapses are always threatening and saddening to the group as a whole, and as with psychiatric relapses, the health of the group must be attended to as well as the needs of the individual patient.

4.8. Termination

Since program participation is based on goal setting and goal attainment, the length of involvement varies tremendously. However, counselors guard against the group becoming a maintenance program. Although maintenance for this clientele is certainly a worthy goal and not to be discounted, a growing caseload limits the program to an active treatment program. Patients and counselors are often mutually aware when the client's progress has reached the point of sufficient goal attainment. Sufficient time is allowed, however, for proper termination. Ideally, the patient will announce a decision to terminate several weeks in advance so that the individual and group can work through the termination process together. Patients are offered Transitions group for graduated program withdrawal, although the option of evening attendance is not acceptable to all. Some have difficulty facing the prospect of engaging with a new group and wish to terminate outright. Some attend Alcoholics Anonymous or other support groups simultaneously and plan to rely on them for continued maintenance. Patients are always invited to re-establish contact should the need arise.

Case Illustration

Janice had diagnoses of dysthymia, dependent personality disorder, and alcohol abuse. After attending group for 1 month as an inpatient and 4 months as an outpatient, she believed she "ought to" be ready to terminate. A young mother of four boys, separated from her husband, she had been an active and empathic group participant and had maintained abstinence. When Janice shared her decision, the group confronted her pattern of avoiding her own feelings and needs to meet some arbitrary expectation. When Janice acknowledged that she was not in fact comfortable with her decision, she was encouraged to remain and work on meeting her own needs. Janice terminated several months later after greatly resolving her burden of unnecessary guilt and unrealistic expectations.

5. Staff Issues and Community Concerns

5.1. Staff Development

Dual programs are inherently multidisciplinary undertakings. Each of the mental health disciplines, and those skilled in work with substance abuse, have unique and essential contributions to make. Ideally, a multidisciplinary staff is a creative and harmonious team; in actuality, the occasions for conflict are many.

Historically, the mental health and substance abuse professions have harbored suspicion and distrust for each other, each viewing the other as limited, nonresponsive to important patient needs, and, sometimes, imperialistic. Our program is located in the State of Minnesota, which has a strong tradition of responsiveness to the needs of substance abusers ("chemically dependent" is the local terminology), and explicit standards concerning who is qualified to provide these services. In recent years, these traditions are increasingly working their way into rules and laws governing service provisions. However, tension still exists between chemical dependency and mental health service providers, with each discipline often believing that its contributions are devalued by other professionals.

Most conflicts have not been of the sort to be resolved once-and-for-all, but continue to require staff awareness. Ongoing staff development to encourage an atmosphere of mutual respect and appreciation of differences (not to mention accurate knowledge of what the differences actually are) have been essential. Staff are careful to insulate vulnerable patients from staff conflict.

We do not mean to paint a pessimistic picture of endless and unrewarding staff turmoil. On the contrary, staff members and patient care have often benefited from the conscious exploration of questions close to the heart of therapeutic endeavors, and a staff acquainted with struggling to resolve its own conflicts is a more confident role model of conflict resolution skills. Yet, forging a multidisciplinary collection of individualistic practitioners into a cohesive team responsive to the full range of dual patient needs is likely to confront a staff forcefully with a number of dilemmas; several of the most pervasive and problematic are described in more detail in the following sections.

5.1.1. The Role of Recovering Persons

Several areas of potential conflict have their roots in the role of recovering individuals who serve as substance abuse counselors in substance abuse treatment. Traditionally, most chemical dependency counselors have themselves been recovering, and this has a broad impact upon the field's understanding of appropriate intervention. Because most counselors have "been there" themselves, identification with clients is seen as the hallmark of empathy. One result is the important role of counselors' self-disclosure in treatment. Another is staff reliance upon directive approaches; since they have "been there" and have found a way back to greater health, counselors are often quite comfortable in telling others how to recover. Additionally, counselors' own recovery programs may interact with their service provision. Program activities may be seen as a part of one's own "Twelfth Step" work, resulting in substantial involvement in the lives of former and current patients.

Mental health professionals cannot so easily assume that they have been where their patients have been, and are less likely to view identification as the best basis for therapy. Also, they are attuned to the vulnerability of patients in psychotherapeutic relationships. For these reasons, mental health professionals rely on different techniques to establish rapport and express empathy. Self-disclosure is infrequent; therapy techniques may be more exploratory than directive, and firmer boundaries are set around involvement with patients outside the office.

Members of a multidisciplinary staff may see each other's behavior as inappropriately familiar or overly remote, too confrontive, or too tolerant, intrusive, or overly reserved. Stylistic differences between practitioners can create conflict for patients also. For instance, patients accustomed to self-disclosure by substance abuse counselors may feel rejected by mental health professionals who reveal less about themselves. Some patients would prefer not to cope with awareness of their counselors' personal histories, and may feel overwhelmed by self-disclosure. Patients who need to test others extensively may perceive staff conflict and attempt to capitalize on it. Staff reactions to patients' puzzlement or distress sometimes adds fuel to the fire.

However, patients need a degree of consistency in the staff's responses to them. In a group setting, consistency across patients is especially important. The broad outlines of appropriate interventions, self-disclosure, and involvement in patients' lives are probably program policy matters, at least at the outset, while the staff works toward a common understanding of patient needs. Although problematic, these areas of concern create opportunity for all staff to broaden their repertoire of therapeutic techniques and to consider together the indications for various approaches.

5.1.2. Staff Polarization

Polarization of the staff over whether the psychiatric or the substance use disorder is primary is another pitfall. Professionals focusing on one disorder may either overreact or underreact to symptoms of other conditions; alcohol and drug withdrawal, suicidal ideation, the threat of relapse to substance use, hallucinations, depression, and so on. Patients would often be relieved to have to accept only one disorder and be able to deny the other. Any indication that staff will align with them in such a maneuver, out of varying comfort levels with different pathologies, can interfere with the patients' attempts to come to grips with the reality of dual disorders.

The program must strive to maintain a balance in responding to both substance use and psychiatric disorders. Conjoint treatment means an effective and integrated response to the multiplicity and complexity of patient needs.

5.1.3. Alternative Recovery Models

Finally, a major issue for the staff is the search for alternative recovery models. Inpatient and outpatient chemical dependency treatment programs are based predominately upon the Alcoholics Anonymous model of recovery, which has indeed proved to be of tremendous value to many recovering individuals (Hoffmann, Harrison, & Belille, 1983). Staff are cautious, in some cases reluctant, to deviate from this program, especially those staff members who owe their own recovery to Alcoholics Anonymous. Dual program patients, however, have encountered a variety of problems with both the AA program and AA participation.

Although AA itself does not, in principle, discourage the use of prescribed psychotropic medications, patients have often encountered vocal opposition to medication maintenance. They are not "chemically free," they are told, and thus are not "working the program." Many patients are reluctant to conceal their medication use because of the honesty mandated by Alcoholics Anonymous. Patients' tolerance for "defending" their medication varies, and some have felt severely shamed.

AA's "one drink—one drunk" precept parallels the dichotomous thinking of many clients, and can act as a self-fulfilling prophecy for some who might not otherwise lose control with their first ingestion. Viewing success or failure in absolute terms can lead to shame and despair, even when a relapse follows a long phase of quality abstinence. Dual counselors encourage patients to take a more positive view that any progress achieved in the recovery process cannot be erased, and can be recaptured with a renewed commitment to sobriety. Total, permanent abstinence is a desirable goal for dual patients as well as for other substance abusers, but it is neither realistic nor humane as a sole measure of the successful recovery of dual patients.

Other difficulties include the natural conflict between AA's suppressive approach to managing life problems and psychotherapeutic techniques stressing self-awareness and choices of behavior options; and the shame patients feel in the face of others' reactions to their inability to apply the AA program successfully to their psychiatric symptoms. (It is difficult to "turn over" major depression or schizophrenia.) Finally, many dual disorder patients are not well equipped to participate in the social atmosphere of Alcoholics Anonymous. This is, perhaps, particularly true of severely paranoid individuals.

Therefore, dual program staff must strike a delicate and careful balance concerning AA. The value of AA is not discounted by staff, and no patient who finds it helpful is discouraged from participating. The dual program itself, however, is not based on AA principles. This stance toward AA is not equally acceptable to all staff, or to any one staff member at all times.

5.2. Community Liaison

Forming and maintaining working alliances with other service providers is a crucial aspect of the program's success. Typically, no one agency can meet all the needs of this multiply impaired population. Like conflict within program staff, there may be controversy among providers who respond to either psychiatric problems or substance abuse problems, but are seldom trained to meet needs on both dimensions.

A particularly acute problem is finding residential care for patients leaving the inpatient phase of the dual program. Some chemical dependency halfway houses expressly forbid the use of prescribed psychotropic medications. In many other chemical dependency halfway houses, overt policy may permit the use of psychiatric medications, but covert policy frequently discourages such use. Patients often report pressure from halfway house counselors and residents to discontinue their medications. Some chemical-dependency facilities seem to persist in the belief that all psychiatric symptomatology can be traced to chemical use, and that sobriety is equivalent to recovery. Patients may feel that they are being told that they cannot recover until they are totally "chemically free," and in fact, that their symptoms are due to their medications.

Halfway houses or board-and-care residences for the mentally ill pose a different problem for dual-disorder patients. In these facilities, less attention is paid to maintaining an abstinent environment than is the case in chemical-dependency halfway houses, and drugs may be freely available in some of them. Some such facilities have explicit policies banning the acceptance of chemically dependent clients until a set period of abstinence has been achieved—often 3 months or more. Although some relax their entrance requirements for a client actively involved in the dual treatment program, few openings are available to these clients.

Patients who have made recent suicide attempts, or who are seen as high suicide risks, are harder yet to place because of the responsibility they place on the facilities that accept them. Consequently, it is often the patient most in need of supportive residential care who is least able to obtain it. Residential care tailored to the specific needs of the dual-program patient is difficult indeed to locate.

Admission requirements for community facilities can be understood as realistic assessments of their own staff capabilities in dealing with patients who have additional disorders that are not their primary focus of training and treatment. However, in spite of major educational efforts, community support for these patients is still woefully lacking. While residential placement provides a clear example of the difficulties involved, similar problems arise with respect to the provision of other necessary services. Dual-program staff devote a great deal of time and attention to interagency linkages, and this will probably be necessary for the foreseeable future.

6. Conclusions and Future Directions

Treatment of a dual population stands to benefit from both basic research and outcome studies. Basic research projects having a direct bearing upon treatment include epidemiological studies, explorations of genetic or developmental relationships between various psychiatric and substance use disorders, and identification of risk factors, among many others.

Program evaluation and outcome research are always complicated endeavors, and perhaps especially so when the presenting problems are multifaceted. Participating staff believe in what they do and frequently hear informal accounts of their former patients' successes. However, observation, anecdotes, and intuition are not adequate substitutes for organized empirical studies, and systematic collection of history, intake, and follow-up data has now begun. The studies in progress are based upon a multidimensional concept of recovery—or improvement. Relevant criteria of success include frequency and duration of psychiatric hospitalizations and chemical-dependency treatments; quality and length of abstinence; general medical care utilization; occupational functioning; financial independence; legal conflicts; satisfaction in interpersonal relationships; self-esteem; and personal and social responsibility.

The lack of definitive outcome studies is, of course, not unique to conjoint treatment of substance abuse and psychiatric disorders, but rather is a common state of affairs in mental health disciplines. The staff has developed a deep appreciation of each other's ability to offer care to this population—with their complex histories and present distress—while tolerating considerable ambiguity about the program's ultimate effectiveness.

Finally, the histories of our patients suggest that many of them had clearly identifiable dual disorders for a substantial period of time before they came to the attention of the dual program. This offers some hope that earlier intervention can be designed to meet the needs of at least some individuals suffering from dual disorders before they reach the level of impairment and dysfunction typical of our current population.

Acknowledgments

The authors wish to express their appreciation to Cindi Claypatch, of the Dual Disorders staff, for her assistance in data collection, and Rosemary Perron for her secretarial skills through countless revisions.

The authors also wish to acknowledge St. Paul-Ramsey Hospital Medical Education and Research Foundation and Dual Disorders staff members W. S. Balcerzak, Michael J. Gross, and Manuel S. Mejia.

7. References

Alterman, A. I., Erdlen, F., & Murphy, E. Alcohol abuse in the psychiatric hospital population. *Addictive Behaviors*, 1981, 6, 69–73.

Carroll, J. F. X. Mental illness and disease: Outmoded concepts in alcohol and drug rehabilitation. *Community Mental Health Journal*, 1975, *11*, 418–429.

Dichter, M., & Eusanio, A. Rationale for a generic based training model for mixed substance abuse–psychiatric populations. Paper presented at the 11th Annual Eagleville Conference, Eagleville, Pennsylvania, May 19, 1978.

Fine, E. W. The syndrome of alcohol dependency and depression. In E. Gottheil, A. T. McLellan, & K. A. Druley (Eds.), *Substance abuse and psychiatric illness*. New York: Pergamon Press, 1980.

Hoffmann, N. G., Harrison, P. A., & Belille, C. A. Alcoholics Anonymous after treatment: Attendance and abstinence. *The International Journal of the Addictions*, 1983, *18*, 311–318.

McLellan, A. T., Druley, K. A., & Carson, J. Evaluation of substance abuse problems in a psychiatric hospital. *Journal of Clinical Psychiatry*, 1978, *39*, 425.

McLellan, A. T., MacGahan, J. A., & Druley, K. A. Changes in drug abuse clients, 1972–1977: Implications for revised treatment. Paper presented at American Psychiatric Association Conference, Atlanta, Georgia, May, 1978.

Pokorny, A. D. The multiple readmission psychiatric patient. *Psychiatric Quarterly*, 1965, *39*, 70–78.

Powell, B. J., Penick, E. C., Othmer, E., Bingham, S. F., & Rice, A. S. Prevalence of additional psychiatric syndromes among male alcoholics. *Journal of Clinical Psychiatry*, 1982, *43*, 404–407.

Tuason, V. B., Fair-Riedesel, P., & Hoffmann, N. G. Effects of deinstitutionalization on acute care psychiatric facilities. *Minnesota Medicine*, 1982, *65*, 697–699.

Ziegler-Driscoll, G., Sax, P., Deal, D., & Ostreicher, P. Selection and training of staff to work with the psychiatrically ill substance abuser. In E. Gottheil, A. T. McLellan, & K. A. Druley (Eds.), *Substance abuse and psychiatric illness*. New York: Pergamon Press, 1980.

Zosa, A. *Psychiatric problems can no longer be automatically used to screen patients out of drug and alcohol rehabilitation programs*. In D. J. Ottenberg, J. F. X. Carroll, & C. Bolognese (Eds.), Proceedings of the Eleventh Annual Eagleville Conference, May 19, 1978.

Index

7973

DATE DUE			